T0338188

Approximate Antenna Analysis for CAD

Approximate Antenna Analysis for CAD

Hubregt J. Visser
Antenna Engineer, The Netherlands

A John Wiley and Sons, Ltd, Publication

This edition first published 2009

Registered office
John Wiley & Sons Ltd, The Atrium, Southern Gate, Chichester, West Sussex, PO19 8SQ, United Kingdom

For details of our global editorial offices, for customer services and for information about how to apply for permission to reuse the copyright material in this book please see our website at www.wiley.com.

Library of Congress Cataloging-in-Publication Data

Visser, Hubregt J.
Approximate antenna analysis for CAD / Hubregt J. Visser.
p. cm.
Originally presented as author's thesis–Ph. D.
Includes bibliographical references and index.
ISBN 978-0-470-51293-7 (cloth)
1. Antennas (Electronics)–Computer-aided design. 2. Electromagnetic fields–Computer simulation.
I. Title.
TK7871.6.V569 2009
621.382'4–dc22

2008041825

A catalogue record for this book is available from the British Library.

ISBN 9780470512937 (H/B)

Set in 10/12pt Times by Sunrise Setting Ltd, Torquay, UK.
Printed in Great Britain by CPI Antony Rowe, Chippenham, Wiltshire

Contents

Preface

In der Beschränkung zeigt sich erst der Meister,[1] wrote Johann Wolfgang von Goethe on 26 June 1802. It is a quote much used in PhD theses to accentuate and justify the compactness of a thesis. For this book, which also serves the purpose of a PhD thesis, this quote is completely unjustified. I have tried to be as elaborate as possible in explaining the approximate antenna models developed.

This book is the result of more than 15 years of work in the field of antenna modeling. After working for a number of years on the full-wave modeling of large phased array antennas, I found that, for a customer, it is very hard to wait till a full-wave computer code has been developed. Therefore I started developing so-called 'engineering' or approximate models in parallel with the full-wave models. These engineering models, which can be produced much faster, but at the cost of reduced accuracy, can give the customer a preview of what will be possible, and may be used to create 'predesigns' to be fine-tuned by applying the full-wave model. Nowadays I focus completely on developing approximate models. Most of the topics encountered in this book were developed over the last few years, but some date back almost 15 years.

The reason for being 'as elaborate as possible' in explaining the approximate models is twofold. First, as a young engineer fresh from university, I found it hard, when starting on a new assignment, to work backwards from a relevant paper and understand all the steps taken in the development of a model. In those days, I would have wanted a book that would have taken me by the hand and explained to me all the necessary steps taken in the development of

[1] 'Constraint is where you show you are a master'.

a model. With this book, I have tried to accommodate this wish. Second, I have always been in the privileged situation of having literature search facilities and a large technical library at my immediate disposal. For those not in this privileged situation, it may be very hard to get access to the necessary references. Therefore, rather than just referring to the sources, I have also written down all of the equations needed for implementing the model into software.

This may have the effect that the book will become a bit dreary for experienced antenna engineers. For the inexperienced antenna engineer, I hope that, referring again to Goethe, the following quote will be appropriate after reading the book: *Das also war des Pudels Kern*[2] [1].

REFERENCE

1. J.W. von Goethe, *Faust: Der Tragödie erster und zweiter Teil. Urfaust*, Beck Verlag, Munich, Germany, 2006.

Hubregt J. Visser
Veldhoven, The Netherlands

[2] 'So this, then, was the kernel of the brute'.

Acknowledgments

This book could not have been written without the help of many individuals whom I would like to thank for their contributions. Chapter 2 is the result of a cooperation between the Electromagnetics Department of the Faculty of Electrical Engineering of Eindhoven University of Technology (TU/e) and the Image Science Institute of the University Medical Center Utrecht (UMC Utrecht), both in The Netherlands. From UMC Utrecht, I would like especially to thank Chris Bakker, Jan-Henry Seppenwoolde and Wilbert Bartels. I would also like to thank my MSc students Nicole Op den Kamp and Marjan Aben for contributing to that chapter. I would like to thank my MSc students Iwan Akkermans and Jeroen Theeuwes for their contributions to Chapter 5. Frank van den Boogaard, from TNO Defence and Safety, is thanked for his kindness in permitting me to use material on waveguide array antenna modeling for Chapter 6. K.K. Chan from Chan Technologies, Inc., Canada, is thanked for his many helpful suggestions and support in developing the model. A word of special thanks is reserved for Anton Tijhuis from TU/e for being my promoter and pushing me forward to finish this work. Also, a word of special thanks is reserved for Guy Vandenbosch from the Catholic University of Leuven, Belgium, for also being my promoter and for keeping faith in me for more than ten years. Ad Reniers is thanked for preparing the many antenna prototypes and performing part of the measurements. Sarah Hinton, Sarah Tilley and Tiina Ruonamaa from Wiley are thanked for their incredible patience and support. Finally, I would like to thank my wife Dianne and daughter Noa for accepting, again, a long period of book-related neglect.

H.J.V.

Acronyms

AC	Alternating Current
BBC	British Broadcasting Corporation
CAT	Computed Axial Tomography
COTS	Commercial Off-the-Shelf
CPS	Coplanar Strip
CPW	Coplanar Waveguide
CT	Computed Tomography
DC	Direct Current
FE	Finite Element
FFT	Fast Fourier Transform
FID	Free Induction Decay
FIT	Finite Integration Technique
FR	Flame Retardant
GA	Genetic Algorithm
GPS	Global Positioning System
GSM	Global System for Mobile Communications; Generalized Scattering Matrix
iMRI	Interventional Magnetic Resonance Imaging

MEN	Multimode Equivalent Network
MIT	Massachusetts Institute of Technology
MoM	Method of Moments
MRI	Magnetic Resonance Imaging
NEC	Numerical Electromagnetic Code
NMI	Nuclear Medicine Imaging
NMRI	Nuclear Magnetic Resonance Imaging
OFDM	Orthogonal Frequency Division Multiplexing
PCB	Printed Circuit Board
PEC	Perfect Electric Conductor
PET	Positron Emission Tomography
RC	Relative Convergence
RF	Radio Frequency
RFID	Radio Frequency Identification
RK	Runge–Kutta Method
SAR	Specific Absorption Rate
SMA	Subminiature Version A
SNR	Signal-to-Noise Ratio
SPECT	Single-Photon-Emission Computed Tomography
TE	Transverse Electric
TEM	Transverse Electromagnetic
TL	Transmission Line
TLM	Transmission Line Matrix
TM	Transverse Magnetic
UWB	Ultrawideband
WAIM	Wide-Angle Impedance Match
WLAN	Wireless Local Area Network

1 Introduction

From the moment that Heinrich Rudolf Hertz experimentally proved the correctness of the Maxwell equations in 1886, antennas have been in use. The fact that Guglielmo Marconi's success depended on the 'finding' of the right antenna in 1895 indicates the importance of antennas and thus of antenna analysis. It was, however, common practice up until the middle of the 1920s to design antennas empirically and produce a theoretical explanation after the successful development of a working antenna. It took a world war to evolve antenna analysis and design into a distinct technical discipline. The end of the war was also the starting point of the development of electronic computers that eventually resulted in the commercial distribution of numerical electromagnetic analysis programs. Notwithstanding the progress in numerical electromagnetic analysis, a need still exists for approximate antenna models. They are needed both in their own right and as part of a synthesis process that also involves full-wave models.

1.1 THE HISTORY OF ANTENNAS AND ANTENNA ANALYSIS

The history of antennas dates back almost entirely to the understanding of electromagnetism and the formulation of the electromagnetic-field equations. In the 1860s, James Clerk Maxwell saw the connection between Ampère's, Faraday's and Gauss's laws. By extending Ampère's law with what he called a *displacement current* term, he united electricity and magnetism into electromagnetism [1]. His monumental work of 1873, *A Treatise on Electricity and Magnetism*, is still in print [2]. With light now described as and proven to be an electromagnetic phenomenon, Maxwell had already predicted the existence of electromagnetic waves at radio frequencies, i.e. at much lower frequencies than light.

Approximate Antenna Analysis for CAD Hubregt J. Visser

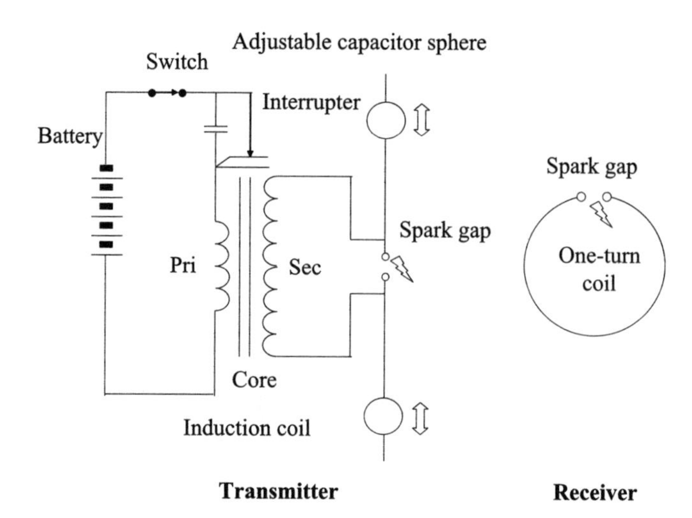

Figure 1.1 Hertz's open resonance system. With the receiving one-turn loop, small sparks could be observed when the transmitter discharged. From [4].

It was not until 1886 that he was proven right by Heinrich Rudolf Hertz, who constructed an *open* resonance system as shown in Figure 1.1 [3, 4]. A spark gap was connected to the secondary windings of an induction coil. A pair of straight wires was connected to this spark gap. These straight wires were equipped with electrically conducting spheres that could slide over the wire segments. By moving the spheres, the capacitance of the circuit could be adjusted for resonance. When the breakdown voltage of air was reached and a spark created over the small air-filled spark gap, the current oscillated at the resonance frequency in the circuit and emitted radio waves at that frequency (Hertz used frequencies of around 50 MHz). A single-turn square or circular loop with a small gap was used as a receiver. Without being fully aware of it, Hertz had created the first radio system, consisting of a transmitter and a receiver.

Guglielmo Marconi grasped the potential of Hertz's equipment and started experimenting with wireless telegraphy. His first experiments – covering the length of the attic of his father's house – were conducted at a frequency of 1.2 GHz, for which he used, like Hertz before him, cylindrical parabolic reflectors, fed at the focal point by half-wave dipole antennas. In 1895, however, he made an important change to his system that suddenly allowed him to transmit and receive over distances that progressively increased up to and beyond 1.5 km [5–7]. In his own words, at the reception for the Nobel Prize for physics in Stockholm in 1909 [7]:

> In August 1895 I hit upon a new arrangement which not only greatly increased the distance over which I could communicate but also seemed to make the transmission independent from the effects of intervening obstacles. This arrangement [Figure 1.2(a)] consisted in connecting one terminal of the Hertzian oscillator or spark producer to earth and the other terminal to a wire or capacity area placed at a height above the ground and in also connecting at the receiver end [Figure 1.2(b)] one terminal of the coherer to earth and the other to an elevated conductor.

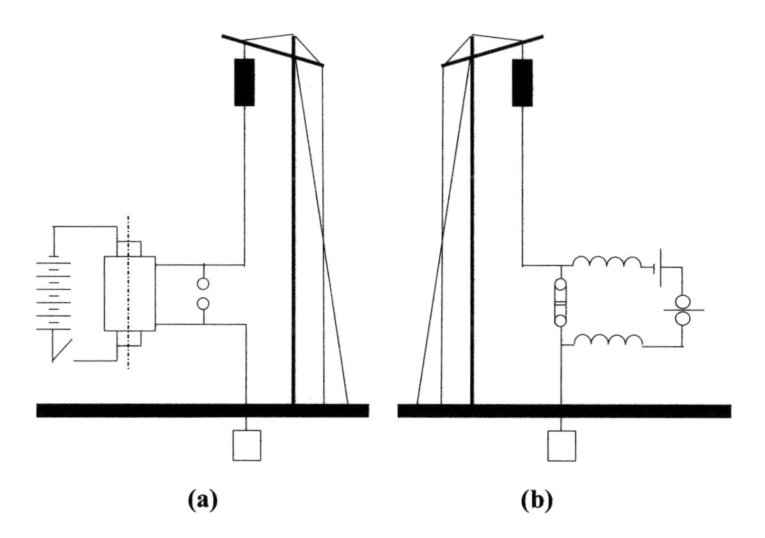

Figure 1.2 Marconi's antennas of 1895. (a) Scheme of the transmitter used by Marconi at Villa Griffone. (b) Scheme of the receiver used by Marconi at Villa Griffone. From [4]. Reproduced, with permission, from Ofir Glazer, Bio-Medical Engineering Department, Tel-Aviv University, Israel. Part of M.Sc. final project, tutored by Dr. Hayit Greenspan.

Marconi had enlarged the antenna. His monopole antenna was resonant at a wavelength much larger than any that had been studied before, and it was this creation of long-wavelength electromagnetic waves that turned out to be the key to his success. It was also Marconi who, in 1909, introduced the term *antenna* for the device that was formerly referred to as an *aerial* or *elevated wire* [7, 8].

The concept of a monopole antenna, forming a dipole antenna together with its image in the ground, was not known by Marconi at the time of his invention. In 1899, the relation between the antenna length and the operational wavelength of the radio system was explained to him by Professor Ascoli, who had calculated that the 'length of the wave radiated [was] four times the length of the vertical conductor' [9].

Up to the middle of the 1920s it was common practice to design antennas empirically and produce a theoretical explanation after the successful development of a working antenna [10]. It was in 1906 that Ambrose Fleming, a professor at University College, London, and consultant to the Marconi Wireless Telegraphy Company, produced a mathematical explanation of a monopole-like antenna[1] based on image theory. This may be considered the first ever antenna design that was accomplished both experimentally and theoretically [10]. The first theoretical description of an antenna may be attributed to H.C. Pocklington, who, in 1897, first formulated the frequency domain integral equation for the total current flowing along a straight, thin wire antenna [11].

[1]This antenna was a *suspended long wire antenna*, nowadays also called an *inverted L antenna* or *ILA*, and used for transatlantic transmissions.

The invention of the *thermionic valve*, or *diode*, by Fleming in 1905 and of the *audion*, or *triode*, by Lee de Forest in 1907 paved the way for the reliable detection, reception and amplification of radio signals. From 1910 onwards, broadcasting experiments were conducted that resulted, in Europe, in the formation in 1922 of the British Broadcasting Corporation (BBC) [12]. The early antennas in the broadcasting business were makeshift antennas, derived from the designs used in point-to-point communication. Later, T-configured antennas were used for transmitters [13], and eventually vertical radiators became standard, owing to their circularly symmetrical coverage (directivity) characteristic [13, 14]. The receiver antennas used by the public were backyard L-structures and T-structures [4].

In the 1930s, a return of interest in the higher end of the radio spectrum took place. This interest intensified with the outbreak of World War II. The need for compact communication equipment as well as compact (airborne) and high-resolution radar made it absolutely necessary to have access to compact, reliable, high-power, high-frequency sources. In early 1940, John Randall and Henry Boot were able to demonstrate the first cavity magnetron, creating 500 kW at 3 GHz and 100 kW at 10 GHz. In that same year, the British Prime Minister, Sir Winston Churchill, sent a technical mission to the United States of America to exchange wartime secrets for production capacity. As a result of this Tizard Mission, named after its leader Sir Henry Tizard, the cavity magnetron was brought to the USA and the MIT Rad Lab (Massachusetts Institute of Technology Radiation Laboratory) was established. At the Rad Lab, scientists were brought together to work on microwave electronics, radar and radio, to aid in the war effort.

The Rad Lab closed on 31 December 1945, but many of the staff members remained for another six months or more to work on the publication of the results of five years of microwave research and development. This resulted in the famous 28 volumes of the Rad Lab series, many of which are still in print today [15–42].

In relation to antenna analysis, we have to mention the volume *Microwave Antenna Theory and Design* by Samuel Silver [26], which may be regarded as one of the first 'classic' antenna theory textbooks. Soon, it was followed by several other, now 'classic' antenna theory textbooks, amongst others *Antennas* by John Kraus in 1950 [43], *Antennas, Theory and Practice* by S.A. Schelkunoff in 1952 [44], *Theory of Linear Antennas* by Ronold W.P. King in 1956 [45], *Antenna Theory and Design* by Robert S. Elliott in 1981 [46] and *Antenna Theory, Analysis and Design* by Constantine A. Balanis in 1982 [47]. Specifically for phased array antennas, we have to mention *Microwave Scanning Antennas* by Robert C. Hansen [48] (1964), *Theory and Analysis of Phased Array Antennas* by N. Amitay, V. Galindo and C.P. Wu [49] (1972), and *Phased Array Antenna Handbook* by Robert J. Mailloux [50] (1980).[2]

At the end of World War II, antenna theory was mature to a level that made the analysis possible of, amongst others, freestanding dipole, horn and reflector antennas, monopole antennas, slots in waveguides and arrays thereof. The end of the war was also the beginning of the development of electronic computers. Roger Harrington saw the potential of electronic computers in electromagnetics [51] and in the 1960s introduced the method of moments (MoM) in electromagnetism [52]. The origin of the MoM dates back to the work of

[2]For the 'classic' antenna theory textbooks mentioned here, we refer to the first editions. Many of these books have by now been reprinted in second or even third editions.

Galerkin in 1915 [53]. The introduction of the IBM PC[3] in 1981 helped considerably in the development of numerical electromagnetic analysis software. The 1980s may be seen as the decade of the development of numerical microwave circuit and planar antenna theory. In this period, the Numerical Electromagnetics Code (NEC) for the analysis of wire antennas was commercially distributed. The 1990s, however, may be seen as the decade of numerical electromagnetic-based design of microwave circuits and (planar, integrated) antennas. In 1989 the distribution of Sonnet started, followed, in 1990, by the HP (now Agilent) High Frequency Structure Simulator (HFSS)[4] [51]. These two numerical electromagnetic analysis tools were followed by Zeland's IE3D, Remcom's XFdtd, Agilent's Momentum, CST's Microwave Studio, FEKO from EM Software & Systems, and others.

Today, we have evolved from the situation in the early 1990s when the general opinion appeared to be 'that numerical electromagnetic analysis cannot be trusted' to a state wherein numerical electromagnetic analysis is considered to be the ultimate truth [51]. The last assumption, however, is as untrue as the first one. Although numerical electromagnetic analysis software has come a long way, incompetent use can easily throw us back a hundred years in history. One only has to browse through some recent volumes of peer-reviewed antenna periodicals to encounter numerous examples of bizarre-looking antenna structures designed by iterative use of commercially off-the-shelf (COTS) numerical electromagnetic analysis software. These reported examples of the modern variant of trial and error, although meeting the design specifications, are often presented without even a hint of a tolerance analysis, let alone a physical explanation of the operation of the antenna.

The advice that James Rautio, founder of Sonnet Software, gave in the beginning of 2003 [51],

> No single EM tool can solve all problems; an informed designer must select the appropriate tool for the appropriate problem,

is still valid today, as a benchmarking of COTS analysis programs showed at the end of 2007 [54, 55]. Apart from the advice to choose the right analysis technique for the right structure to be analyzed, these recent studies also indicate the importance of being careful in the choice of the feeding model and the mesh for the design to be analyzed. So, notwithstanding the evolution of numerical electromagnetic analysis software, it still takes an experienced antenna engineer, preferably one having a PhD in electromagnetism or RF technology, to operate the software in a justifiable manner and to interpret the outcomes of the analyses.

Having said this, we may now proceed with a discussion of how to use full-wave analysis software for antenna synthesis.

1.2 ANTENNA SYNTHESIS

Antenna synthesis should make use of a manual or automated iterative use of analysis steps. The analysis techniques occupy a broad time consumption 'spectrum' from quick physical

[3] 4.77 MHz, 16 kB RAM, no hard drive.
[4] Currently Ansoft HFSS.

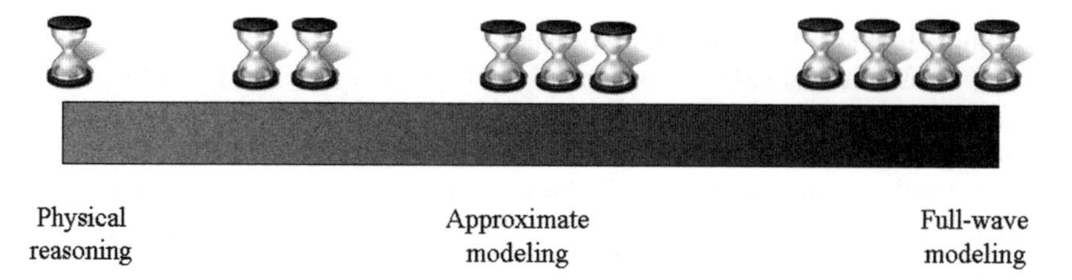

Figure 1.3 Analysis techniques ordered according to calculation time involved.

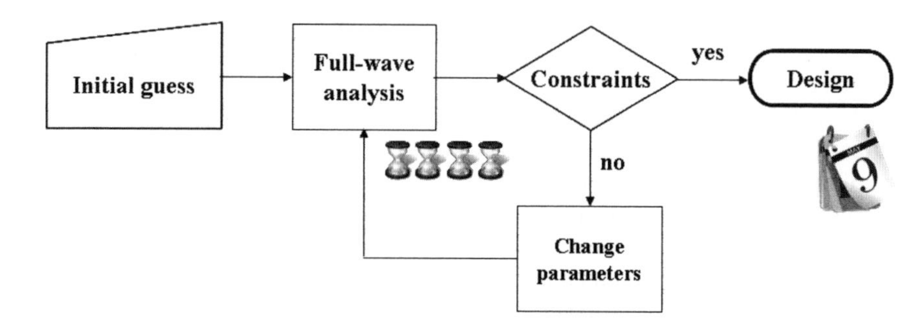

Figure 1.4 Stochastic optimization based on iteration of full-wave analysis is a (too) time-consuming process.

reasoning ('the length of a monopole-like antenna should be about a quarter of the operational wavelength') to lengthy (in general) full-wave numerical electromagnetic analysis. The 'spectrum' of analysis techniques is shown in Figure 1.3, where the hourglasses indicate symbolically the time involved in applying the various analysis techniques.

For an automated synthesis, starting with mechanical and electromagnetical constraints and possibly an initial guess,[5] we have to rely on stochastic optimization. Since stochastic optimization needs a (very) large number of function evaluations or analysis steps, such an optimization scheme based on full-wave analysis (Figure 1.4) is not a good idea.

Therefore, we propose a two-stage approach [56], where, first, a stochastic optimization is used in combination with an approximate analysis and, second, line search techniques are combined with full-wave modeling (Figure 1.5). Since one of the key features of the approximate analysis model needs to be that its implementation in software is fast while still sufficiently accurate, we may employ many approximate analysis iterations and therefore use a stochastic optimization to get a predesign. This predesign may then be fine-tuned using a limited number of iterations using line-search techniques. Owing to the limited number of iterations, we may now – in the final synthesis stage – employ a full-wave analysis model.

Using an approximate but still sufficiently accurate model, the automated design – using stochastic optimization – may be sped up considerably. The output at this stage of the synthesis process is a preliminary design. Depending on the accuracy of this design and

[5] An initial guess may be created by randomly choosing the design variables.

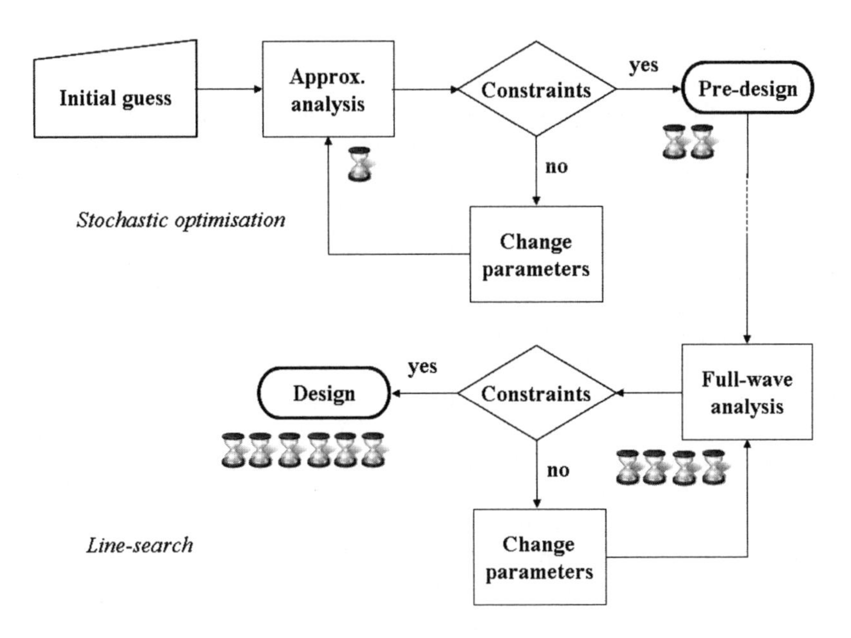

Figure 1.5 Antenna synthesis based on stochastic optimization in combination with an approximate model and line search with a full-wave model.

the design constraints, it is very well possible that the design process could end here; see for example [56]. If a higher accuracy is required or if the design requirements are not fully reached, this preliminary design could be used as an input for a line search optimization in combination with a full-wave model. For the complete synthesis process using both approximate and full-wave models (Figure 1.5), the time consumption will drop with respect to a synthesis process involving only a full-wave model. The reason is that the most time-consuming part of the process, i.e. when the solution space is randomly sampled, is now conducted with a fast, approximate, reduced-accuracy model. The question that remains is what may be considered to be 'sufficiently accurate'.

1.3 APPROXIMATE ANTENNA MODELING

From the point of view of synthesis, approximate antenna models are a necessity. They need to be combined with a full-wave analysis program, but if – depending on the application – the accuracy of the approximation is sufficient, the approximate model alone will suffice. In [51, 54], the use of (at least) two full-wave simulators is advised, but not many companies or universities can afford to purchase or lease multiple full-wave analysis programs. For many companies that do not specialize in antenna design, even the purchase or lease of one full-wave analysis program may be a budgetary burden. Therefore the availability of approximate, sufficiently accurate antenna models is required not only for the full synthesis process. It is also valuable for anyone needing an antenna not yet covered in the standard antenna textbooks who does not have access to a full-wave analysis program.

The purpose of the approximate and full-wave models is to replace the realization and characterization of prototypes, thus speeding up the design process. This does not mean, however, that prototypes should not be realized at all. At least one prototype should be realized to verify the (pre)design. A range of slightly different prototypes could be produced as a replacement for the fine-tuning that employs line search techniques in combination with full-wave modeling.

A question that still remains with respect to the approximate modeling is what may be considered 'sufficiently accurate'. This question cannot be answered unambiguously. It depends on the application; the requirements for civil and medical communication antennas, for example, are much less stringent than those for military radar antennas. If we look at a communication antenna to be matched to a standard 50 Ω transmission line, we should not look at the antenna input impedance but rather at the reflection level. In general, any reflection level below −10 dB over the frequency range of interest is considered to be satisfactory. This means that, if we assume the input impedance to be real-valued, we may tolerate a relative error in the input impedance of up to 100%. For low-power, integrated solutions, working with a 50 Ω standard for interconnects may not be the best solution. A conjugate matching may be more efficient. If we are looking at antennas to be conjugately matched to a complex transmitter or receiver front-end impedance, however, we cannot tolerate the aforementioned large impedance errors. In general, we may say that we consider an approximate antenna model sufficiently accurate if it predicts a parameter of interest to within a few percent relative to the measured value or the (verified) full-wave analysis result. Such an accuracy also prevents the answer drifting away during the stochastic optimization.

Another question is when to develop an approximate model. The answer to this question is dictated both by the resources available and a company's long-term strategy. If neither a full-wave analysis program for the problem at hand nor an existing approximate model is available, then one can resort to trial and error or develop an approximate model or a combination of both, where the outputs of experiments dictate the path of the development of the model. If a full-wave analysis program is available and the antenna to be designed is a one-of-a-kind antenna or time is really critical, one can resort to an educated software variant of design by trial and error, meaning that the task should be performed by an antenna expert. When the antenna to be designed can be considered to belong to a class of antennas, meaning that similar designs are foreseen for the future, but for different materials and other frequency bands or for use in other environments, it is beneficial to develop a dedicated approximate model. The additional effort put into the development of the model for the first design will be compensated for in the subsequent antenna designs. An antenna design may also be created by generating a database of substructure analyses, employing a full-wave analysis model. Then, a smart combining of these preanalyzed substructures results in the desired design. The generation of the database will be very time-consuming but once this task has been accomplished, the remainder of the design process will be very time-efficient.

The last question is how to develop an approximate model. First of all, the approximate model should be tailored to the antenna class at hand. To achieve that, the antenna structure should be broken down into components for which analytical equations have been derived in the past, in the precomputer era, or for which analytical equations may be derived.

By distinguishing between main and secondary effects, approximations may be applied with different degrees of accuracy, thus speeding up computation time. It appears that much of the work performed in the 1950s, 1960s and 1970s that seems to have been forgotten is extremely useful for this task. In this book, we have followed this approach for a few classes of antennas. For each class of antennas, we have taken a generic antenna structure and decomposed it into substructures, such as sections of transmission line, dipoles and equivalent electrical circuits. For these substructures and for the combined substructures, approximate analysis methods have been selected or developed. The main constraints in developing approximate antenna models were the desired accuracy in the antenna parameter to be evaluated (the amplitude of the input reflection coefficient or the value of the complex input impedance) and the computation time for the software implementation of the model. Examples of the development of approximate models will be given in the following chapters.

1.4 ORGANIZATION OF THE BOOK

In Chapter 2, we start with the development of an approximate model for intravascular antennas, i.e. loops and solenoids embedded in blood (Figure 1.6). A reason for undertaking this development was the unavailability of a full-wave analysis program fit for the task at the time of development. But even if such a program had been available, it would have taken too much time to be of practical value in designing intravascular antennas.

The antennas were meant as receiving antennas in a magnetic resonance imaging (MRI) system, either for visualizing catheter tips during interventional MRI or for obtaining detailed information about the inside of the artery wall. The figure shows that the quasi-static model developed here may be used in a stochastic optimization process. The optimization times were of the order of minutes.

In Chapter 4, we describe an example of the use of a full-wave analysis program for designing a printed ultrawideband (UWB) monopole antenna, the reason being that this antenna was a 'one-of-a-kind' design. We begin with physical reasoning about how the proposed antenna operates. In the design process, it becomes clear that it may be beneficial to use or develop approximate models for parts of the structure, such as filtering structures in the feeding line. Next, an approximate model is developed for a non-UWB printed monopole antenna (Figure 1.7) that is considered to belong to a class of antennas. The model is based on an equivalent-radius dipole antenna with a magnetic covering.

Then, in Chapter 5, we discuss folded-dipole antennas and some means to control the input impedance of these antennas. The envisaged application is in the field of radio frequency identification (RFID), where the antenna needs to be conjugately matched to the RFID chip impedance, which will, in general, be some complex value different from 50 Ω. An approximate model based on dipole antenna analysis and transmission line analysis is applied to both thin-wire folded-dipole structures and folded-dipole structures consisting of strips on a dielectric slab. Also, arrays of reentrant folded dipoles will be analyzed, as shown in Figure 1.8.

Pursuing the modeling of 'non-50 Ω' antennas, in Chapter 6 we discuss an efficient, approximate but accurate modeling of a *rectenna*, i.e. an antenna connected to a rectifying element (diode), meant for collecting RF energy and transforming it to usable DC energy.

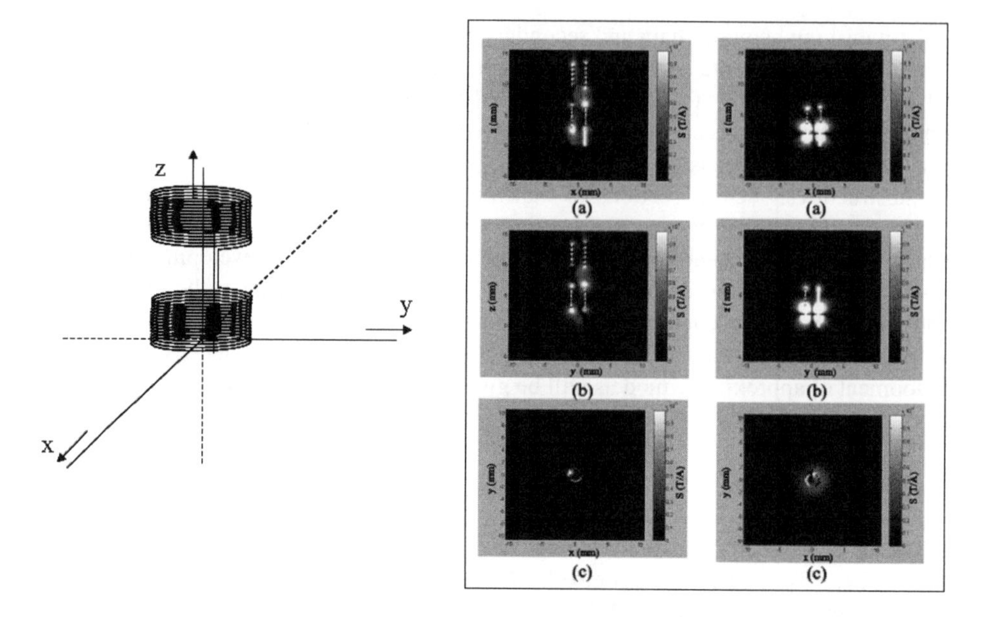

Figure 1.6 Intravascular antenna, and optimization results. Left: antenna. Right: magnetic field intensity calculated after optimization for local antenna 'visibility' (left), and calculated after optimization for maximum magnetic field intensity at the position of the artery wall (right) for different planes through the antenna.

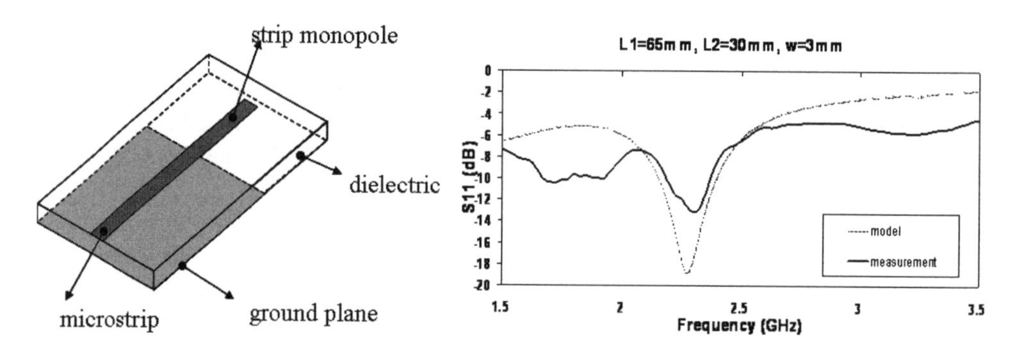

Figure 1.7 Printed monopole antenna and results of analysis by an approximate model. Left: antenna configuration. Right: calculated and measured return loss as a function of frequency for a particular configuration.

We start by modeling the rectifying circuit with the aid of a large-signal equivalent model. Once the input impedance of this circuit has been determined, we use a modified cavity model for a rectangular microstrip patch antenna to find the complex conjugate impedance value. Thus we may directly match the antenna and the rectifying circuit. To complete the chapter, we discuss a means of using antennas for power and data exchange simultaneously, based on the concept of the Wilkinson power combiner (Figure 1.9).

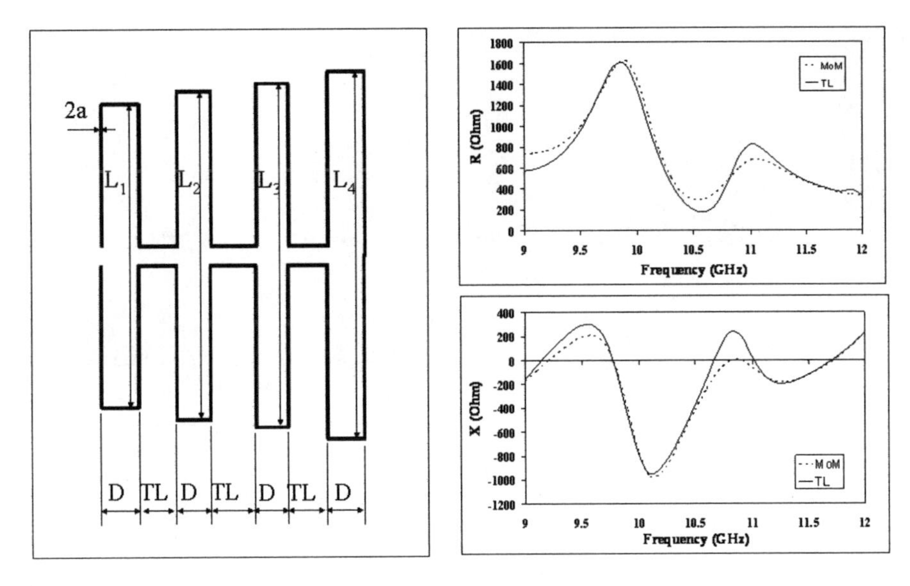

Figure 1.8 Linear array of reentrant folded dipoles. Left: array configuration. Right: real and imaginary parts of the array input impedance as a function of frequency, calculated with the approximate model and with the method of moments.

Chapter 7 deals with 'approximation' in a different way. In this chapter, we use an approximation for large, planar array antennas. The approximation consists of considering the array antenna to be infinite in two directions in the transverse plane. This approximation allows us, for an array of identical radiating elements positioned in a regular lattice, to consider the array to be periodic and uniformly excited, and therefore we only have to analyze a single unit cell (Figure 1.10) that contains all of the information about the mutual couplings with the (infinite) environment. The approximation is applied to an array consisting of open-ended waveguide radiators with or without obstructions in the waveguides and with or without dielectric sheets in front of the waveguide apertures. The infinite-array approximation works best for very large array antennas where the majority of the elements experience an environment identical to that of an element in an infinite array. In practice, even arrays consisting of a few tens of elements may be approximated in this way.

Although the material in this chapter dates back to the mid 1990s and a lot of work on this type of array antennas has been performed since [57–60], we find it appropriate to present a 'classic' mode-matching approach. The material here may aid in understanding new developments and may be relatively easy implemented in software for analyzing rectangular waveguide structures and infinite arrays of open-ended waveguides.

Since the different chapters may be read independently, we have opted for a form where conclusions and references are given per chapter. Throughout the book, we indicate vectors by boldface characters, for example, $\boldsymbol{A}$ and $\boldsymbol{b}$. Unit vectors are further denoted by hats, for example, $\hat{\boldsymbol{u}}_x$, $\hat{\boldsymbol{u}}_y$ and $\hat{\boldsymbol{u}}_z$. The dB scale is defined as $10^{10}\log|x|$, where x is a normalized power. The definition $20^{10}\log|x|$ is used when x is a normalized amplitude (electric field, voltage, magnetic field, current, etc.); $20^{10}\log|x| = 10^{10}\log|x^2|$. The natural numbers $\mathbb{N}$

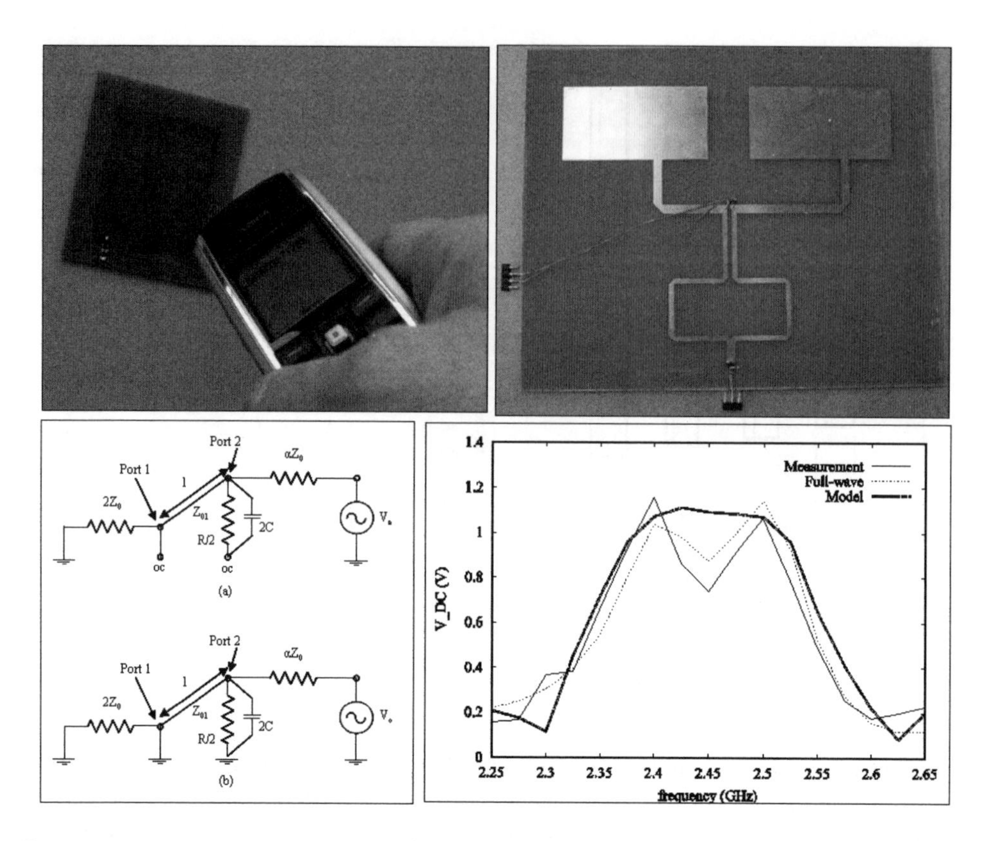

Figure 1.9 Rectennas. Top left: rectenna feeding an LED, wirelessly powered by a GSM phone. Top right: antenna and power-combining network for simultaneously receiving power and data. Bottom left: even–odd mode analysis for power combiner with rectifying element. Bottom right: calculated and measured open-source voltage as a function of frequency across the rectifying element in the power combiner shown in the top right of the figure.

are the set $\{1, 2, 3, \ldots\}$ or $\{0, 1, 2, 3, \ldots\}$. The inclusion of zero is a matter of definition [61]. Here we define $\mathbb{N}$ to include zero. Finally, a superscript number placed after a word indicates a footnote, for example, 'example[1]'.

1.5 SUMMARY

Notwithstanding the progress in numerical electromagnetic analysis, the automated design of integrated antennas based on full-wave analysis is not yet feasible. In a two-stage approach, where stochastic optimization techniques are used in combination with approximate models to generate predesigns and these predesigns are used as input for line search optimization in combination with full-wave modeling, automated antenna design is feasible. Therefore, a need exists for approximate antenna models for different classes of antennas.

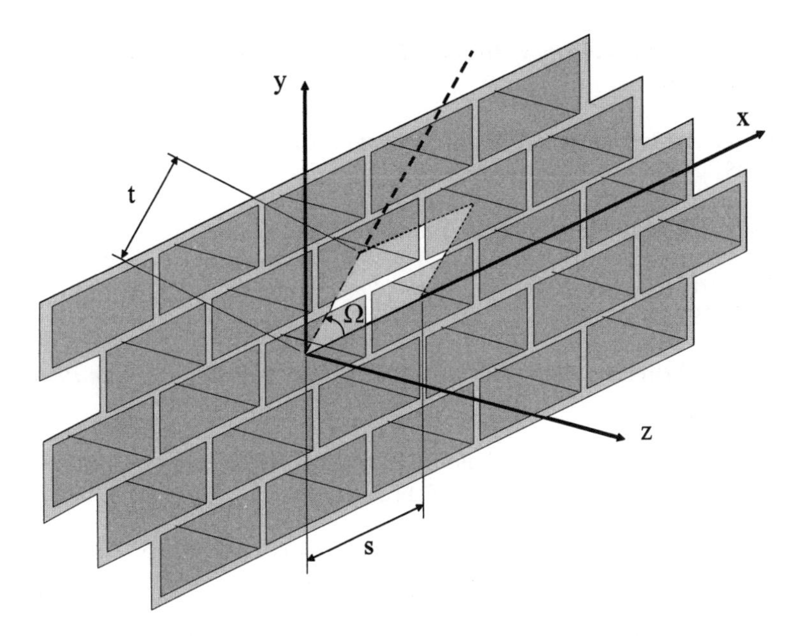

Figure 1.10 Planar, infinite, open-ended waveguide array antenna with the radiators arranged into a triangular grating, plus an indication of a single unit cell.

For one-of-a-kind antenna designs, the iterative, manual use of a full-wave analysis program is advised. So, today, not only are full-wave models needed but also there still exists a need for approximate models. That both full-wave and approximate models are needed cannot be said more eloquently than Ronold W.P. King did in 2004 [62]:

> At this age of powerful computers, there are those who believe that numerical methods have made analytical formulas obsolete. Actually, the two approaches are not mutually exclusive but rather complementary. Numerical methods can provide accurate results within the resolution determined by the size of the subdivisions. Analytical formulas provide unrestricted resolution. Numerical results are a set of numbers for a specific set of parameters and variables. Analytical formulas constitute general relations that exhibit functional relationships among all relevant parameters and variables. They provide the broad insight into the relevant physical phenomena that is the basis of new knowledge. They permit correct frequency and dimensional scaling. Computer technology and mathematical physics are a powerful team in the creation of new knowledge.

REFERENCES

1. J.C. Maxwell, 'A dynamical theory of the electromagnetic field', *Royal Society Transactions*, Vol. 155, p. 459, 1865.

2. J.C. Maxwell, *A Treatise on Electricity and Magnetism*, Dover Publications, New York, 1954.

3. R.S. Elliot, *Electromagnetics: History, Theory and Applications*, John Wiley & Sons, New York, 1999.

4. H.J. Visser, *Array and Phased Array Antenna Basics*, John Wiley & Sons, Chichester, UK, 2005.

5. G. Masini, *Marconi*, Marsilio, New York, 1995.

6. G.C. Corazza, 'Marconi's history', *Proceedings of the IEEE*, Vol. 86, No. 7, pp. 1307–1311, July 1998.

7. G. Marconi, 'Wireless telegraphic communications', *Nobel Lectures in Physics*, 1901–21, Elsevier, 1967.

8. G. Pelosi, S. Selleri and B.A. Valotti, 'Antennae', *IEEE Antennas and Propagation Magazine*, Vol. 42, No. 1, pp. 61–63, February 2000.

9. B.A. Austin, 'Wireless in the Boer War', *100 Years of Radio*, 5–7 September 1995, Conference Publication 411, IEE, pp. 44–50, 1995.

10. A.D. Olver, 'Trends in antenna design over 100 years', *100 Years of Radio*, 5–7 September 1995, Conference Publication 411, IEE, pp. 83–88, 1995.

11. H.C. Pocklington, 'Electrical oscillations in wires', *Proceedings of the Cambridge Philosophical Society*, pp. 324–332, 25 October 1897.

12. J. Hamilton (ed.), *They Made Our World; Five Centuries of Great Scientists and Inventors*, Broadside Books, London, pp. 125–132, 1990.

13. W.F. Crosswell, 'Some aspects of the genesis of radio engineering', *IEEE Antennas and Propagation Magazine*, Vol. 35, No. 6, pp. 29–33, December 1993.

14. J. Ramsay, 'Highlights of antenna history', *IEEE Communications Magazine*, pp. 4–16, September 1981.

15. L.N. Ridenour, *Radar System Engineering*, Vol. 1 of MIT Radiation Laboratory Series, McGraw-Hill, New York, 1947.

16. J.S. Hall, *Radar Aids to Navigation*, Vol. 2 of MIT Radiation Laboratory Series, McGraw-Hill, New York, 1947.

17. A.R. Roberts, *Radar Beacons*, Vol. 3 of MIT Radiation Laboratory Series, McGraw-Hill, New York, 1947.

18. J.A. Pierce, A.A. McKenzie and R.H. Woodward, *Loran*, Vol. 4 of MIT Radiation Laboratory Series, McGraw-Hill, New York, 1948.

19. G.N. Glasoe and J.V. Lebacqz, *Pulse Generators*, Vol. 5 of MIT Radiation Laboratory Series, McGraw-Hill, New York, 1948.

20. G.B. Collins, *Microwave Magnetrons*, Vol. 6 of MIT Radiation Laboratory Series, McGraw-Hill, New York, 1948.

21. D.R. Hamilton, J.K. Knipp and J.B. Horner Kuper, *Klystrons and Microwave Triodes*, Vol. 7 of MIT Radiation Laboratory Series, McGraw-Hill, New York, 1948.

22. C.G. Montgomery, R.H. Dicke and E.M. Purcell, *Principles of Microwave Circuits*, Vol. 8 of MIT Radiation Laboratory Series, McGraw-Hill, New York, 1948.

23. G.L. Ragan, *Microwave Transmission Circuits*, Vol. 9 of MIT Radiation Laboratory Series, McGraw-Hill, New York, 1948.

24. N. Marcuvitz, *Waveguide Handbook*, Vol. 10 of MIT Radiation Laboratory Series, McGraw-Hill, New York, 1951.

25. C.G. Montgomery, *Technique of Microwave Measurements*, Vol. 11 of MIT Radiation Laboratory Series, McGraw-Hill, New York, 1947.

26. S. Silver, *Microwave Antenna Theory and Design*, Vol. 12 of MIT Radiation Laboratory Series, McGraw-Hill, New York, 1949.

27. D.E. Kerr, *Propagation of Short Radio Waves*, Vol. 13 of MIT Radiation Laboratory Series, McGraw-Hill, New York, 1951.

28. L.D. Smullin and C.G. Montgomery, *Microwave Duplexers*, Vol. 14 of MIT Radiation Laboratory Series, McGraw-Hill, New York, 1948.

29. H.C. Torrey and C.A. Whitmer, *Crystal Rectifiers*, Vol. 15 of MIT Radiation Laboratory Series, McGraw-Hill, New York, 1948.

30. R.V. Pound, *Microwave Mixers*, Vol. 16 of MIT Radiation Laboratory Series, McGraw-Hill, New York, 1948.

31. J.F. Blackburn, *Components Handbook*, Vol. 17 of MIT Radiation Laboratory Series, McGraw-Hill, New York, 1949.

32. G.E. Valley Jr. and H. Wallman, *Vacuum Tube Amplifiers*, Vol. 18 of MIT Radiation Laboratory Series, McGraw-Hill, New York, 1948.

33. B. Chance, V. Hughes, E.F. MacNichol Jr., D. Sayre and F.C. Williams, *Waveforms*, Vol. 19 of MIT Radiation Laboratory Series, McGraw-Hill, New York, 1949.

34. B. Chance, R.I. Hulsizer, E.F. MacNichol, Jr. and F.C. Williams, *Electronic Time Measurements*, Vol. 20 of MIT Radiation Laboratory Series, McGraw-Hill, New York, 1949.

35. I.A. Greenwood Jr., J.V. Holdam Jr. and D. MacRae Jr., *Electronic Instruments*, Vol. 21 of MIT Radiation Laboratory Series, McGraw-Hill, New York, 1948.

36. T. Soller, M.A. Star and G.E. Valley Jr., *Cathode Ray Tube Displays*, Vol. 22 of MIT Radiation Laboratory Series, McGraw-Hill, New York, 1948.

37. S.N. Van Voorhis, *Microwave Receivers*, Vol. 23 of MIT Radiation Laboratory Series, McGraw-Hill, New York, 1948.

38. J.L. Lawson and G.E. Uhlenbeck, *Threshold Signals*, Vol. 24 of MIT Radiation Laboratory Series, McGraw-Hill, New York, 1950.

39. H.M. James, N.B. Nichols and R.S. Phillips, *Theory of Servomechanisms*, Vol. 25 of MIT Radiation Laboratory Series, McGraw-Hill, New York, 1947.

40. W.M. Cady, M.B. Karelitz and L.A. Turner, *Radar Scanners and Radomes*, Vol. 26 of MIT Radiation Laboratory Series, McGraw-Hill, New York, 1948.

41. A. Svoboda, *Computing Mechanisms and Linkages*, Vol. 27 of MIT Radiation Laboratory Series, McGraw-Hill, New York, 1948.

42. K. Henney (ed.), *Index*, Vol. 28 of MIT Radiation Laboratory Series, McGraw-Hill, New York, 1953.

43. J. Kraus, *Antennas*, McGraw-Hill, New York, 1950.

44. S.A. Schelkunoff, *Antennas, Theory and Practice*, John Wiley & Sons, London, 1952.

45. R.W.P. King, *Theory of Linear Antennas*, Harvard University Press, Cambridge, MA, 1956.

46. R.S. Elliott, *Antenna Theory and Design*, Prentice Hall, Englewood Cliffs, 1981.

47. C.A. Balanis, *Antenna Theory, Analysis and Design*, John Wiley & Sons, New York, 1982.

48. R.C. Hansen, *Microwave Scanning Antennas*, Academic Press, New York, Vols. 1 and 2, 1964, Vol. 3, 1966.

49. N. Amitay, V. Galindo and C.P. Wu, *Theory and Analysis of Phased Array Antennas*, John Wiley & Sons, New York, 1972.

50. R.J. Mailloux, *Phased Array Antenna Handbook*, Artech House, 1980.

51. J.C. Rautio, 'Planar electromagnetic analysis', *IEEE Microwave Magazine*, pp. 35–41, March 2003.

52. R.F. Harrington, *Field Computation by Moment Methods*, Macmillan, New York, 1986.

53. R. Harrington, 'Origin and developments of the method of moments for field computation', *IEEE Antennas and Propagation Magazine*, Vol. 32, No. 3, pp. 31–35, June 1990.

54. A. Vasylchenko, Y. Schols, W. De Raedt and G.A.E. Vandenbosch, 'A benchmarking of six software packages for full-wave analysis of microstrip antennas', *Proceedings of the 2nd European Conference on Antennas and Propagation, EuCAP2007*, November 2007, Edinburgh, UK.

55. A. Vasylchenko, Y. Schols, W. De Raedt and G.A.E. Vandenbosch, 'Challenges in full wave electromagnetic simulation of very compact planar antennas', *Proceedings of the 2nd European Conference on Antennas and Propagation, EuCAP2007*, November 2007, Edinburgh, UK.

56. A.G. Tijhuis, M.C. van Beurden, B.P. de Hon and H.J. Visser, 'From engineering electromagnetics to electromagnetic engineering: Using computational electromagnetics for synthesis problems', *Elektrik, Turkish Journal of Electrical Engineering and Computer Sciences*, Vol. 16, No. 1, pp. 7–19, 2008.

57. H.J. Visser and M. Guglielmi, 'CAD of waveguide array antennas based on "filter" concepts', *IEEE Transactions on Antennas and Propagation*, Vol. 47, No. 3, pp. 542–548, March 1999.

58. D. Bakers, *Finite Array Antennas: An Eigencurrent Approach*, PhD thesis, Eindhoven University of Technology, 2004.

59. B. Morsink, *Fast Modeling of Electromagnetic Fields for the Design of Phased Array Antennas in Radar Systems*, PhD thesis, Eindhoven University of Technology, 2005.

60. S. Monni, *Frequency Selective Surfaces Integrated with Phased Array Antennas: Analysis and Design Using Multimode Equivalent Networks*, PhD thesis, Eindhoven University of Technology, 2005.

61. T.C. Collocot and A.B. Dobson (eds.), *Dictionary of Science and Technology*. Revised edition, Chambers, Edinburgh, UK, 1982.

62. R.W.P. King, 'A review of analytically determined electric fields and currents induced in the human body when exposed to 50–60-Hz electromagnetic fields', *IEEE Transactions on Antennas and Propagation*, Vol. 52, No. 5, pp. 1186–1191, May 2004.

2

Intravascular MR Antennas: Loops and Solenoids[1]

The rapid developments in the field of (nuclear) magnetic resonance imaging ((N)MRI), especially the fast growth in temporal efficiency, and the development of 'open' MRI systems have contributed significantly to the feasibility of interventional MRI (iMRI). In this context, a need exists for intravascular MR antennas, to be used for either tracking of guide wires and catheters through blood vessels during surgery or for obtaining high-resolution images of vessel walls, images that cannot be obtained by conventional MRI operation. Although various intravascular MR antenna concepts have already been investigated, an electromagnetic model – leading to fast calculations when implemented in a computer code – to quantitatively compare such concepts or even synthesize optimum antennas is needed. An approximate model, based on a quasi-static magnetic-field computation, is developed here and thoroughly compared with exact solutions to assess its range of validity. With the thus verified approximate model, various antenna concepts for tracking and imaging are quantitatively compared and a selection of the 'best' antenna concepts is made. Next, *in vitro* tests are described, confirming the results obtained theoretically. Finally, we describe optimization using a genetic algorithm based on the approximate model, to synthesize antenna designs.

[1]Parts of this chapter are the result of a cooperation between the Electromagnetics Department of the Faculty of Electrical Engineering of Eindhoven University of Technology (TU/e) and the Image Science Institute of the University Medical Center Utrecht (UMC Utrecht), both in The Netherlands. Within this cooperation, two students from TU/e performed MSc thesis projects on intravascular MR antennas at UMC Utrecht, supervised by representatives of both universities.

Approximate Antenna Analysis for CAD Hubregt J. Visser

2.1 INTRODUCTION

This chapter addresses a recent development in medical imaging: *magnetic resonance imaging* (MRI). More specifically, it addresses means to expand the applications of MRI by intravascular collection of measurement data. MRI is one of many medical imaging techniques. Medical imaging (MI) is the process by which parts of the body, not normally visible, are examined and diagnosed, preferably by visualizing those parts.

The best-known imaging technique – skipping the obvious 'tapping, feeling and interpreting' of a physician – is that of radiology, employing *X-rays*. The classical X-ray image, which can show bone fractures and pathological changes in the lungs, is a shadow image resulting from the attenuation of X-ray photons by (parts of) the body. An extension of this technique is found in *computed tomography* (CT) and *computed axial tomography* (CAT). In a CT scan or CAT scan, many X-ray images of a slice of the body are taken from different angles. These X-ray images are then mathematically processed to produce a comprehensive image of the slice. A major disadvantage of these radiology techniques is that the use of ionizing radiation imposes a limit on the image acquisition time, especially for children.

Ultrasound is a widely used, sound-based technique. Waves of high-frequency (2–10 MHz) acoustic energy are radiated into the body. These waves are reflected by tissue to varying degrees, detected by an acoustic transducer and transformed into an image. These images are produced in real time, which is one of the advantages of the technique. Another advantage of ultrasound is that it is safe to use, as ultrasound does not seem to harm the patient. The major drawback is that an ultrasound image shows less detailed information than a CT or CAT scan. The resolution is directly related to the wavelength used. For X-rays, the wavelengths are of the order of 0.01 nm. For ultrasound, the frequencies are of the order of 4.5 Hz, but since the wave velocity is of the order of 1.5×10^3 m s^{-1}, the wavelengths are of the order of 0.3 mm.

In *nuclear medicine imaging* (NMI), a radioactive source is injected into the patient. This radioactive source functions as a tracer and is 'designed' to tag molecules that seek specific sites in the body. A detector is positioned next to or around the patient and the radiation emitted from the body is measured. The technique is very similar to that of a CT or CAT scan, but with the difference that the radiation source is now internal and its distribution is unknown. The two most commonly employed types of NMI are *single-photon-emission computed tomography* (SPECT), which uses radiotracers that emit photons when decaying, and *positron emission tomography* (PET), which uses radiotracers that produce positron–electron pairs. The drawbacks of these techniques are the ones mentioned above for techniques employing ionizing radiation.

Magnetic resonance imaging (MRI) does not employ ionizing radiation. A patient is positioned in a high-intensity static magnetic field. The magnetic field causes the spin-possessing molecules in the body to align their magnetic moments with this field. When a radio frequency (RF) pulse is emitted, causing the main magnetic field to deflect, the molecules absorb energy, which is reradiated after the RF pulse has ceased to exist. This reradiation induces a current in a receiver coil, and this received signal is a measure of the tissue being excited. By applying a position-dependent gradient in the main magnetic field, it is possible to identify the spatial location of reemitted RF energy. As in a CT or CAT scan image, slices of patients are produced, but the image contrast that can be achieved in soft

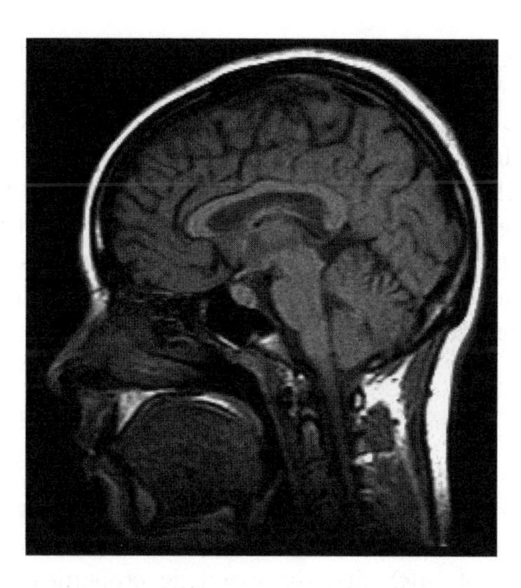

Figure 2.1 MR image of the human brain. Courtesy NASA/JPL-Caltech.

matter by MRI is superior [1]. The radiation involved is nonionizing and roughly in the range 30–120 MHz. An example of an MR image is shown in Figure 2.1.

The fast growth in the temporal efficiency of MRI systems has contributed significantly to the feasibility of interventional MRI (iMRI). In this context, a need exists for intravascular MR antennas, to be used for either tracking of guide wires and catheters through blood vessels during surgery and even for obtaining high-resolution images of vessel walls, images that cannot be obtained by conventional MRI operation. For MRI operation, receiver coils are employed to detect the reradiated RF energy that is absorbed by molecules in the tissue when excited by an external RF pulse. The resolution of an image that can be formed is directly related to the signal-to-noise ratio (SNR). When we wish to obtain detailed information about blood vessel walls, the commonly used receiver coils cannot produce the desired SNR. The employment of local receiver coils instead of surface coils to increase the SNR has been successful, for example, in the diagnosis of prostate cancer (endorectal coils, e.g. [2]) and in the detection of tumours of less than 1 cm^3 volume (endovaginal coils, e.g. [3]).

A logical next step would be the employment of intravascular coils or antennas for detecting areas of stenosis, dissection, aneurysm or other vascular pathology. It should be noted, however, that the use of these intravascular antennas will only have practical value in combination with endovascular intervention, when an arterial puncture has already been made and the risks involved in endovascular intervention have already been assessed as being acceptable.

Before moving on to the topic of intravascular antennas, we shall first give a brief overview of the basics of MRI. Subsequently, we shall present an overview of existing intravascular-antenna concepts for tracking and imaging purposes and we shall compare these concepts in a qualitative way. The development of a static electromagnetic model for intravascular MR antennas will be discussed next. To assess the validity of the static model, comparisons will be made with results obtained from a dynamic, small-loop, uniform-current antenna model. Since our reference is an approximate model, the validity of this approximate model is investigated first. Then, results obtained with the static model are compared with results obtained with the dynamic model for a loop antenna immersed in blood. After the model has been verified for a single-loop antenna, length restrictions on a multiturn loop are derived. After this model has been verified, the antenna concepts will be compared again, but now in a quantitative way. Test results for realized intravascular antennas are the next subject, followed by a discussion of synthesis of intravascular MR antennas. Then, patient safety issues related to the use of intravascular MR antennas are discussed, after which the conclusions of this chapter are presented.

2.2 MRI

MRI, *magnetic resonance imaging*, formerly known as *nuclear magnetic resonance imaging* (NMRI),[2] uses magnetic properties of tissue to create internal anatomical images of people. To understand the basics of MRI, first the magnetic properties of atom nuclei must be understood [1,4].

2.2.1 Magnetic Properties of Atomic Nuclei

An atom may be envisaged as a nucleus consisting of positively charged protons and neutral neutrons, surrounded by negatively charged electrons that travel around the nucleus in orbitals. Every particle possesses a property called *spin*, a (fast) rotational motion around its axis. Any nucleus with either an odd atomic number or an odd atomic weight has a net spin, and we may regard that nucleus as a charged, spinning sphere (Figure 2.2).

Since the nucleus has a net charge, the rotation induces a magnetic field or magnetic *moment*, with an axis that corresponds to the axis of spin, as shown in Figure 2.2. The amplitude of the magnetic moment is μ_{M}.[3] For medical MRI, the most important nucleus that has a net spin is the hydrogen nucleus, or proton, owing to its ample occurrence in the human body.

The magnetic moments of the nuclei in any volume of matter are oriented randomly. When an external static magnetic field $\boldsymbol{B}_0$ is applied, the magnetic moments will tend to align with this external field.[4] The alignment, however, is not perfect. In the presence of an external static magnetic field, the nuclei experience a torque which causes the magnetic moments of these nuclei to rotate around the axis of the external field. This *precession* is analogous to

[2] MRI has its roots in the chemical world, where *nuclear magnetic resonance* (NMR) has become the most important analytical technique for the structural analysis of molecules in solution [1]. NMR *imaging* brought NMR technology into the medical world, and although the word 'nuclear' in NMR has nothing to do with radioactivity, this emotion-laden word has been dropped by the medical community to avoid confusion or fear in (potential) patients.

[3] The subscript M has been introduced to avoid confusion with the symbol μ that is used to represent the electromagnetic permeability.

[4] Some of the nuclei will align with the external field and some will align against it. These alignments correspond to quantum-mechanical energy states, the one aligned with the external field corresponding to a lower energy state. Thus, for a large enough sample, there will be a net alignment *with* the external field. The amplitude of this net alignment is proportional to the field strength.

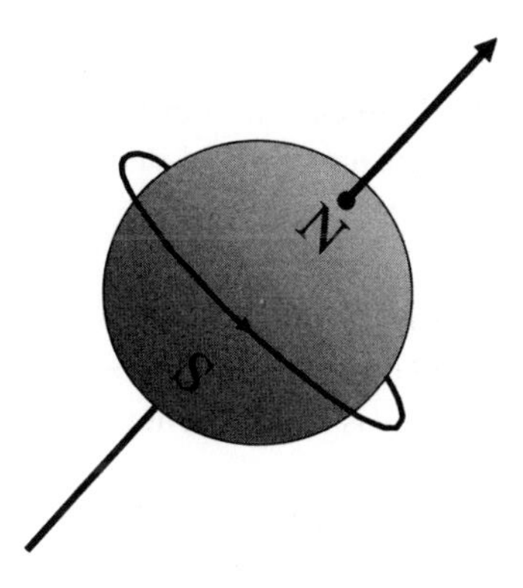

Figure 2.2 Spinning nucleus, where N and S indicate the magnetic north pole and south pole, respectively.

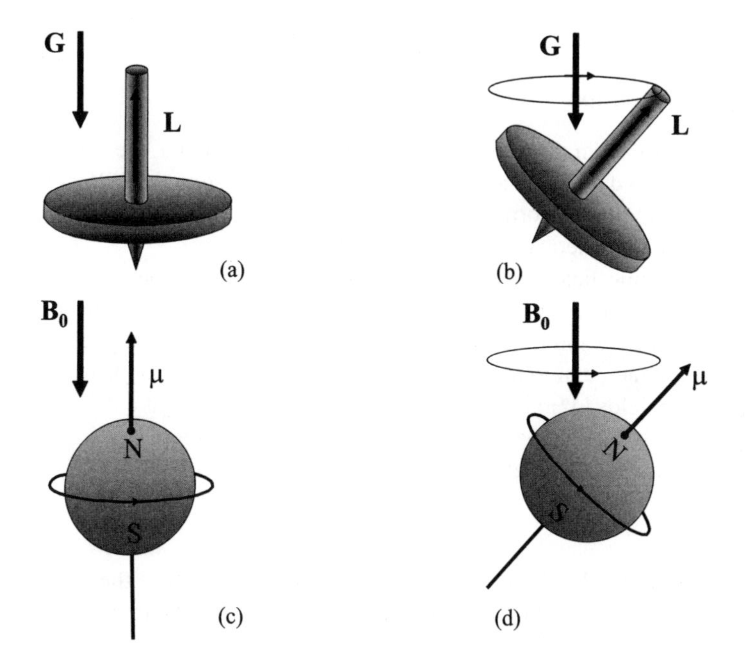

Figure 2.3 Analogy between a spinning top in a gravitational field and a spinning nucleus in an external magnetic field. (a) Spinning top with angular momentum aligned with the gravitational field. (b) Spinning top with angular momentum not aligned with the gravitational field. (c) Spinning nucleus with magnetic moment aligned with the external static magnetic field. (d) Spinning nucleus with magnetic moment not aligned with the external static magnetic field.

the motion of a spinning top with angular momentum L in a gravitational field G, where the spinning top is *not* aligned with the gravitational field (Figure 2.3) [1].

In the situation depicted in Figure 2.3(b), the top precesses around the direction of the gravitational field with an angular frequency ω. For the situation depicted in Figure 2.3(d), the rate of precession is isotope- and magnetic-field-dependent. The precession frequency ω_L is given by (e.g. [1])

$$\omega_L = \gamma B_0, \tag{2.1}$$

where $B_0 = |\boldsymbol{B}_0|$ is the strength of the external static magnetic field (in T), ω_L (in MHz) is known as the *Larmor frequency*, and the *gyromagnetic ratio* γ is a constant for any particular nuclear isotope. The Larmor frequency is the frequency at which atomic nuclei respond when interrogated by RF radiation.

So far, we have described the MR process on the atomic or microscopic level. On a macroscopic level, we only have to deal with the net results. So, on a macroscopic level, we observe – at equilibrium[5] – a net magnetization aligned with the external static magnetic field. Let us assume that the direction of the external static field is the z direction of a Cartesian coordinate system. At equilibrium, the net magnetization is $\boldsymbol{M} = M_z\hat{\boldsymbol{u}}_z$, where $\hat{\boldsymbol{u}}_z$ is the unit vector in the z direction.

2.2.2 Signal Detection

To record a response from the net magnetic moment of the nuclei, the static magnetic field $M_z\hat{\boldsymbol{u}}_z$ is distorted by a dynamic magnetic field $\boldsymbol{M}_T$ with a direction that differs from that of the static field. Owing to this 'distortion' component, the nuclei will precess around the direction of the newly formed magnetization $\boldsymbol{M}$ (Figure 2.4).

When the dynamic field ceases to exist, the net magnetization will be restored. This change in magnetic field, from the deflected field back to the z-directed magnetic field, may be recorded by virtue of induced currents in RF receiver coils. These receiver coils, which are positioned perpendicular to the direction of the static magnetic field, are sensitive only to dynamic magnetic fields ($\boldsymbol{M}_T$) in the transverse plane shown in Figure 2.4.

The angle ϑ between the net magnetization $\boldsymbol{M}$ and the main field $\boldsymbol{B}_0$, known as the *RF flip angle* or *RF pulse angle*, is given by [1]

$$\vartheta = \gamma B_1 t, \tag{2.2}$$

where B_1 is the amplitude of the disturbing magnetic field and t is the duration of the RF pulse, i.e. the time for which the field $\boldsymbol{M}_T$ has been turned on. Maximum signal reception is achieved for $\vartheta = \pi/2$.

When looking at a single nucleus, we see – at equilibrium – the magnetic moment precessing around the direction of the main magnetic field. On a macroscopic scale, we see a magnetization vector directed parallel to the main magnetic field; the individual transverse components cancel each other. When a distorting field (RF pulse) $\boldsymbol{M}_T$ is now applied, deflecting the main magnetic field to $\vartheta = \pi/2$, we see the precession axis move from the

[5]For materials with atomic nuclei that possess a property called *spin*, this spin makes the nuclei behave as small magnets. Applying a strong external magnetic field to these nuclei results in the precessing of the magnetic moments of the nuclei around the direction of the external field and thus in the forming of a *net* magnetization in the same direction as that of the external field. It takes a time denoted by T_1 to develop this steady-state net magnetization.

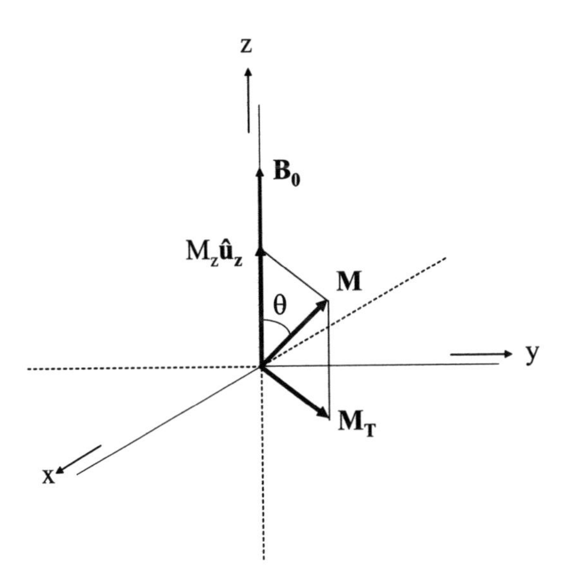

Figure 2.4 Magnetization.

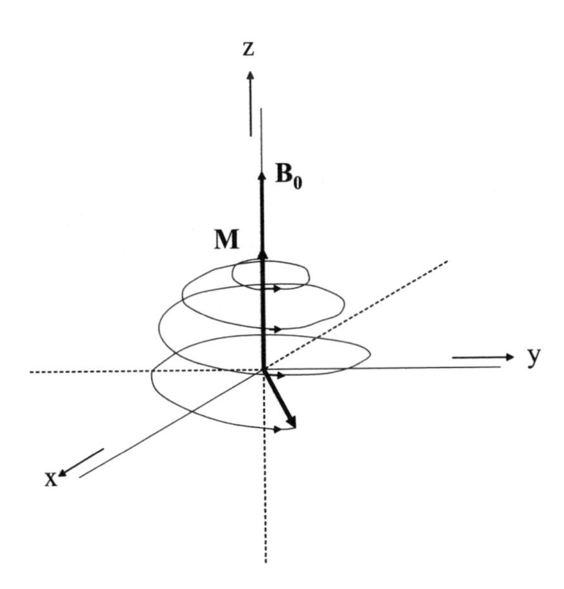

Figure 2.5 Precession of net magnetization during an RF pulse having a flip angle $\vartheta = \pi/2$.

z direction to a transverse direction and, to an observer in the external *laboratory frame* of reference, the magnetization vector spirals down – the precession axis following the net magnetization – towards the *xy* plane (Figure 2.5). In a *rotating frame* of reference, the net magnetization vector simply rotates from the *z* direction to the transverse direction.

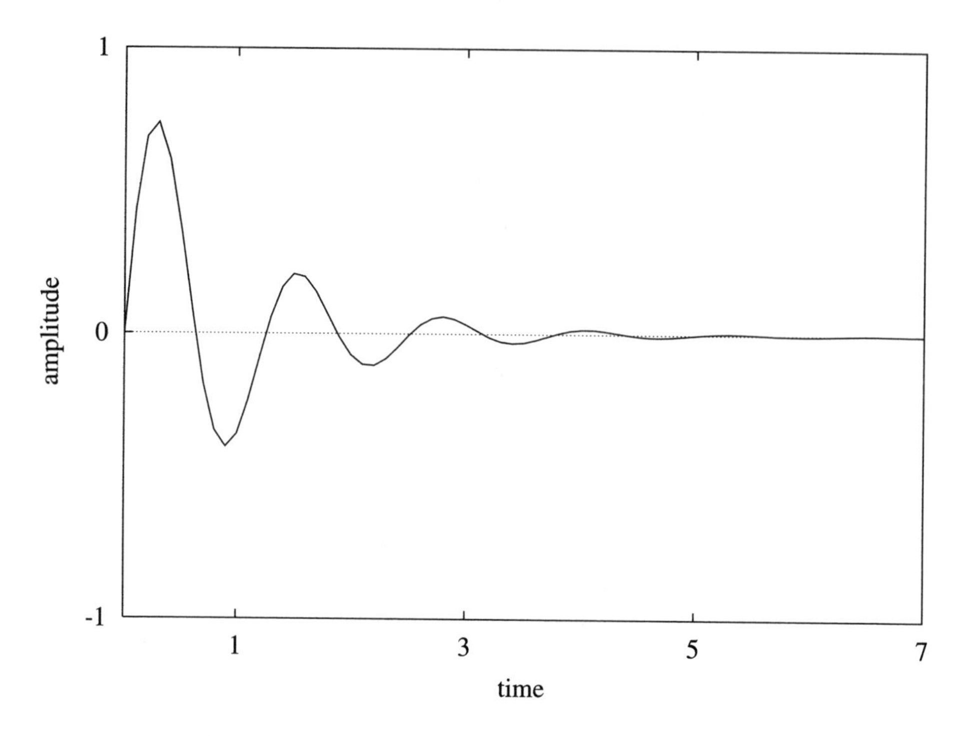

Figure 2.6 FID signal. Time and amplitude normalized.

When the RF pulse (at the Larmor frequency) has been transmitted, the RF energy absorbed by the protons (having made them jump to higher energy states) is retransmitted (again at the Larmor frequency). The magnetization starts to return to equilibrium and the protons begin to dephase.

The recovery of the magnetization to its thermodynamic equilibrium value M_0 with time is described by [4]

$$M_z = M_0(1 - \mathrm{e}^{-t/T_1}). \tag{2.3}$$

The time constant T_1 is called the *spin–lattice relaxation time*.[6] Spin–lattice relaxation is the process whereby the energy absorbed by excited protons or spins is released back into the surrounding lattice, reestablishing thermodynamic equilibrium.

The dephasing is the result of proton–magnetic-field interactions, also known as *spin–spin interactions*. The result of the dephasing is a decay in the magnitude of the transverse component of the net magnetization. The return of the transverse magnetization M_{xy} to its equilibrium value M_{xy_0} with time is described by [4]

$$M_{xy} = M_{xy_0}\mathrm{e}^{-t/T_2}. \tag{2.4}$$

[6]The exponential recovery is characterized by a time constant T_1, at which 63.2% of the magnetization has recovered its alignment with the main magnetic field. The value of T_1 is unique to every tissue.

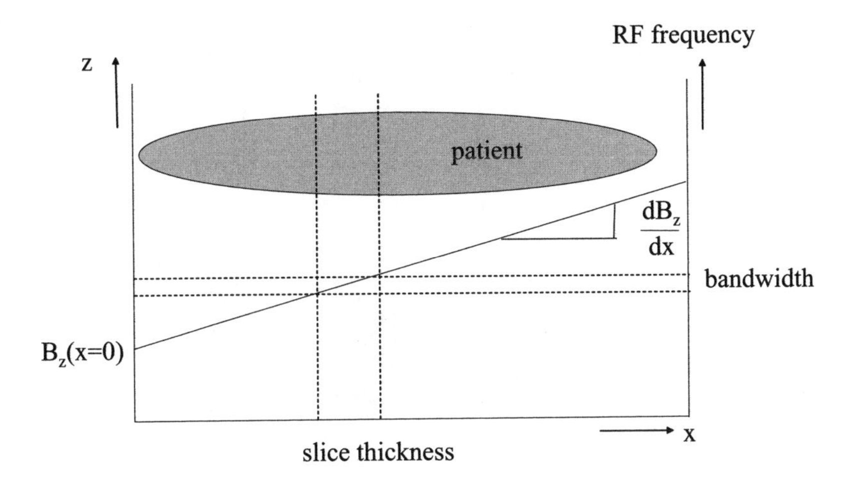

Figure 2.7 Magnetic field gradient for slice-selective excitation. The figure shows, schematically, the projection of a patient on the xz plane. When an x-dependent magnetic field gradient is added to the static magnetic field, the Larmor frequency becomes linearly dependent on x. Therefore, every frequency bandwidth (see the right vertical axis) selected in the received signal corresponds to a 'slice' of the patient. By choosing the central frequency, the position of the slice can be selected. The slice thickness may be decreased by selecting a smaller frequency bandwidth.

The time constant T_2 is known as the *spin–spin relaxation time*.[7] Spin–spin relaxation is a temporary, random interaction between two excited spins that causes a cumulative loss in phase, resulting in an overall loss of signal. In the absence of any gradient in the main magnetic field, the received MR signal (Figure 2.6), is known as the *free induction decay* (FID). The oscillation of the FID signal is due to the Larmor precession of the net magnetization around $\boldsymbol{B}_0$.

Since the Larmor frequency is directly related to the strength of the main magnetic field, adding a field gradient that depends on a direction orthogonal to the main field direction opens up the possibility to select slices of the patient to be imaged (Figure 2.7). Every slice will now return signals that correspond to a different Larmor frequency. The intensity will give information about the concentration of protons.

2.3 INTRAVASCULAR MR ANTENNAS

The application of intravascular receiver coils or, put more generally, intravascular antennas, will lead to an increase in SNR compared with the use of surface receiver coils. Intravascular antennas can be placed in close proximity to the specific target tissue, which results in a

[7]The value of T_2 is the time after excitation when the signal amplitude has been reduced to 36.8% of its original value. The value of T_2 is unique to every tissue.

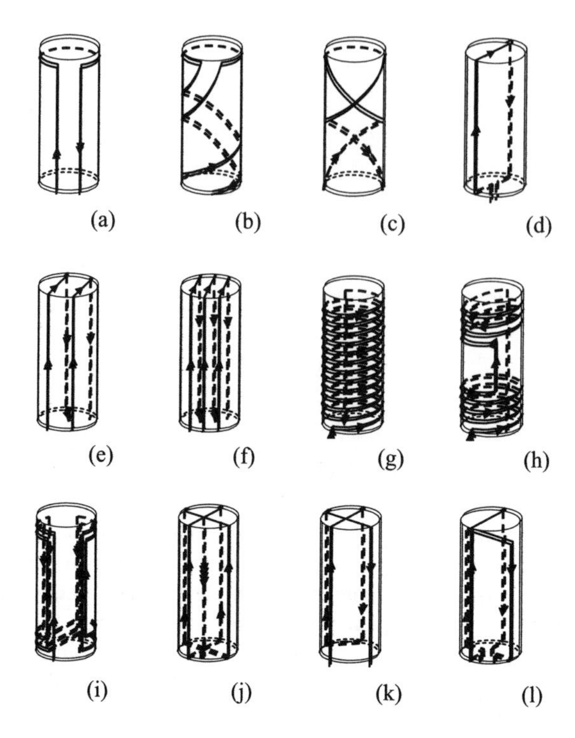

Figure 2.8 Intravascular-antenna concepts. (a) Antiparallel wires [5]. (b) Double-helix wire [5]. (c) Opposed double-helix wire [5]. (d) Single loop [6]. (e) Double loop. (f) Triple loop. (g) Solenoid [7]. (h) Dual-opposed solenoids [7]. (i) Saddle coil. (j) Four-wire center return. (k) Four-wire birdcage. (l) Quadrature coil [8].

considerable reduction in the amount of received noise. An intravascular antenna can be used for imaging artery walls and may also be employed for tracking purposes, i.e. locating the position and/or orientation of a catheter or catheter tip.

2.3.1 Antenna Designs for Tracking

In Figure 2.8, we show some intravascular-antenna concepts reported in the literature. Some of these antenna concepts are more suited for imaging purposes, and some are more suited for tracking purposes.

For tracking purposes, the intravascular antenna needs to aid in visualizing the catheter. This may be accomplished semiactively [9], using resonant antennas to locally add gain to the RF magnetic field, or actively, where the antenna is detuned during the RF pulse with external circuitry. The antenna may, for example, be a loop mounted along the *complete length* of the catheter, used to induce locally, along the catheter, a magnetic-field distortion. Examples of such antennas are the antiparallel-wire antenna (Figure 2.8(a)), the double-helix wire antenna (Figure 2.8(b)) and the opposed-double-helix wire antenna (Figure 2.8(c)).

Table 2.1 Qualitative comparison of intravascular-antenna concepts for tracking [9, 12].

Concept	Mechanics	Signal sensitivity	Orientation	Safety
Resonant antenna	+	++	−	−
Antiparallel wires	+	+	−−	−
Double helix	+	+	++	−
Opposed double helix	+	++	++	−
Guide wire	+	+/−	−	−
Three dual-opposed solenoids	+	++	−	−
Perpendicular coils	+	++	++	−

The main magnetic-field distortion gives rise to a local dephasing, which will be visible as a deviation in the MRI image and thus acts as a position indicator for the catheter.

The *tip* of a catheter may be detected by placing a small resonant antenna or coil at the tip of the catheter and using the detected MR frequency to steer the three orthogonal main-field gradients to locally code the Larmor frequency of the protons. Thus the catheter tip *location* can be determined in three dimensions and may be projected onto MRI images [10]. If the catheter *orientation* needs to be determined as well, the single-coil antenna may be replaced by multiple coils. This provides multiple high-signal locations. Instead of using the 'along-the-catheter loop' antenna mentioned above for detecting the catheter, the guide wire may also be used as a dipole antenna [11].

On the basis of intravascular-antenna results reported in the open literature, a qualitative comparison of antenna concepts for imaging is given in Table 2.1 [9, 12]. Passive tracking methods such as using contrast agents or adding magnetic rings to the catheter are not considered here. These methods lack the possibility of dynamically compensating for signal loss for different catheter orientations as is possible with the application of intravascular antennas.

First of all, the table shows that none of the antenna concepts is safe. An intravascular MR antenna is safe when it does not present an additional risk to the patient. The largest risk is presented not by the antenna itself, but by its electrical leads. Leads that have a length equal to or longer than half a wavelength (in the surrounding medium) may act as linear antennas and become resonant. This may result in heating of (parts of) the leads to temperature in excess of 70°C [10]. At such levels, the surrounding tissue will be destroyed. In section 2.8, we shall address safety issues more thoroughly. Furthermore, the table indicates that the antenna concepts to be studied in more detail are the *opposed-double-helix antenna*, the *three-dual-opposed-solenoids antenna* (for determining catheter *orientation*) and the *perpendicular-coils antenna*. The double-helix antenna will not be considered for further investigation, since it is outperformed by the opposed-double-helix antenna. The resonant antenna will not be considered, since this antenna is part of a semiactive tracking system that has the same drawbacks as these mentioned for passive tracking methods.

The perpendicular-coils antenna needs some explanation. Originally developed as a fiducial marker [13], i.e. not electrically connected to the MRI scanner hardware, but applied here as an antenna for active tracking [12], the perpendicular-coils antenna consists of two coils, wound on top of each other, the first coil making an angle α with the catheter axis, and the second making an angle $\pi - \alpha$ with the catheter axis (Figure 2.9). When the angle α is

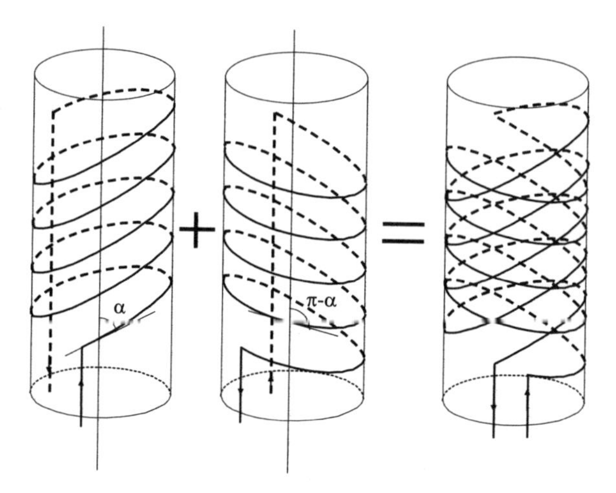

Figure 2.9 Perpendicular-coils antenna developed from two skewed coils.

chosen to be $\pi/4$, the two skewed coils are perpendicular to each other. In this configuration, the net magnetic field is concentrated in the center of the two orthogonal coils and is directed in the radial direction.

2.3.2 Antenna Designs for Imaging

The main requirements on an antenna for arterial-wall imaging are a high sensitivity outside the catheter boundaries up to and beyond the artery wall, a radially homogeneous sensitivity pattern and, preferably, a large longitudinal coverage for multislice imaging. On the basis of results reported in the open literature, a qualitative comparison of antenna concepts for imaging is given in Table 2.2 [9]. In this table, *mechanics* stands for size, rigidity and complexity, *orientation* stands for sensitivity to the antenna orientation and *safety* relates to the length of the antenna.

In selecting antenna concepts worthwhile to be further investigated, we shall start by omitting all candidates that show a double minus sign in one or more column entries. That leaves us with the *double-loop antenna*, the *triple-loop antenna* and the *dual-opposed-solenoids antenna* to be evaluated in more detail. Although the *saddle-coil antenna* must be discarded owing to its complexity, we shall still investigate this concept, assuming a feasible construction method.

2.4 MR ANTENNA MODEL

To compare the different antenna concepts without actually constructing prototypes and performing *in vitro* and *in vivo* experiments[8] with an MRI scanner, the availability of an

[8] *In vitro* – Latin for 'in glass' – is an experimental technique where the experiments are performed outside a living organism. Here, it means that experiments are performed within a *phantom*. *In vivo* – Latin for 'in the living' – indicates that the experiments are performed in the presence of a living organism.

Table 2.2 Qualitative comparison of intravascular-antenna concepts for imaging [9].

Concept	Mechanics	Radial sensitivity	Axial sensitivity	Orientation	Safety
Single loop	+	−	++	−−	+/−
Single loop Multiturn	−	−	+/−	−	+/−
Double loop	+	+	++	−	+/−
Triple loop	−	+	++	−	+/−
Dipole	−	+/−	++	−	−−
Twin lead	++	+	+	−	−−
Dual opposed solenoids	++	++	+/−	−	−
Saddle coil	−−	+	++	−	−
Center return	−−	+	−−	−	−
Birdcage	−−	+	−−	−	−
Quadrature coil	−−	−	++	−	−

electromagnetic model for calculating the fields is desirable. To this end, commercial-off-the-shelf (COTS), three-dimensional, full-wave electromagnetic analysis software may be applied, for example [14] (finite element method), [15] (transmission line matrix method) and [16] (method of moments). To obtain analysis results very rapidly and possibly optimize antenna designs through repeated analyses, however, we prefer to develop an approximate model; this is feasible for antennas that are small compared with the wavelength [16].

Using equation (2.1) for protons (^{1}H), which have a gyromagnetic ratio of 42.58 MHz T^{-1} [1], an MRI scanner that produces a 1.5 T strong static magnetic field [9, 12], gives us a Larmor frequency of 63.87 MHz. Owing to the field gradient applied, the frequency will vary around this value and, for convenience, we shall therefore assume, from now on, a resonance frequency $f = 64$ MHz.

The medium that surrounds the intravascular antenna will be mainly blood, which is characterized by a relative permeability $\mu_r \approx 1$, a relative permittivity $\varepsilon_r \approx 80$ and a conductance $\sigma \approx 8$ S m^{-1} at a temperature of 37°C [17, 18]. The wavelength may thus be calculated as

$$\lambda = \frac{c_0}{\sqrt{\mu_r \varepsilon_r}} \frac{1}{f} = \frac{1}{f\sqrt{\mu_0 \mu_r \varepsilon_0 \varepsilon_r}} = 0.52 \text{ m}. \tag{2.5}$$

The large blood vessels for which MRI antennas are needed have a diameter between 4 and 6 mm [9, 18]. Therefore an antenna diameter of about 2 mm is anticipated, allowing the catheter to be maneuvered through the vascular system and preventing complete blood flow blockage.[9] The far field for a small antenna [19] starts at a distance of $\lambda/2\pi = 82.8$ mm from the antenna. With the stated dimensions of intravascular antennas and vessels, the vessel wall position will be in the near field of the antenna.

[9]The first prototypes, used for *in vitro* experiments, were made a little larger. A diameter of 4 mm was dictated by the materials and construction facilities available at the time [9]. Later, *in vitro* experiments dedicated to intravascular antennas for tracking purposes were performed with antennas positioned on 5F catheters, which have a diameter of 1.67 mm [12].

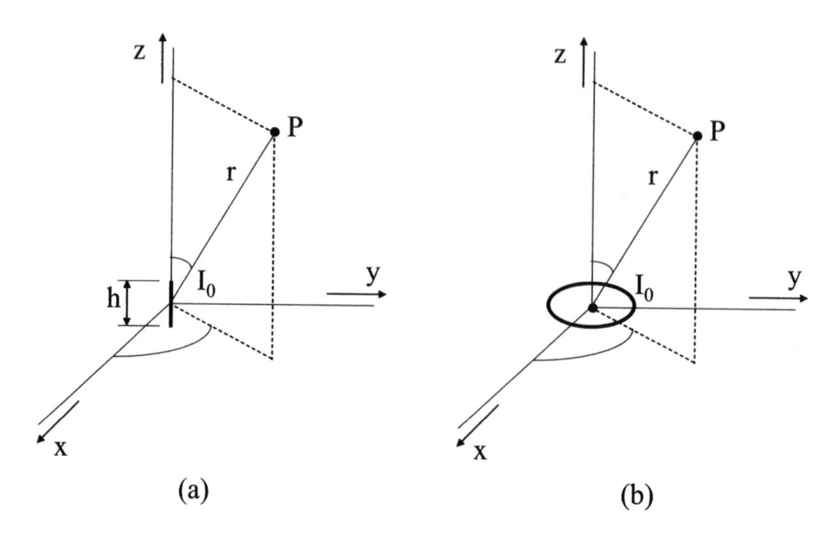

Figure 2.10 Elementary radiators. (a) Elementary, or Hertzian, dipole. (b) Elementary loop, or magnetic dipole.

The radiated fields of an elementary, or Hertzian, dipole[10] at an observation position $P(r, \vartheta, \varphi)$ (Figure 2.10(a)) may be calculated as follows [20]:

$$H_{\varphi} = \mathrm{j}\frac{kI_0 h \sin(\vartheta)}{4\pi r}\left(1 + \frac{1}{\mathrm{j}kr}\right)\mathrm{e}^{-\mathrm{j}kr}, \tag{2.6}$$

$$E_r = \eta\frac{I_0 h \cos(\vartheta)}{2\pi r^2}\left(1 + \frac{1}{\mathrm{j}kr}\right)\mathrm{e}^{-\mathrm{j}kr}, \tag{2.7}$$

$$E_{\vartheta} = \mathrm{j}\eta\frac{kI_0 h \sin(\vartheta)}{4\pi r}\left(1 + \frac{1}{\mathrm{j}kr} - \frac{1}{(kr)^2}\right)\mathrm{e}^{-\mathrm{j}kr}, \tag{2.8}$$

$$H_r = H_{\vartheta} = E_{\varphi} = 0, \tag{2.9}$$

where I_0 is the amplitude of the uniform current, h is the length of the dipole, $\omega = 2\pi f$ is the radial frequency, η is the characteristic impedance of free space, $\varepsilon = \varepsilon_0\varepsilon_{\mathrm{r}}$, $\mu = \mu_0\mu_{\mathrm{r}}$ and k is the wave number.

The radiated fields of an elementary magnetic dipole[11] at an observation position $P(r, \vartheta, \varphi)$ (Figure 2.10(b)), may be obtained from the results for an elementary dipole by

[10] A Hertzian dipole is a dipole antenna with a length much smaller than the wavelength, in fact so small that the current may be considered to be uniform over the length.

[11] An elementary magnetic dipole can be realized as an electric-current loop with a circumference that is so much smaller than the wavelength that the current may be considered to be uniform over the loop.

virtue of duality[12] [20], or may be calculated as follows [21]:

$$E_\varphi = \eta \frac{I_0 (ka)^2 \sin(\vartheta)}{4r} \left(1 + \frac{1}{\mathrm{j}kr}\right) \mathrm{e}^{-\mathrm{j}kr}, \tag{2.10}$$

$$H_r = \mathrm{j} \frac{I_0 k a^2 \cos(\vartheta)}{2r^2} \left(1 + \frac{1}{\mathrm{j}kr}\right) \mathrm{e}^{-\mathrm{j}kr}, \tag{2.11}$$

$$H_\vartheta = -\frac{I_0 (ka)^2 \sin(\vartheta)}{4r} \left(1 + \frac{1}{\mathrm{j}kr} - \frac{1}{(kr)^2}\right) \mathrm{e}^{-\mathrm{j}kr}, \tag{2.12}$$

$$E_r = E_\vartheta = H_\varphi = 0, \tag{2.13}$$

where a is the radius of the loop.

Far away from these elementary antennas, the r^{-1} terms dominate; very close to the antenna, the r^{-3} terms are dominant. In between, the r^{-2} terms are dominant. If, for a very small but not elementary antenna, we can identify a region where the r^{-2} terms are clearly dominant and if the artery wall is in this region, then it is likely that we may be able to approximate the magnetic field in this region of interest by a quasi-static magnetic field. The *Biot–Savart law* [20] tells us that the static magnetic field produced by a steady current I_0 shows an $I_0 r^{-2}$ dependence. Upon inspection of equations (2.8)–(2.11), we see that for dominating r^{-2} terms, the dynamic magnetic field also shows an $I_0 r^{-2}$ dependence.

To verify this hypothesis, we shall calculate the magnetic field of a small loop antenna of radius a by employing equations (2.9) and (2.10) and compare these results with those obtained by applying the Biot–Savart law to a direct current in a loop. The loop will be approximated by a finite number of straight wire segments. We have specifically chosen a small loop antenna, since most concepts for intravascular antennas that have been demonstrated are based upon small loops.

Before we start this comparison, we first need to verify our reference, i.e. the small-loop approximation, based on a constant direct current, which gave rise to equations (2.10)–(2.13). In [22] it was shown that, first of all, the small-loop approximations resulting in equations (2.10)–(2.13) may be obtained as a limiting case of general exact series representations for a uniform current loop. Secondly, it was demonstrated in [22] that the field component H_ϑ in the plane of the loop, at a distance of half a wavelength from the loop center, as calculated by equation (2.12), is less than 5% in error compared with the exact solution for a uniform current for loop radii up to 0.11λ and less than 10% in error for loop radii up to 0.15λ. For practical purposes, therefore, we need to find up to what radius a loop antenna may be regarded as supporting a uniform current. In [23], it was demonstrated that for loop circumferences larger than a wavelength, the current may not be treated as uniform. To find limiting values for the loop radius, we shall look at the admittance of a loop.

[12] The concept of duality states that exchanging $\boldsymbol{H}$ for $\boldsymbol{E}$, $\boldsymbol{E}$ for $\boldsymbol{H}$, μ for ε and ε for μ leaves Maxwell's curl equations for source-free regions unchanged. Thus solutions for a problem with an electric source can be adapted to problems with a magnetic source [20].

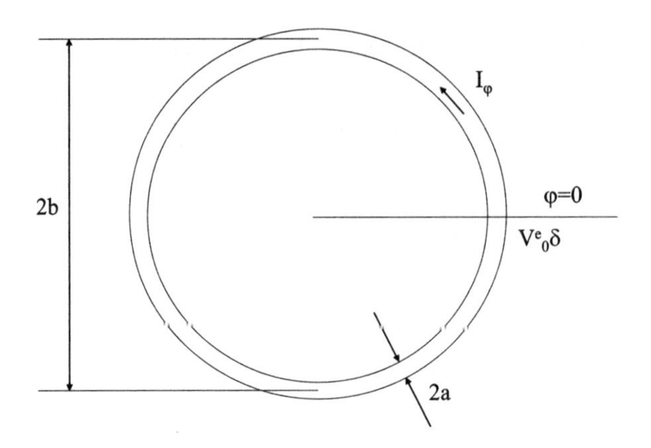

Figure 2.11 Circular loop antenna.

2.4.1 Admittance of a Loop

We start by considering a loop in air. The loop has a radius b and is made of a wire having a circular cross section defined by a radius a. The loop is excited at $\varphi = 0$ with a voltage delta-gap generator $V_0^{\mathrm{e}}\delta(\varphi)$ and carries a current I_φ (Figure 2.11).

Expanding the current in a Fourier series [24–26],

$$I_\varphi = -\mathrm{j}\frac{V_0^{\mathrm{e}}}{\eta_0\pi}\left[\frac{1}{a_0} + 2\sum_{n=1}^{\infty}\frac{\cos(n\varphi)}{a_n}\right], \tag{2.14}$$

results in an input impedance

$$Y_{\text{in}} = \frac{I(0)}{V_0^{\mathrm{e}}} = -\mathrm{j}\frac{1}{\eta_0\pi}\left[\frac{1}{a_0} + 2\sum_{n=1}^{\infty}\frac{1}{a_n}\right], \tag{2.15}$$

where $\eta_0 = \sqrt{\mu_0/\varepsilon_0}$ is the characteristic impedance of free space and

$$a_n = \frac{k_0 b}{2}(F_{n+1} + F_{n-1}) - \frac{n^2}{k_0 b}F_n. \tag{2.16}$$

Here $k_0 = \omega\sqrt{\varepsilon_0\mu_0}$ is the free-space wave number,

$$F_0 = \frac{1}{\pi}\ln\left(\frac{8b}{a}\right) - \frac{1}{2}\int_0^{2k_0 b}[\Omega_0(x) + \mathrm{j}J_0(x)]\,dx \tag{2.17}$$

and

$$F_n = F_{-n} = \frac{1}{\pi}\left[K_0\left(\frac{na}{b}\right)I_0\left(\frac{na}{b}\right) + C_n\right] - \frac{1}{2}\int_0^{2k_0 b}[\Omega_{2n}(x) + \mathrm{j}J_{2n}(x)]\,dx,\ n \neq 0. \tag{2.18}$$

In the above, $J_0(x)$ is the Bessel function of the first kind and order zero with argument x, $I_0(x)$ is the modified Bessel function of the first kind and order zero with argument x, $K_0(x)$

is the modified Bessel function of the second kind and order zero with argument x,

$$C_n = \gamma - 2\sum_{m=0}^{n-1}(2m+1)^{-1} + \ln(4), \tag{2.19}$$

where $\gamma = 0.5772\ldots$ is Euler's constant, and

$$\Omega_m(x) = \frac{1}{\pi}\int_0^{\pi} \sin[x\sin(\vartheta) - m\vartheta]\, d\vartheta, \tag{2.20}$$

is the Lommel–Weber function of order m with argument x.

For a loop antenna immersed in a dissipative medium (such as blood), the equations stated above still apply, but we need to replace ε_0 by $\varepsilon - \mathrm{j}\sigma/\omega$, μ_0 by μ and k_0 by k [24–26], where

$$k = \beta - \mathrm{j}\alpha = \omega\sqrt{\mu\varepsilon}\sqrt{1-\mathrm{j}p}. \tag{2.21}$$

Here $p = \sigma/\omega\varepsilon$ and

$$\sqrt{1-\mathrm{j}p} = \cosh\left(\frac{1}{2}\sinh^{-1}(p)\right) - \mathrm{j}\sinh\left(\frac{1}{2}\sinh^{-1}(p)\right). \tag{2.22}$$

Furthermore, $\varepsilon = \varepsilon_0\varepsilon_r$, where, for blood and at a frequency of 64 MHz, $\varepsilon_r \approx 80$ [17, 18] and $\sigma \approx 8\ \mathrm{S\,m^{-1}}$ [17, 18]. The input admittance is then found as

$$Y_{\mathrm{in}} = -\mathrm{j}\Delta\frac{(1-\mathrm{j}\alpha/\beta)}{\pi\eta_0}\left[\frac{1}{a_0} + 2\sum_{n=1}^{\infty}\frac{1}{a_n}\right], \tag{2.23}$$

where

$$\alpha = \omega\sqrt{\mu\varepsilon}\sinh\left[\frac{1}{2}\sinh^{-1}(p)\right], \tag{2.24}$$

$$\beta = \omega\sqrt{\mu\varepsilon}\cosh\left[\frac{1}{2}\sinh^{-1}(p)\right], \tag{2.25}$$

$$\Delta = \sqrt{\frac{\varepsilon_r}{\mu_r}}\cosh\left[\frac{1}{2}\sinh^{-1}(p)\right] \tag{2.26}$$

and

$$p = \frac{\sigma}{\omega\varepsilon}. \tag{2.27}$$

With the use of equations (2.15)–(2.20), the input impedance of a loop antenna in air with a thickness parameter $\Omega = 2\ln(2\pi b/a) = 10$ has been calculated as a function of the loop radius expressed in wavelengths. For the analysis, 20 Fourier terms were used. As shown in [25], that number of Fourier terms leads to convergent impedance values for loop radii satisfying $\beta b \leq 0.5$, or $b \leq 0.08\lambda$. The results for the resistance are shown in Figure 2.12, and the results for the modulus of the reactance are shown in Figure 2.13. The input impedance of a uniform current loop antenna as a function of the loop radius was also calculated using the same equations, but taking only the a_0^{-1} term into account. The results for this small-loop

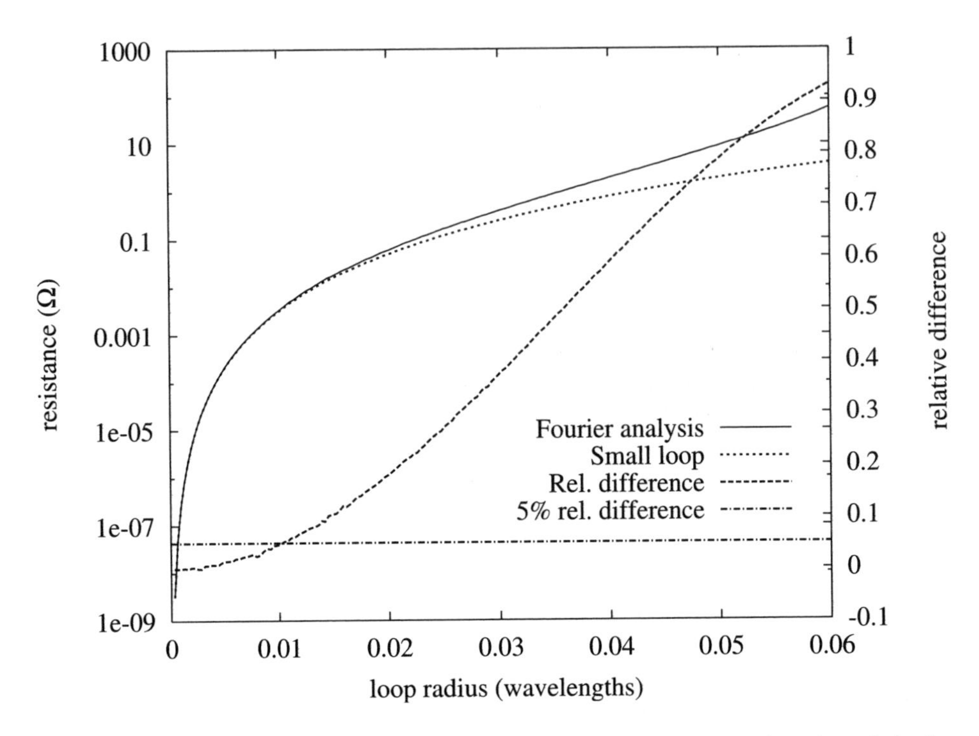

Figure 2.12 Real part of the input impedance of a loop antenna as a function of the loop radius.

approximation are shown in the same two figures. The figures agree with the results presented in [19], both for the 20-term Fourier analysis and for the small-loop approximation.

In Figures 2.12 and 2.13, the relative difference between the 20-term Fourier analysis and the small-loop approximation is also shown as a function of loop radius. Accepting a 5% difference between the approximate and exact results and also taking account of the fact that the reactance dominates over the resistance leads to the commonly quoted rule of thumb, [19, 21,27], that the circumference of a loop antenna should be smaller than approximately a tenth of a wavelength for the uniform-current approximation to be valid.

Having thus established the validity of the analysis methods in air, we took the final step of immersing the loop antenna in a dissipative medium, more specifically blood, for which we have used $\varepsilon_r = 80$ and $\sigma = 8$. Applying equation (2.23) would result in replacing the real integration limits in equations (2.17) and (2.18) by complex ones and having to deal with Bessel and Weber–Lommel functions of complex arguments. Since solving this problem would be beyond the scope of establishing the validity of our reference, we shall follow [28] and use a power series expansion for a small loop antenna.

Taking only the first two terms in equation (2.23) into account, developing power series for the functions F_n (equations (2.17) and (2.18)) for complex $k = \beta(1 - j\alpha/\beta)$ and using

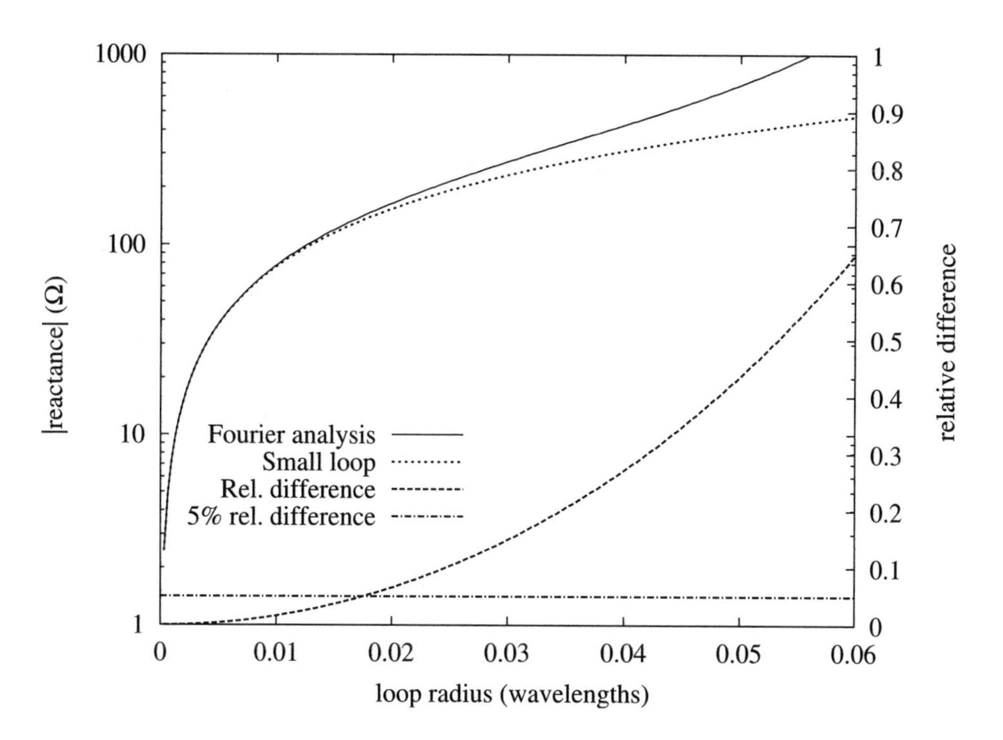

Figure 2.13 Imaginary part of the input impedance of a loop antenna as a function of the loop radius.

only the first few terms of these power series leads to [28]

$$
\begin{aligned}
Y_{\text{in}} &= -\mathrm{j}\frac{\Delta(1-\mathrm{j}\alpha/\beta)}{\pi\eta_0}\frac{1}{kb}\left[\frac{1}{a_0}+\frac{2}{a_1}\right] \\
&= -\mathrm{j}\Delta\frac{1}{120\pi^2}\frac{1}{\beta b}\left[\frac{1}{F_1}+\frac{4}{F_0+F_2-(2)/((kb)^2)F_1}\right],
\end{aligned}
\tag{2.28}
$$

where

$$
F_0 = \frac{1}{\pi}\ln\left(\frac{8b}{a}\right)-\frac{1}{\pi}\left[2(kb)^2-\frac{4}{9}(kb)^4+\frac{32}{675}(kb)^6\right]-\mathrm{j}\left[kb-\frac{1}{3}(kb)^3+\frac{1}{20}(kb)^5\right],
\tag{2.29}
$$

$$
F_1 = \frac{1}{\pi}\ln\left(\frac{8b}{a}\right)-\frac{1}{\pi}\left[2-\frac{2}{3}(kb)^2+\frac{4}{15}(kb)^4-\frac{160}{4725}(kb)^6\right]-\mathrm{j}\left[\frac{1}{6}(kb)^3-\frac{1}{30}(kb)^5\right],
\tag{2.30}
$$

$$
\begin{aligned}
F_2 = {}&\frac{1}{\pi}\ln\left(\frac{8b}{a}\right)-\frac{1}{\pi}\left[\frac{8}{3}-\frac{2}{15}(kb)^2-\frac{4}{105}(kb)^4+0.0104(kb)^6\right] \\
&-\mathrm{j}\left[\frac{1}{120}(kb)^5-\frac{1}{840}(kb)^7\right],
\end{aligned}
\tag{2.31}
$$

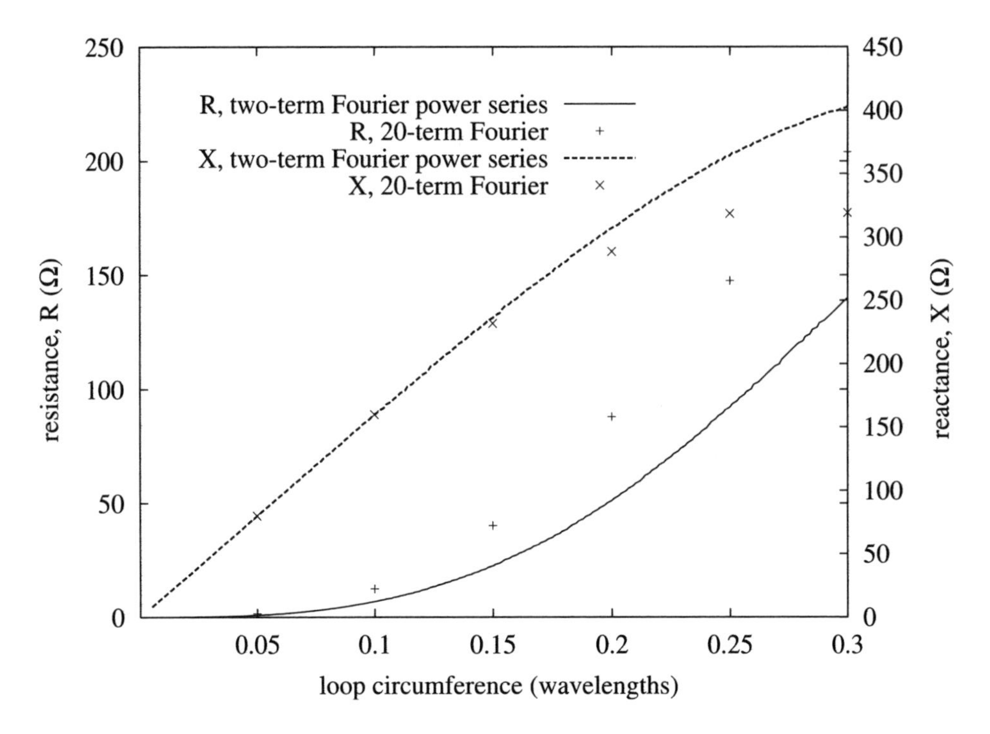

Figure 2.14 Normalized impedance of a circular loop antenna in a dissipative medium for which $\alpha/\beta = 1$. Exact and approximate analysis.

and where use has been made of

$$a_0 = kbF_1 \tag{2.32}$$

and

$$a_1 = kb\left[\frac{F_0 + F_2}{2}\right] - \frac{1}{kb}F_1. \tag{2.33}$$

Δ is given in equation (2.26).

In Figure 2.14, we show the normalized input impedance thus calculated as a function of loop circumference over the wavelength for $\alpha/\beta = 1$. The normalization is with respect to Δ. In the same figure, we show the results of a 20-term Fourier analysis, taken from [24,25]. The results apply to the problem at hand, since for blood ($\varepsilon_r = 80$ and $\sigma = 8$), the ratio of α and β is equal to 0.98.

The figure shows that, up to relatively large loop sizes, the loop reactance is well modeled by the first two terms of the Fourier analysis. The approximation for the loop resistance, however, starts to deviate seriously from the exact value for loop circumferences exceeding 0.05 wavelengths. So, to find the loop radius limit for which a uniform current may be assumed, we must restrict our analysis to loop circumferences $2\pi b \leq 0.05\lambda$. In Figure 2.15, we show again the results of a 20-term Fourier analysis and also an approximate two-term Fourier analysis for the input impedance of a small loop antenna over a smaller circumference-over-wavelength range than that in Figure 2.14. In this figure, we show also

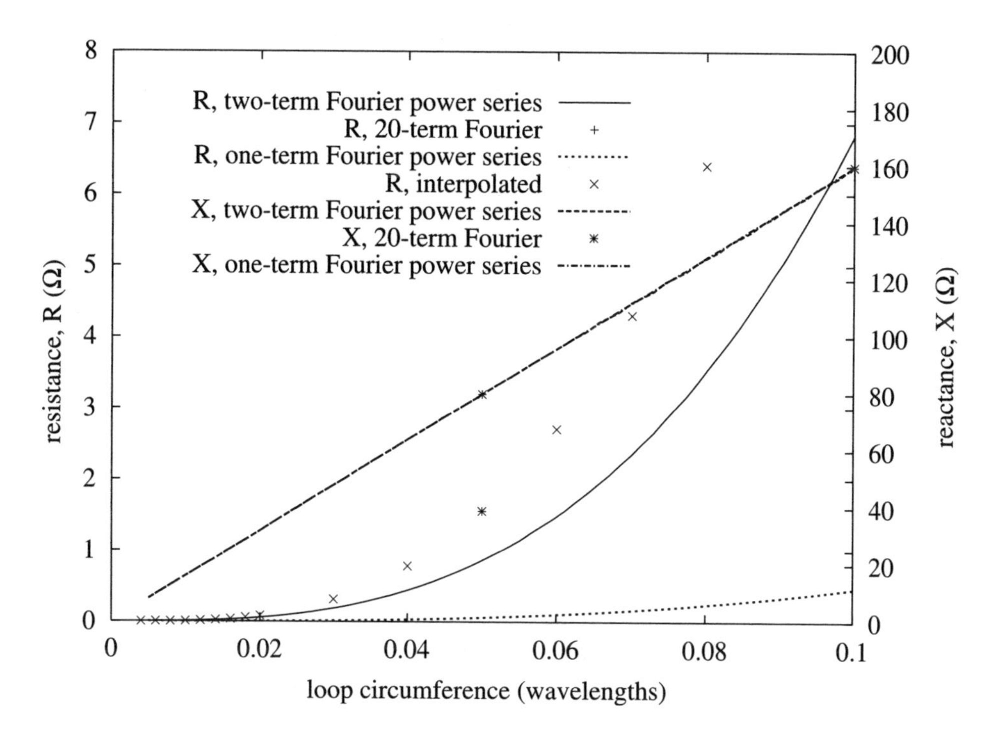

Figure 2.15 Normalized impedance of a circular loop antenna in a dissipative medium for which $\alpha/\beta = 1$. Exact, approximate and uniform-current analysis.

the results of a single-term Fourier analysis, i.e. the results of a uniform-current analysis. Finally, the figure shows the result of a rational-function interpolation [29] based on the two smallest two-term Fourier analysis results and the four smallest 20-term Fourier analysis results, bridging the gap between these two analyses.

The figure again shows fair agreement over the entire circumference range between all simulation results for the reactance of the loop, but poor agreement between the resistance simulation results. For small loop circumferences or radii, the agreement between the exact reactance and the reactance based on a uniform-current approximation is excellent. The resistance for a uniform-current approximation approaches that for the two-term Fourier analysis when the circumference becomes infinitely small. The resistance value, though, for small loops is outweighed by the reactance value.[13] Taking this into account, we start by determining the 5% deviation between the one- and two-term Fourier analysis results for the loop reactance as a function of the circumference. As Figure 2.16 shows, this restricts us to loops of circumference smaller than 0.2λ.

Next, somewhat arbitrarily but aiming at a fair assessment, we determine the circumference value below which the reactance is two to three orders of magnitude larger than the resistance, and take this value as our maximum circumference value that allows a

[13] A small loop antenna carrying a uniform current may be considered as a radiating inductor [19].

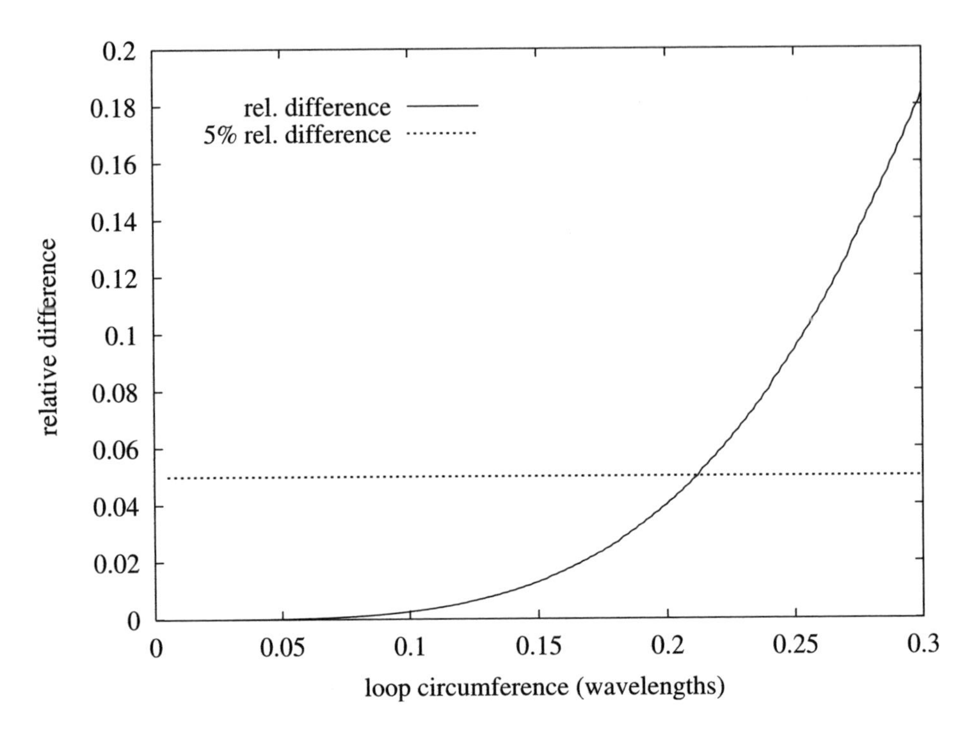

Figure 2.16 Relative difference between the approximate and the uniform-current analysis of the reactance of a circular loop antenna in a dissipative medium for which $\alpha/\beta = 1$.

uniform-current approximation. Following this line of reasoning, we put a restriction on the circumference of a loop antenna in blood of 0.02 wavelengths. This means that the radius should not exceed 1.7 mm.

Thus, we may consider a loop antenna immersed in blood and subject to a 1.5 T main MR magnetic field to carry a uniform current if the radius of the loop does not exceed 1.7 mm.

2.4.2 Sensitivity

In the remainder of this section, we shall calculate *sensitivity patterns*. A sensitivity pattern displays the sensitivity of an intravascular antenna to the magnetic field as a function of position (in the near field). Owing to the reciprocity of a passive antenna, this three-dimensional pattern or two-dimensional sections of this pattern may be calculated from the magnetic-field amplitude as a function of the near-field position, transmitted by the antenna when a current I flows through the wire. The sensitivity is defined as [9]

$$S = \frac{1}{I}\sqrt{B_x^2 + B_y^2}\,(\mathrm{T\,A^{-1}}) \tag{2.34}$$

where $B_i = \mu H_i$, $i = x, y$. This definition is based on the argument that magnetic fields will be measured only in the transverse (xy) plane in an MR scanner where the main magnetic field is z-directed.

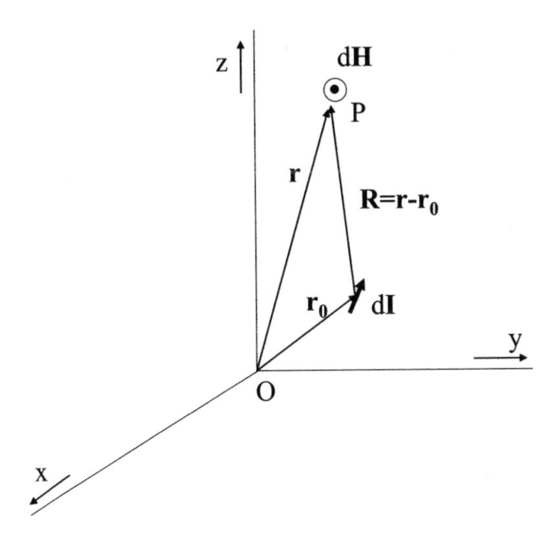

Figure 2.17 Magnetic field induced by an infinitesimal straight wire segment.

2.4.3 Biot–Savart Law

Ampère's law relates the induced magnetic field of a general but stationary current path to that current path. Before Ampere formulated this relation, Biot and Savart derived a quantitative relation for the special case of a straight wire [20].

For a current element $\boldsymbol{dI}$ at position $\boldsymbol{r}_0$ relative to a chosen origin, the induced magnetic field $d\boldsymbol{H}$ at a position $P = P(\boldsymbol{r})$ relative to the same origin (Figure 2.17), is given by

$$d\boldsymbol{H}(\boldsymbol{r}) = \frac{\boldsymbol{dI} \times \boldsymbol{R}}{4\pi R^3} = I(\boldsymbol{r}_0)\frac{\boldsymbol{dl} \times \boldsymbol{R}}{4\pi R^3}, \tag{2.35}$$

where $\boldsymbol{dI} = I(\boldsymbol{r}_0)\boldsymbol{dl}$. Equation (2.35) is known as the *Biot–Savart law*, although, for the reasons mentioned above, it is also referred to as *Ampère's law* [20].[14] The total magnetic field $\boldsymbol{H}(\boldsymbol{r})$ of the current elements around a current path C is obtained by integrating this equation over the path:

$$\boldsymbol{H}(\boldsymbol{r}) = \int_C I(\boldsymbol{r}_0)\frac{\boldsymbol{dl} \times \boldsymbol{R}}{4\pi R^3}. \tag{2.36}$$

This integral is readily evaluated for observation positions on the axis of a circular loop, but off axis and for shapes more complex than a circular loop, this is difficult or impossible [31]. In order to use the Biot–Savart law with more complicated wire structures, it is necessary to subdivide the structure into segments that result in integrals that can be evaluated in closed form. To that end, it is desirable that the equation of such a line segment may be expressed in terms of a single parameter ζ [31],

$$\boldsymbol{l} = \boldsymbol{l}(\zeta) = \hat{\boldsymbol{u}}_x x(\zeta) + \hat{\boldsymbol{u}}_y y(\zeta) + \hat{\boldsymbol{u}}_z z(\zeta), \tag{2.37}$$

[14]The Biot–Savart law is postulated in [20] and derived in [30] for the static situation, and may be derived for the quasi-static situation as well, as is demonstrated in Appendix 2.A.

where the $\hat{\boldsymbol{u}}_i$, $i = x, y, z$, are unit vectors in the x, y and z directions, respectively, of a Cartesian coordinate system. The infinitesimal segment in equation (2.36) is then

$$\boldsymbol{dl} = \frac{\boldsymbol{dl}(\zeta)}{d\zeta}\,d\zeta = \left[\hat{\boldsymbol{u}}_x \frac{dx(\zeta)}{d\zeta} + \hat{\boldsymbol{u}}_y \frac{dy(\zeta)}{d\zeta} + \hat{\boldsymbol{u}}_z \frac{dz(\zeta)}{d\zeta}\right] d\zeta. \tag{2.38}$$

The vector $\boldsymbol{R}$ in equation (2.36) is given by (Figure 2.17)

$$\boldsymbol{R} = \hat{\boldsymbol{u}}_x[x(\zeta) - x] + \hat{\boldsymbol{u}}_y[y(\zeta) - y] + \hat{\boldsymbol{u}}_z[z(\zeta) - z], \tag{2.39}$$

where it is understood that $P = P(x, y, z)$.

For a straight wire segment between the positions (x_1, y_1, z_1) and (x_2, y_2, z_2), the functions $x(\zeta)$, $y(\zeta)$ and $z(\zeta)$ are simply $(x_1 + (x_2 - x_1)\zeta)$, $(y_1 + (y_2 - y_1)\zeta)$ and $(z_1 + (z_2 - z_1)\zeta)$, respectively, with $0 \le \zeta \le 1$. So, equations (2.38) and (2.39), for this straight wire segment, become

$$\boldsymbol{dl} = [\hat{\boldsymbol{u}}_x(x_2 - x_1) + \hat{\boldsymbol{u}}_y(y_2 - y_1) + \hat{\boldsymbol{u}}_z(z_2 - z_1)]\,d\zeta \tag{2.40}$$

and

$$\begin{aligned}\boldsymbol{R} = {} & \hat{\boldsymbol{u}}_x[(x_1 - x) + (x_2 - x_1)\zeta] + \hat{\boldsymbol{u}}_y[(y_1 - y) + (y_2 - y_1)\zeta] \\ & + \hat{\boldsymbol{u}}_z[(z_1 - z) + (z_2 - z_1)\zeta].\end{aligned} \tag{2.41}$$

The cross product of $\boldsymbol{dl}$ and $\boldsymbol{R}$ may then be computed as

$$\boldsymbol{dl} \times \boldsymbol{R} = \begin{vmatrix} \hat{\boldsymbol{u}}_x & \hat{\boldsymbol{u}}_y & \hat{\boldsymbol{u}}_z \\ (x_2 - x_1)\,d\zeta & (y_2 - y_1)\,d\zeta & (z_2 - z_1)\,d\zeta \\ (x_1 - x) + (x_2 - x_1)\zeta & (y_1 - y) + (y_2 - y_1)\zeta & (z_1 - z) + (z_2 - z_1)\zeta \end{vmatrix}, \tag{2.42}$$

and

$$\begin{aligned}R^3 = {} & \{[(x_1 - x) + (x_2 - x_1)\zeta]^2 + [(y_1 - y) + (y_2 - y_1)\zeta]^2 \\ & + [(z_1 - z) + (z_2 - z_1)\zeta]^2\}^{3/2}.\end{aligned} \tag{2.43}$$

With the use of equation (2.36), the magnetic field at position $P = P(x, y, z)$ due to a unit current $I(\boldsymbol{r}_0) = 1$ flowing in a straight wire segment between the points (x_1, y_1, z_1) and (x_2, y_2, z_2) is given by [31]

$$\begin{aligned}\boldsymbol{H}(\boldsymbol{r})|_{I(\boldsymbol{r}_0)=1} &= \int_C \frac{\boldsymbol{dl} \times \boldsymbol{R}}{4\pi R^3} \\ &= \hat{\boldsymbol{u}}_x \int_{\zeta=0}^{1} \frac{D_x}{(A + B\zeta + C\zeta^2)^{3/2}}\,d\zeta + \hat{\boldsymbol{u}}_y \int_{\zeta=0}^{1} \frac{D_y}{(A + B\zeta + C\zeta^2)^{3/2}}\,d\zeta \\ &\quad + \hat{\boldsymbol{u}}_z \int_{\zeta=0}^{1} \frac{D_z}{(A + B\zeta + C\zeta^2)^{3/2}}\,d\zeta,\end{aligned} \tag{2.44}$$

where

$$A = (x_1 - x)^2 + (y_1 - y)^2 + (z_1 - z)^2, \tag{2.45}$$

$$B = 2[(x_1 - x)(x_2 - x_1) + (y_1 - y)(y_2 - y_1) + (z_1 - z)(z_2 - z_1)], \tag{2.46}$$

$$C = (x_2 - x_1)^2 + (y_2 - y_1)^2 + (z_2 - z_1)^2, \tag{2.47}$$

$$D_x = (y_2 - y_1)(z_1 - z) - (z_2 - z_1)(y_1 - y), \tag{2.48}$$

$$D_y = (z_2 - z_1)(x_1 - x) - (x_2 - x_1)(z_1 - z) \tag{2.49}$$

and

$$D_z = (x_2 - x_1)(y_1 - y) - (y_2 - y_1)(x_1 - x). \tag{2.50}$$

The magnetic field at position P produced by multiple straight wire segments is the sum of the contributions calculated for the isolated wire segments.

2.4.4 Model Verification

To validate the applicability of the Biot–Savart model thus derived for our intravascular MR antennas subject to a 1.5 T main magnetic field, we shall compare the results obtained from the Biot–Savart model with results that can be obtained analytically for a small loop antenna. We have seen that, for the above static MR magnetic field, a small loop antenna immersed in blood may be considered to carry a uniform current as long as the radius of the loop does not exceed 1.7 mm. The radiated fields of such a loop antenna are obtained from those derived for a loop in air, stated in equations (2.10)–(2.13). The adaptation for the surrounding dispersive medium is accomplished by substituting $\tilde{k}$ for k, $\tilde{\eta}$ for η and $\tilde{\varepsilon}$ for ε in these equations, where [28, 32]

$$\tilde{k} = \beta\left(1 - \mathrm{j}\frac{\alpha}{\beta}\right), \tag{2.51}$$

$$\tilde{\eta} = \frac{\sqrt{\mu_0/\varepsilon_r\varepsilon_0}(1/f(p))}{1 - \mathrm{j}\alpha/\beta}, \tag{2.52}$$

$$\tilde{\varepsilon} = \varepsilon_r\varepsilon_0 - \mathrm{j}\frac{\sigma}{\omega}, \tag{2.53}$$

with

$$p = \frac{\sigma}{\omega\varepsilon}, \tag{2.54}$$

$$f(p) = \cosh\left(\frac{1}{2}\sinh^{-1}(p)\right) \tag{2.55}$$

and

$$\frac{\alpha}{\beta} = \tanh\left(\frac{1}{2}\sinh^{-1}(p)\right). \tag{2.56}$$

For blood, $p = 28.09$, $f(p) = 3.81$, $\lambda = 0.52$ m and $\alpha/\beta = 0.97$.

To verify the model, single-loop antennas of radii 0.5 mm, 1.0 mm and 1.5 mm were placed at an angle ϑ relative to the z axis of a rectangular coordinate system. The sensitivity

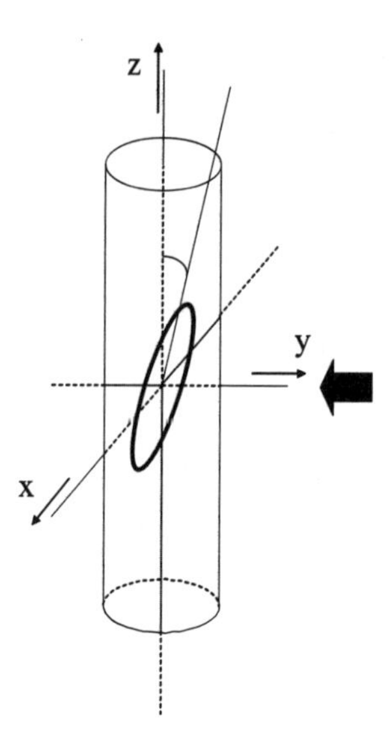

Figure 2.18 Single-loop antenna, positioned in a z-directed artery.

was calculated for different positions y relative to the loop center (Figure 2.18). In this figure, the position of the vascular wall is indicated by a circular cylinder, with its central axis along the z direction.

For these special situations, we may equivalently keep the loop positioned parallel to and in the xy plane, and calculate the sensitivity for different rotation angles ϑ and for a unit current as $S = \sqrt{B_r^2 + B_\varphi^2} = B_r$ (Figure 2.19). With the aid of [21], the sensitivity along the y axis ($\varphi = \pi/2$) can be obtained from the Cartesian components of the magnetic flux density as $S = B_r = \sin(\vartheta)B_y + \cos(\vartheta)B_z$.

For $\vartheta = 0$, the sensitivity is equal to the magnetic flux density on the axis of the loop antenna. For a loop carrying a uniform current I_0, the static magnetic flux density on the axis may be calculated in closed form as (Figure 2.18) [20]

$$S = B = B_y = \mu_0 \frac{I_0 a^2}{2(a^2 + y^2)^3/2}, \tag{2.57}$$

where a is the radius of the loop and y is the distance between the observation point on the axis of the loop and the center of the loop.

In Figures 2.20–2.22, the sensitivities calculated using the dynamic loop model and using the Biot–Savart model for a segmented loop are shown as a function of the distance from the center of the loop. The results are shown for three different loop radii, all loops being at an angle $\vartheta = 0$ with respect to the z axis. The analytic results for the static sensitivity on the loop axis (equation (2.51)) are also shown in these figures. For the calculation based on the Biot–Savart model, 40 straight segments were used to approximate the loop; this is large enough to represent a circular loop accurately [9].

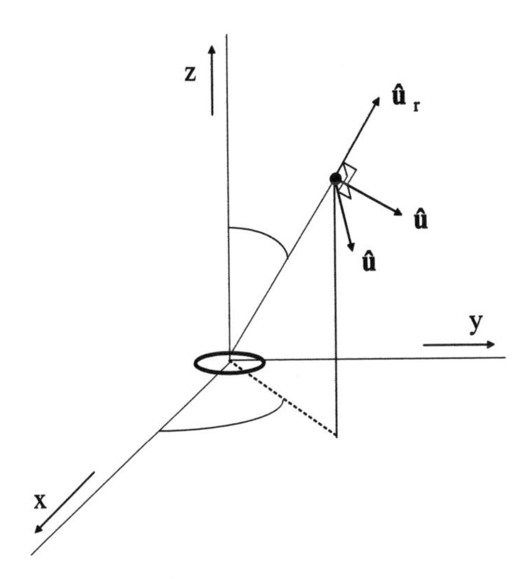

Figure 2.19 Alternative for calculating the sensitivity of a single-loop antenna rotated ϑ from the z axis.

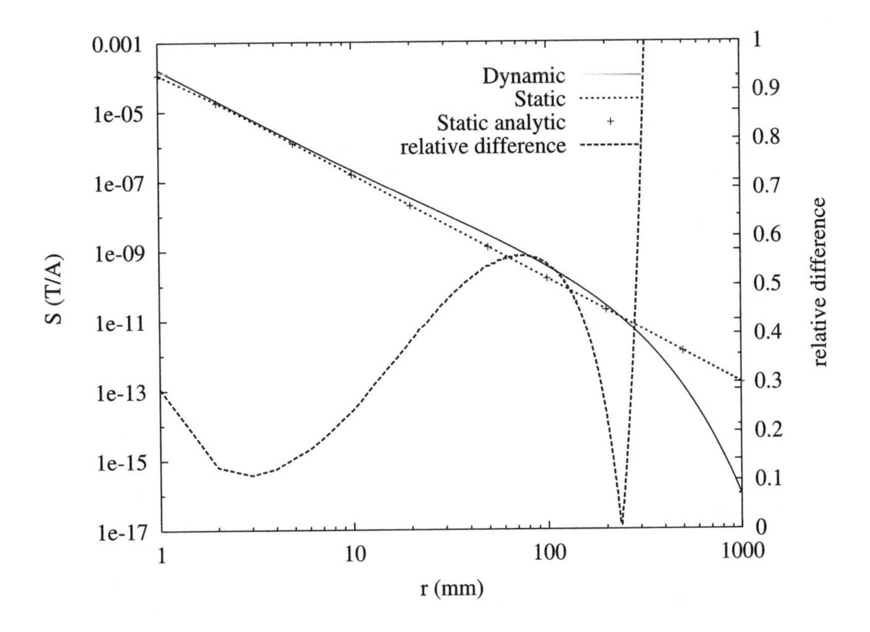

Figure 2.20 Sensitivity versus distance from loop center for $\vartheta = 0$, $a = 0.5$ mm.

Also, in the same figures, the relative difference δ between the dynamic sensitivity S_{dyn} and the static sensitivity S_{stat} is shown ($\delta = (S_{\text{dyn}} - S_{\text{stat}})/S_{\text{dyn}}$). Note that the sensitivity and the distance are displayed on a logarithmic scale.

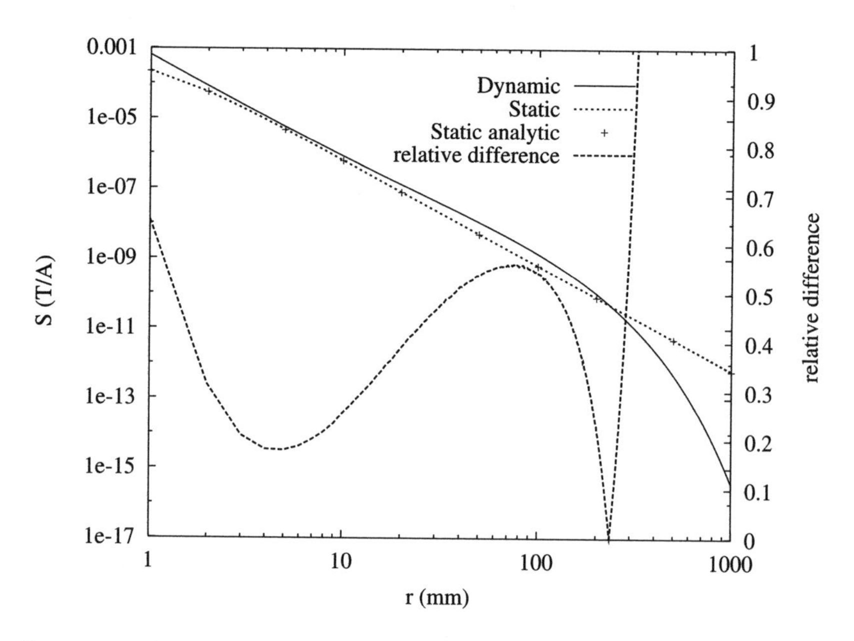

Figure 2.21 Sensitivity versus distance from loop center for $\vartheta = 0$, $a = 1.0$ mm.

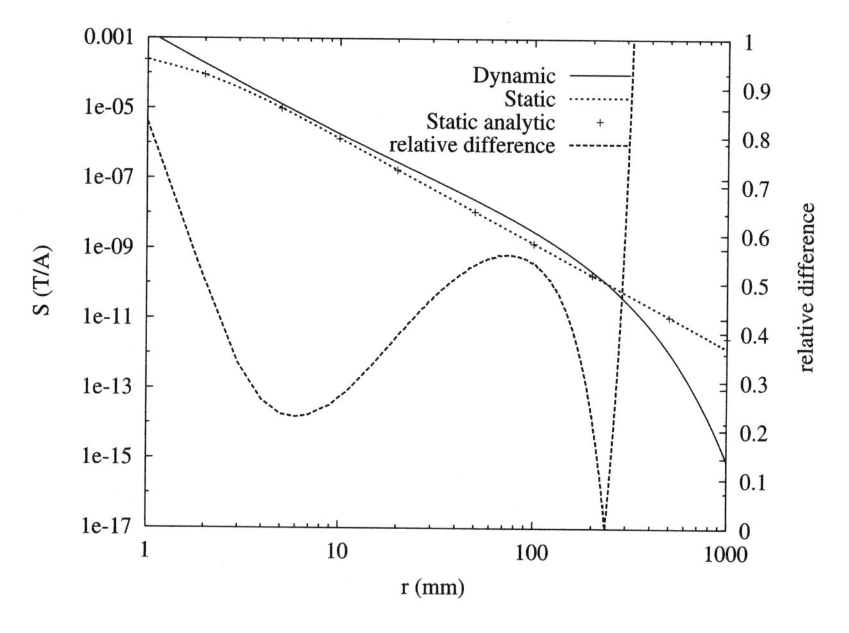

Figure 2.22 Sensitivity versus distance from loop center for $\vartheta = 0$, $a = 1.5$ mm.

These figures clearly show that the sensitivity calculated by the Biot–Savart model for a segmented loop coincides with the static analytical on-axis results for all distance values.

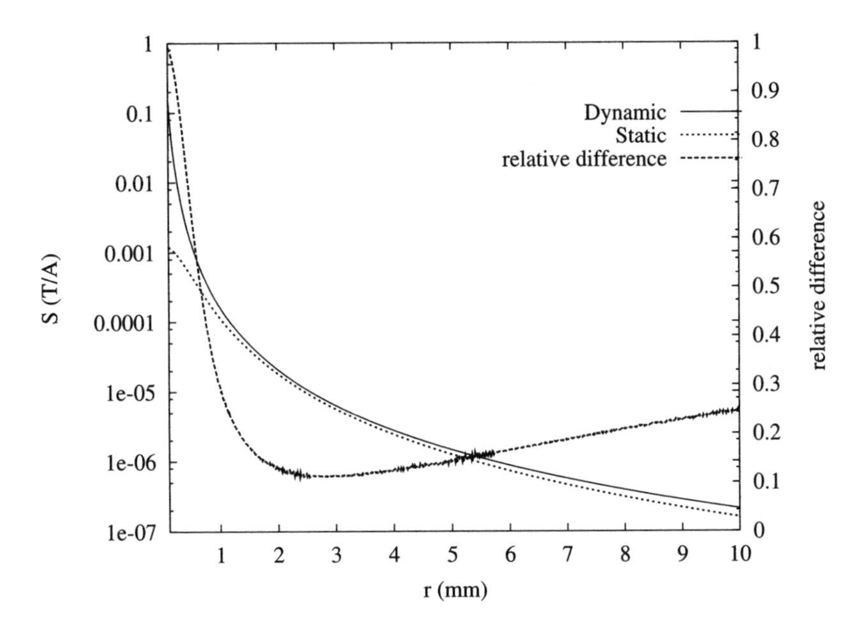

Figure 2.23 Sensitivity versus distance from loop center for $\vartheta = 0$, $a = 0.5$ mm.

The figures also show that very close to the loop, where the fields show an r^{-3} dependence, the dynamic sensitivity differs substantially from the static value. This difference is larger for loops with a larger radius. Far away from the loop, where the fields show an r^{-1} dependence, the dynamic sensitivity starts to differ substantially from the static value again, and very far away from the loop, the difference between the dynamic and the static sensitivity becomes very large, owing to the attenuation of the dynamic fields, which show an $e^{-\alpha r}$ dependence, α being the attenuation constant. This attenuation is accounted for in the dynamic model by the multiplication e^{-jkr} in equations (2.11) and (2.12) but is not taken into account in the static model.[15] In between these areas of substantially different sensitivities, we observe an area of minimum relative difference. This area, where the dynamic (and static) fields show an r^{-2} dependence, seems to coincide with our area of interest: the position of the artery wall, which, for medium and large arteries, varies between 1.0 mm and 3.0 mm [18].

Figures 2.23–2.25 show the same dynamic and static sensitivities and relative differences between these two sensitivities, but now over a smaller distance range. We observe that, on axis, for a well-chosen loop radius, i.e. 0.5 mm or less, the static model approximates the dynamic sensitivity with less than 30% deviation in the region of interest. If we concentrate on large arteries only (radii between 2.0 mm and 3.0 mm), this deviation is less than 13%.

[15]The same attenuation could be introduced into the sensitivity parameter calculated with the Biot–Savart model. This would make the model more realistic but, owing to the r^{-1} behavior of the dynamic fields, the dynamic and static sensitivities would still diverge with the distance r. At distances not very far away from the loop center, the influence of the attenuation is negligible. Therefore the attenuation was not taken into account in the Biot–Savart model.

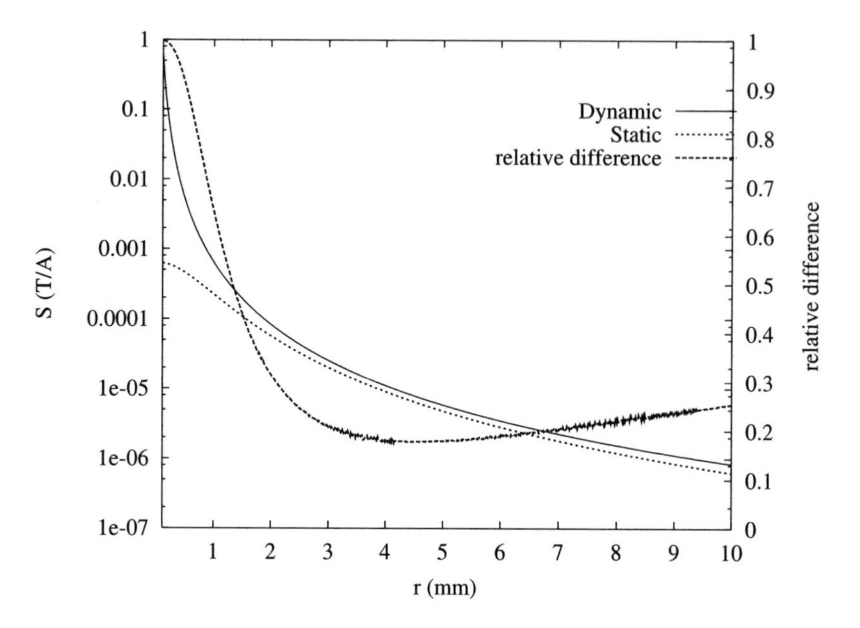

Figure 2.24 Sensitivity versus distance from loop center for $\vartheta = 0$, $a = 1.0$ mm.

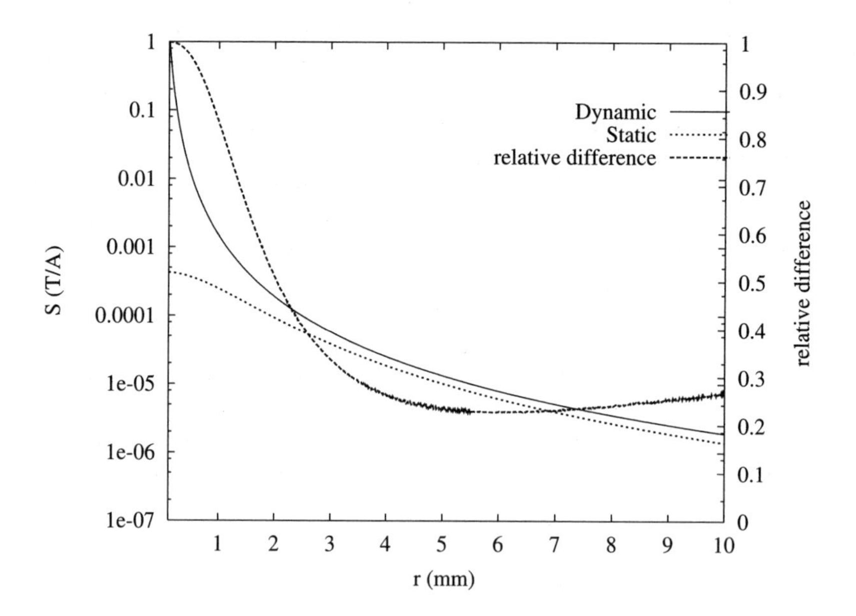

Figure 2.25 Sensitivity versus distance from loop center for $\vartheta = 0$, $a = 1.5$ mm.

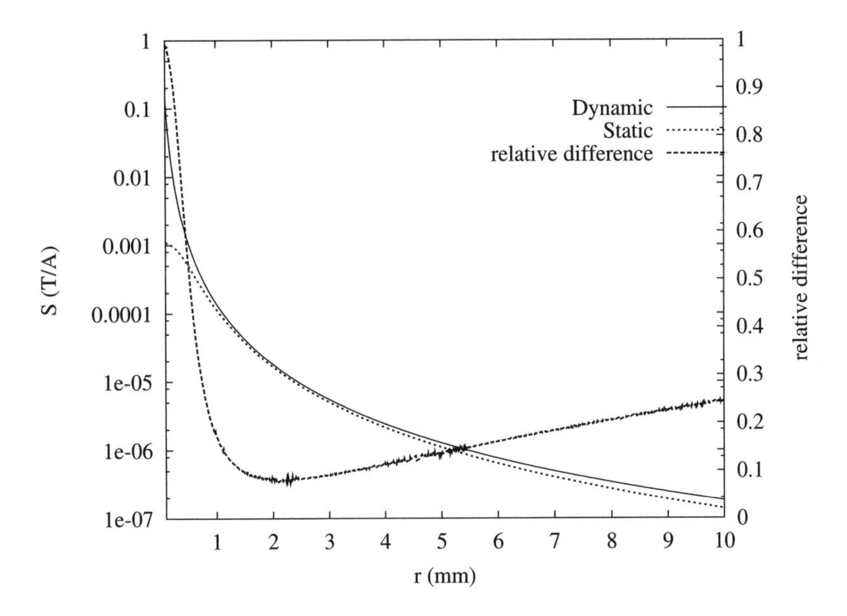

Figure 2.26 Sensitivity versus distance from loop center for $\vartheta = 30°$, $a = 0.5$ mm.

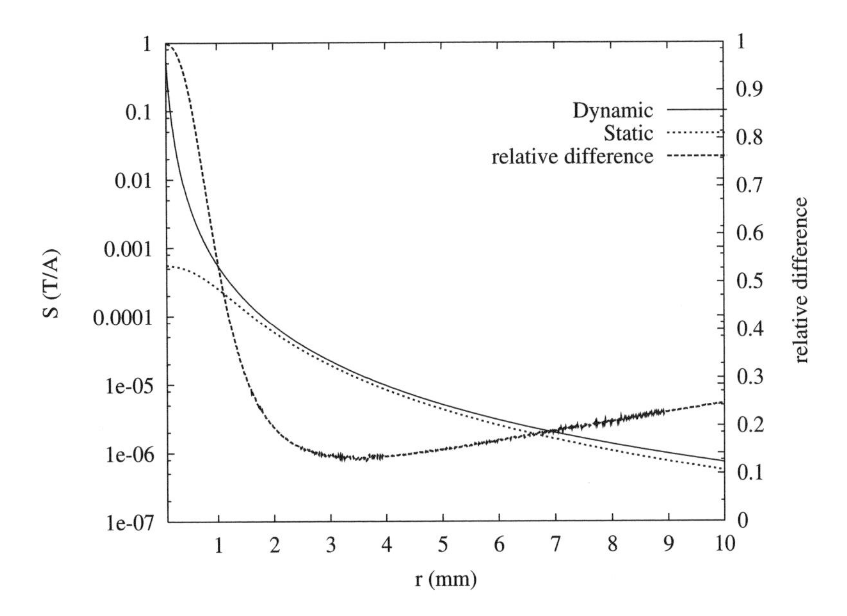

Figure 2.27 Sensitivity versus distance from loop center for $\vartheta = 30°$, $a = 1.0$ mm.

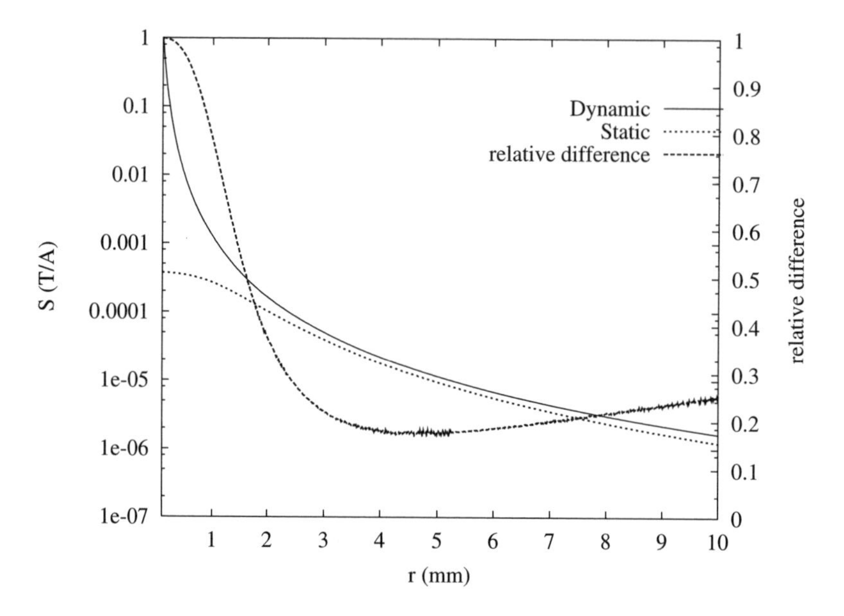

Figure 2.28 Sensitivity versus distance from loop center for $\vartheta = 30°$, $a = 1.5$ mm.

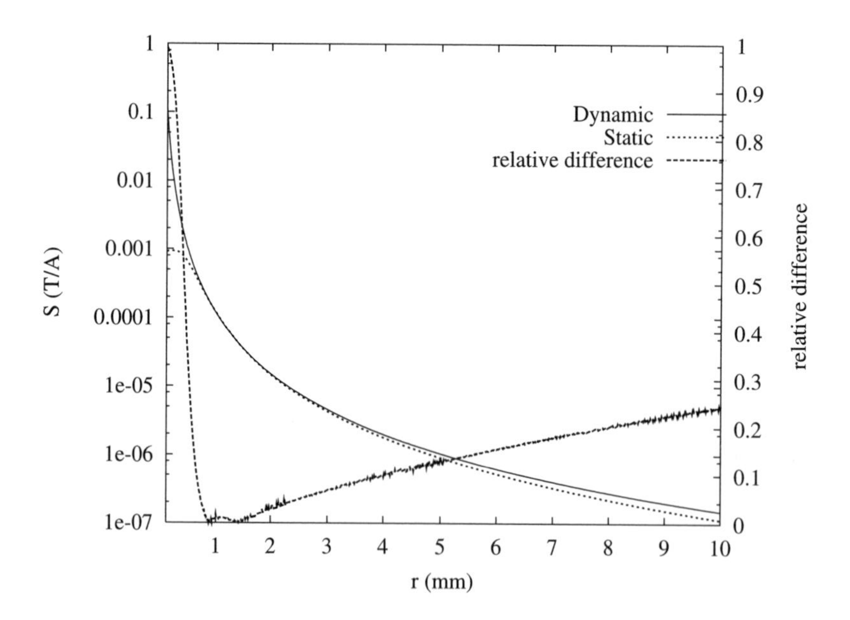

Figure 2.29 Sensitivity versus distance from loop center for $\vartheta = 45°$, $a = 0.5$ mm.

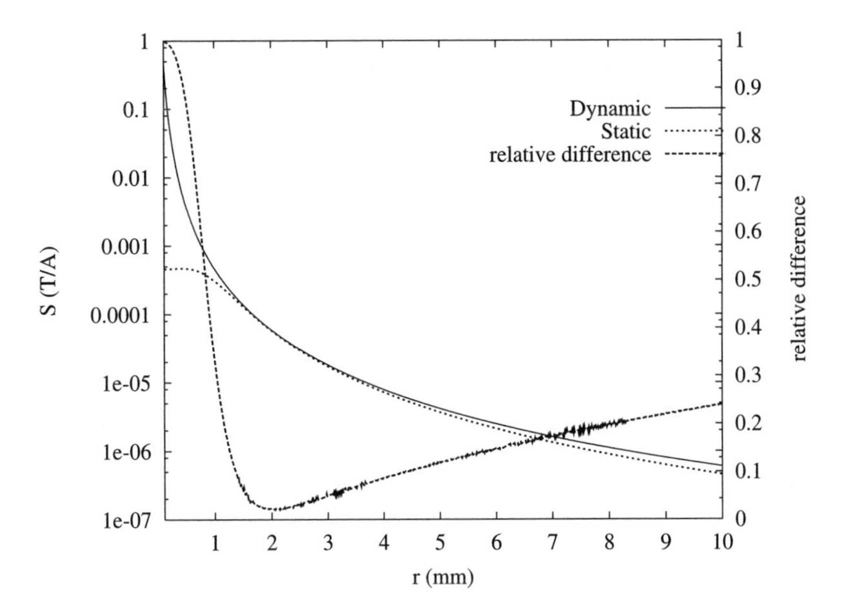

Figure 2.30 Sensitivity versus distance from loop center for $\vartheta = 45°$, $a = 1.0$ mm.

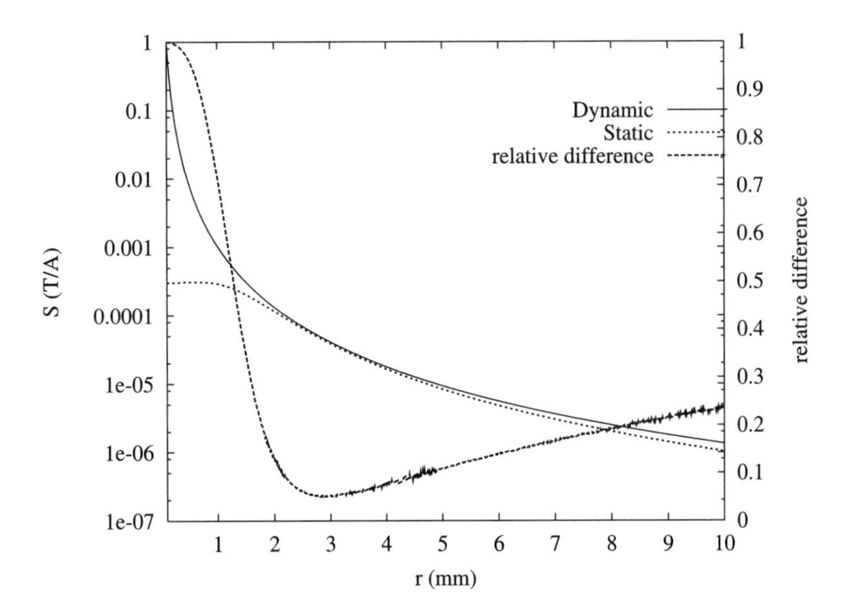

Figure 2.31 Sensitivity versus distance from loop center for $\vartheta = 45°$, $a = 1.5$ mm.

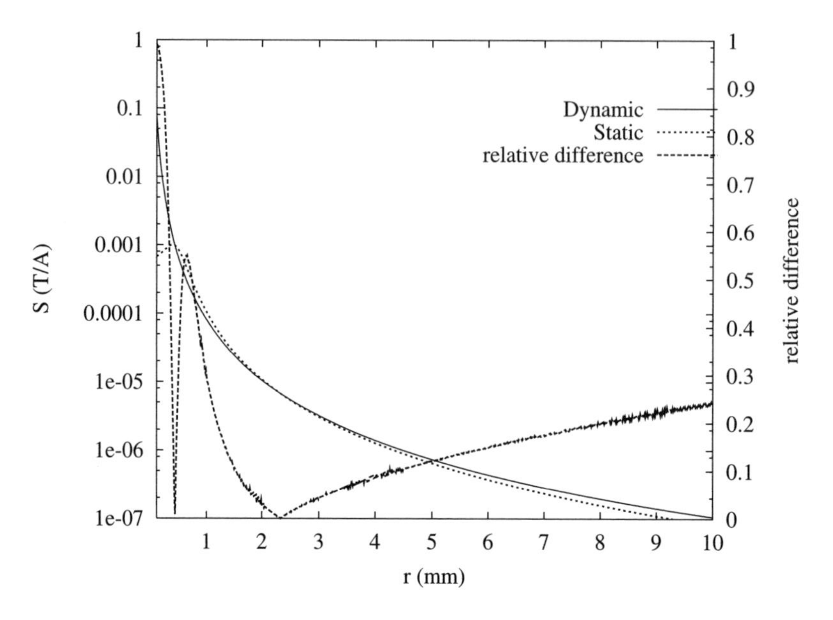

Figure 2.32 Sensitivity versus distance from loop center for $\vartheta = 60°$, $a = 0.5$ mm.

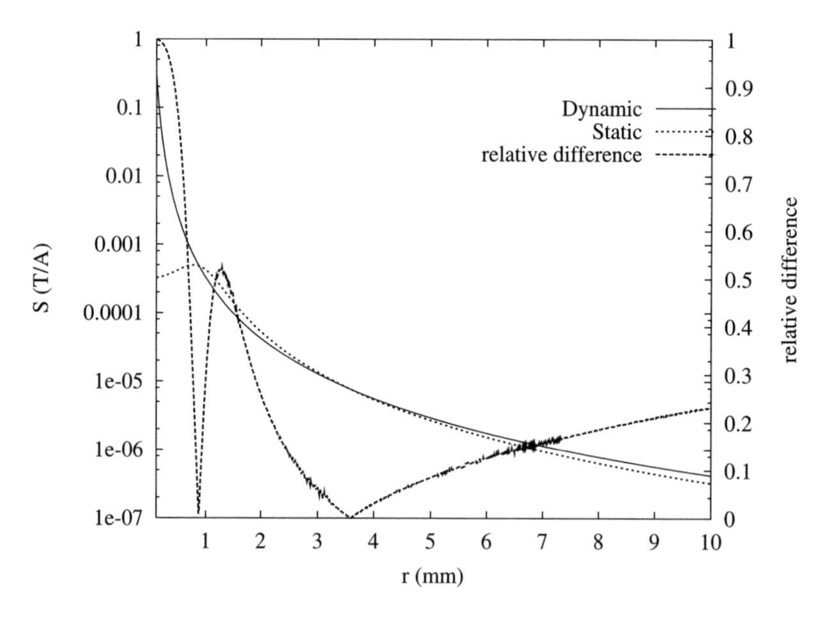

Figure 2.33 Sensitivity versus distance from loop center for $\vartheta = 60°$, $a = 1.0$ mm.

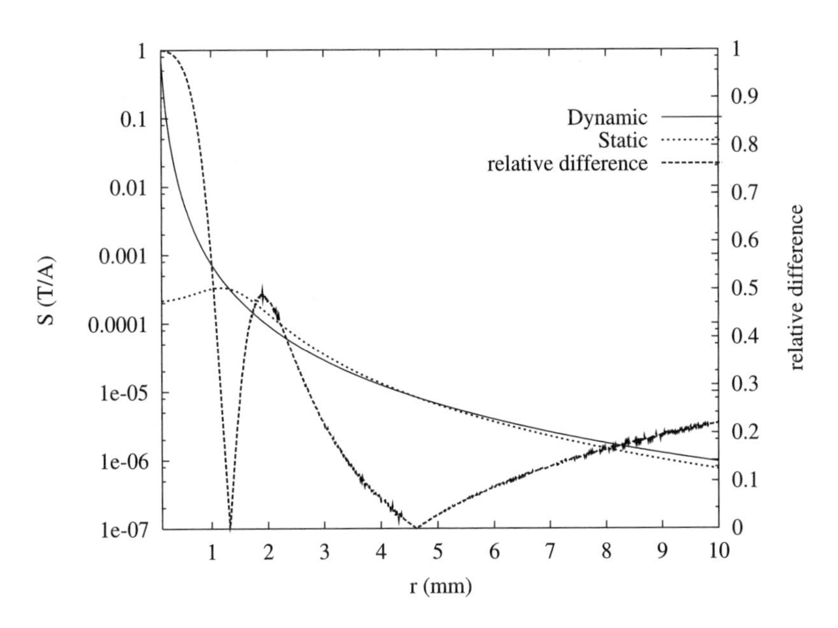

Figure 2.34 Sensitivity versus distance from loop center for $\vartheta = 60°$, $a = 1.5$ mm.

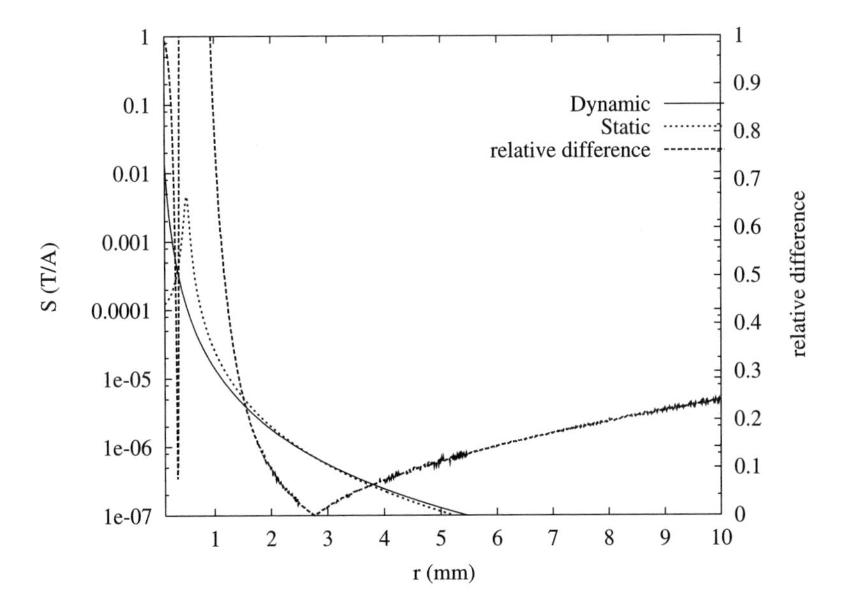

Figure 2.35 Sensitivity versus distance from loop center for $\vartheta = 85°$, $a = 0.5$ mm.

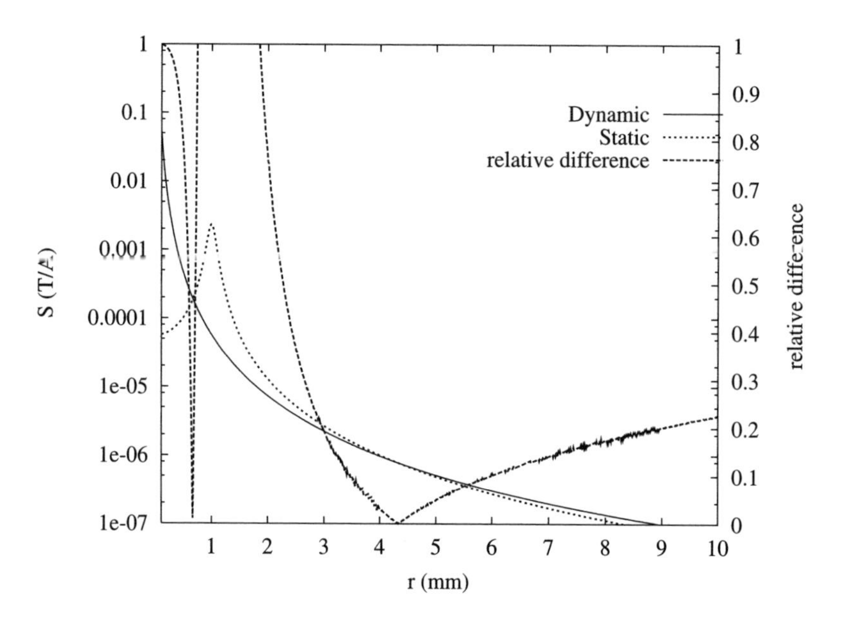

Figure 2.36 Sensitivity versus distance from loop center for $\vartheta = 85°$, $a = 1.0$ mm.

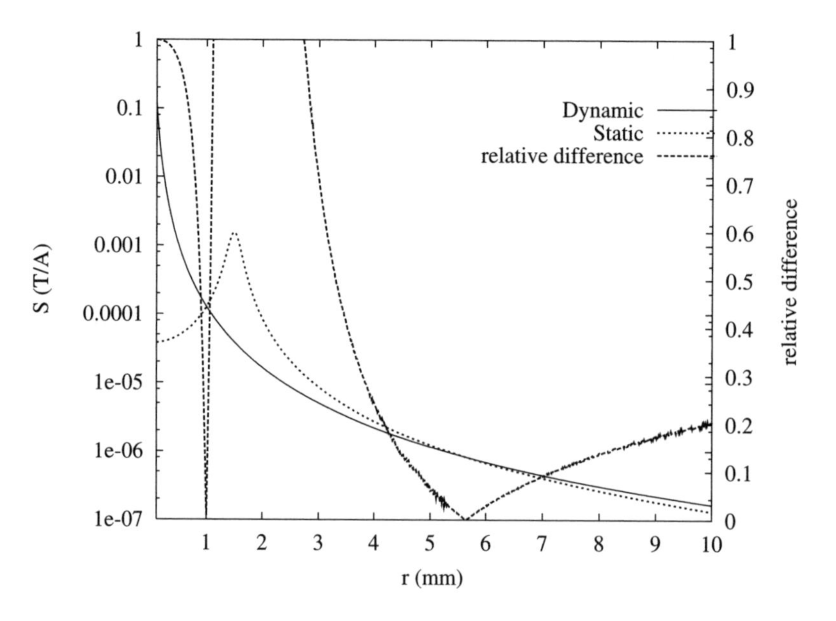

Figure 2.37 Sensitivity versus distance from loop center for $\vartheta = 85°$, $a = 1.5$ mm.

For angles $\vartheta = 30°, 45°, 60°$ and $85°$, the dynamic and static sensitivities as a function of y (Figure 2.18) and the relative difference between these two values are shown for three loop radii, in Figures 2.26–2.28, 2.29–2.31, 2.32–2.34 and 2.35–2.37, respectively. For $\vartheta = 90°$, both the dynamic and static sensitivity are calculated to be zero.

These figures reveal that, especially for large arteries (radii between 2 mm and 3 mm [18]) and small loop antennas ($a = 0.5$ mm), sensitivities that deviate by less than 13% from the exact values may be calculated by the static method. Moreover, the static and dynamic sensitivities in the area of interest (i.e. in and around the position of the vascular wall) show similar behavior as a function of distance from the center of the (tilted) loop antenna. This means that, under well-defined conditions, the static model may be used to predict the absolute value of the sensitivity of a loop antenna with a reasonable accuracy, but – more importantly – the static model may be used to compare different designs with respect to sensitivity profiles.

Now that we have shown the validity of the static model for single-loop antennas immersed in blood, the next question to be answered concerns the validity of employing this model for multiple loops, or, more generally, wire antennas where the length of the wire is larger than that in a single loop of radius 1.7 mm. Restricting ourselves for convenience, for the purpose of this discussion, to multiple-loop antennas, two issues need to be examined. The first is whether the uniform-current assumption still applies and, if not, what the consequences are; the second is the mutual influence of closely spaced turns.

We start with the issue of current uniformity. For single and multiturn loops in air, the restriction on the circumference mentioned in the literature (e.g. [19]) applies to the total length of the wire antenna. So $N2\pi \leq 0.1\lambda$, where N is the number of turns. We have seen that this restriction may be translated into a maximum error of 5% in both the real and the imaginary part of the input impedance of the loop with respect to the exact value. It may be expected, however, that the restrictions on the radiated fields can be relaxed, owing to the averaging effect of the current integrations involved. Moreover, we may recognize that in a multiturn loop antenna, in which the conductor loss may be neglected and every individual loop satisfies the circumference restriction, every loop may be regarded as carrying a uniform current. However, phase differences exist between different turns. Since the turn spacing will be very small in terms of the wavelength, applying array theory to the multiturn loop will result in effectively having N turns at the same position. Of course, this reasoning is only valid if mutual coupling between the turns may be neglected, which, in air, especially for closely packed turns, is not true [19,33]. For multiturn loop antennas in air, one could resort to numerical methods, as explained in, for example, [34] for circular loops or, as explained in [35], approximate methods for rectangular loops.

For loops immersed in blood, the situation is different. Since a current now also flows into the medium, the resistive part of the input impedance increases. Therefore, as we have already observed in section 2.4.1, the maximum allowable loop radius that justifies a uniform-current approach will be smaller. The mutual coupling between two widely spaced small loops immersed in blood will be negligible compared with the self-coupling. When the loops are brought closer together, the mutual coupling increases, but up to short distances the mutual coupling is still negligible compared with the self-coupling [33]. In this respect, the situation differs from that for loops in air. Extrapolating the results stated in [33], the mutual coupling for two small loops immersed in blood, when brought very close together, will increase to

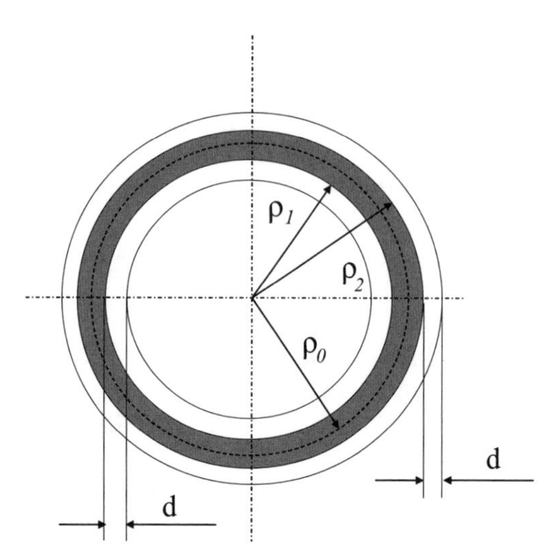

Figure 2.38 Geometrical parameters of an insulated loop.

such a level that it will outweigh the self-coupling by far. This is consistent with our earlier observation that a current flows into the medium surrounding the loops. So, for bare-wire multiturn loop antennas contained in a small volume, the quasi-static approach will fail.

Fortunately, in practice, insulated wire is used to construct intravascular antennas and the insulation does not necessarily compromise our earlier theoretical derivations. For insulated wire, the conducting medium will now act as a shield, thus reducing the mutual coupling between the wires or turns of a multiturn loop [36].

In [37] it was demonstrated that the effect of adding a *thin* layer of insulation to a loop antenna is that the uniform current flow is maintained when the loop antenna is immersed in a conducting medium such as blood. As long as the insulation layer is thin, the impedance of the loop is equal to that of the bare loop immersed in the conducting medium [37]. A practical value for the layer thickness d is given by [37]

$$d \approx 0.2(\rho_2 - \rho_1), \tag{2.58}$$

where d, ρ_1 and ρ_2 are defined in Figure 2.38.

Figures 2.39 and 2.40 [37] show the real and imaginary parts of the admittance of a loop for two loop radii ρ_0 as a function of insulation thickness. These figures show the admittance for a one-term current approximation, Y_0, and the admittance for a two-term current approximation, Y_1. Judging from these figures, a better estimate for the insulation thickness is given by $d \approx x(\rho_2 - \rho_1)$, where $0.4 \leq x \leq 0.5$.

Using the data presented in [37], we shall now look at the situation in which $\Omega = 10$, $\varepsilon_{r3} = 0.2\varepsilon_{r4}$ and $k_3\rho_0 = 0.1$. The relative permittivities ε_{r3} and ε_{r4} are those of the insulating layer and of the surrounding medium, respectively. The wave number k_3 is the wave number in the insulating layer. The graphs in [37] show that in this situation, a uniform-current approach and an unchanged impedance apply for a highly conducting medium and the insulation thickness given in equation (2.58).

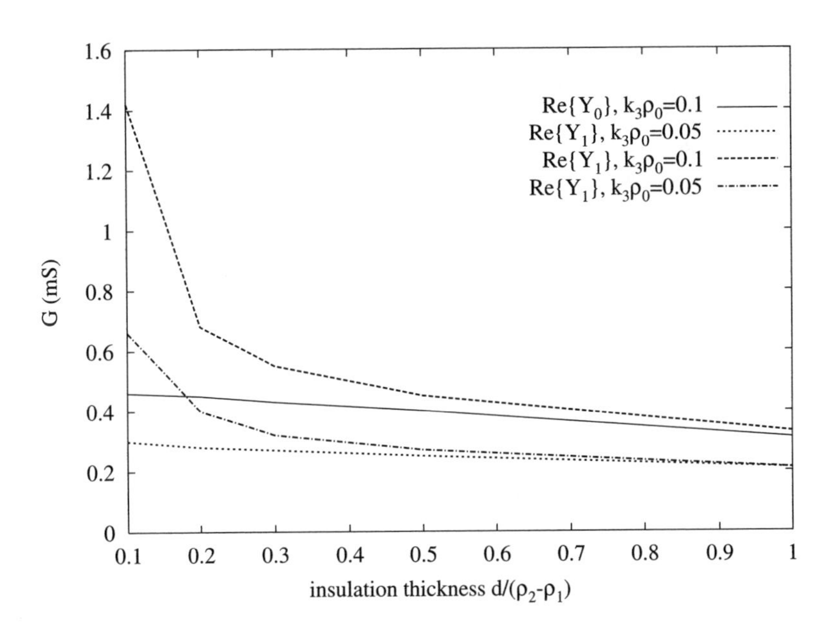

Figure 2.39 Real part of the loop admittance, G, versus insulation thickness.

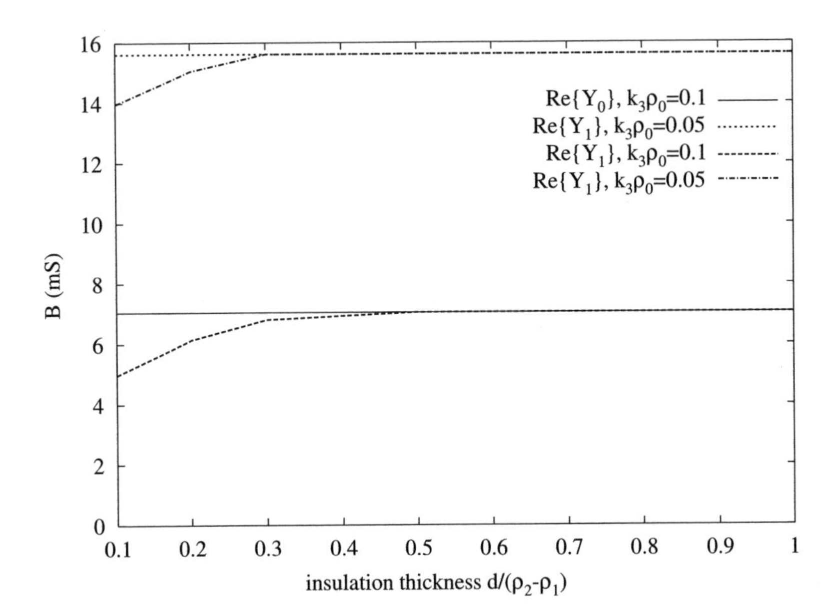

Figure 2.40 Imaginary part of the loop admittance, B, versus insulation thickness.

Substituting the relative permittivity of blood for ε_{r4} (i.e. 80) and assuming the insulating layer to be lossless, so that $k_3 = 2\pi\sqrt{\varepsilon_{r3}}/\lambda_0$, λ_0 being the free-space wavelength, yields

$\rho_0 = 18.65$ mm for the maximum loop radius. Provided that the mutual coupling between the turns of a multiturn loop may be neglected, this value puts a limit on the total length of a multiturn loop. This means that for turns with a radius $a = 0.5$ mm, we may employ up to 37 turns and still assume a uniform current to flow through the wire.

When a uniform current flows through a loop, we have seen that the resistive part of the impedance vanishes and is outweighed by the reactive part. Therefore we now only have to compare the self-inductance and mutual inductance between two coaxial loops as a function of the distance c between them. The two loops have the same loop radius R and wire radius a. The self-inductance L_{11} and mutual inductance L_{12} are given by [20, 38]

$$L_{11} = \mu_0 R\left[\ln\left(\frac{8R}{a}\right) - 2\right] \tag{2.59}$$

and

$$L_{12} = \mu_0 R\left[\left(\frac{2}{k} - k\right)K(k) - \frac{2}{k}E(k)\right], \tag{2.60}$$

where

$$k^2 = \frac{4R^2}{4R^2 + c^2}, \tag{2.61}$$

and [38, 39]

$$K(k^2) = \int_0^{\pi/2} \frac{d\phi}{\sqrt{1 - k^2 \sin^2(\phi)}} \tag{2.62}$$

is the complete elliptic integral of the first kind and

$$E(k^2) = \int_0^{\pi/2} \sqrt{1 - k^2 \sin^2(\phi)}\, d\phi \tag{2.63}$$

is the complete elliptic integral of the second kind.

The ratio L_{12}/L_{11} as a function of the distance between the coaxial loops is shown in Figure 2.41 for $\Omega = 10$ and loops for which $R = 0.5$ mm and $R = 1.0$ mm. This figure shows that, for very small loop separations, the mutual inductance becomes comparable to the self-inductance and may not be neglected. To be able to neglect the mutual coupling effects, an order-of-magnitude difference between the mutual and the self-inductance is advisable. The loops should therefore be separated by at least the radius of the loop.

2.5 ANTENNA EVALUATION

Now that we have demonstrated the validity of the static model, we may employ this model to compare different antenna concepts quantitatively. In section 2.3, we have already conducted a qualitative comparison, based on information collected from various literature sources. As in that section, we shall again separate the antenna concepts into antennas intended for active tracking and antennas intended for imaging. We shall compare the various antenna concepts on the basis of *sensitivity profiles*, i.e. two-dimensional sections through the three-dimensional sensitivity patterns (see section 2.4.2), calculated for antennas positioned along

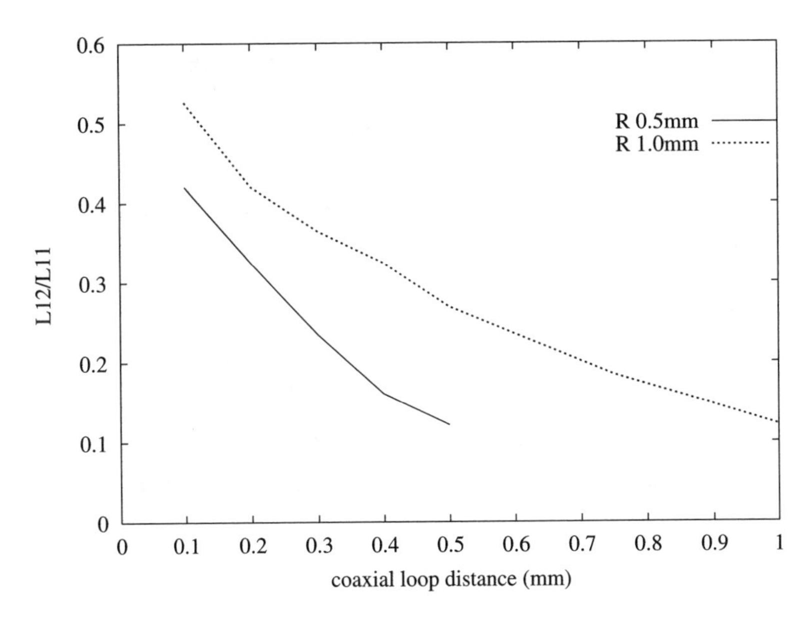

Figure 2.41 Mutual inductance normalized to the self-inductance versus distance between coaxial loops for $R = 0.5$ mm, $R = 1.0$ mm and $\Omega = 10$.

the direction of the main magnetic field of the MR scanner. None of the antennas evaluated was optimized for tracking or imaging purposes. However, all antennas were dimensioned such that they may be mounted on a circular cylinder of radius $R = 1$ mm. The heights of the antennas were 10 mm, and all coils, whether used as the antenna or as part of the antenna, had a height of 3 mm.

2.5.1 Antennas for Active Tracking

For active-tracking purposes, an antenna – mounted on a catheter – needs to be detectable with a high degree of positional accuracy. Therefore, intravascular MR antennas meant for active-tracking purposes need to have a very inhomogeneous sensitivity pattern, with the peak values at or very near the antenna position. In Figures 2.42–2.46, we consider, respectively, the antiparallel-wire antenna, the double-helix antenna, the opposed-double-helix antenna, the center return antenna and the perpendicular-coils antenna. Each figure shows the antenna geometry and three perpendicular sensitivity profiles. The center return antenna originally evaluated for imaging purposes, has been added to the list of antenna concepts for tracking purposes (Figure 2.45). The specifics of the antennas and the positions of the sensitivity profiles are stated in the figure captions.

With the exception of the antiparallel-wire antenna, all antennas demonstrate a localized sensitivity pattern. To obtain better insight into the behavior of the sensitivity as a function of the perpendicular distance from the cylindrical antenna body, Figures 2.47 and 2.48 show the

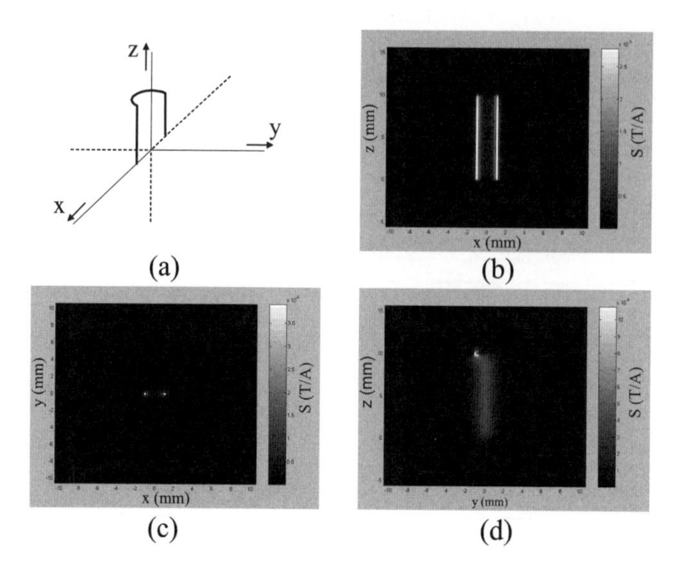

Figure 2.42 Antiparallel-wire antenna: geometry and sensitivity patterns in xz, xy and yz planes. Height 10 mm, radius 1 mm, wire separation 2 mm, 8 segments per circumference. (a) Antenna geometry. (b) Sensitivity pattern in xz plane, $y = 0.05$ mm. (c) Sensitivity pattern in xy plane, $z = 5$ mm. (d) Sensitivity pattern in yz plane, $x = 0.05$ mm.

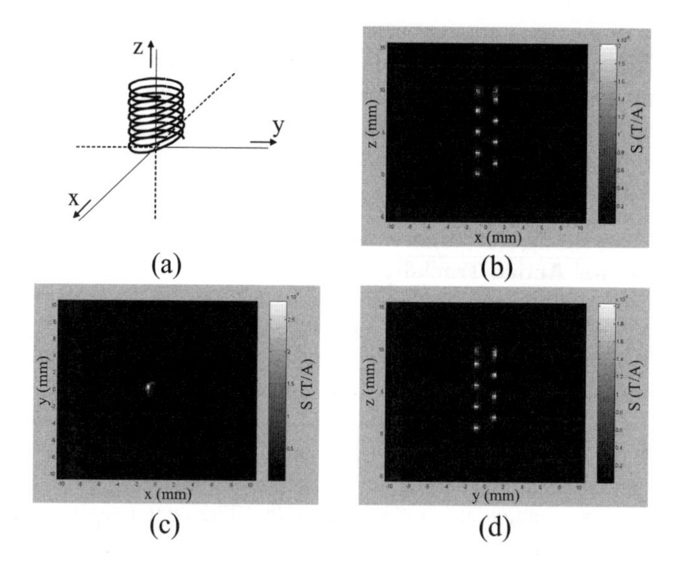

Figure 2.43 Double-helix antenna: geometry and sensitivity patterns in xz, xy and yz planes. Height 10 mm, radius 1 mm, 4 turns up and 4 turns down, 8 segments per circumference. (a) Antenna geometry. (b) Sensitivity pattern in xz plane, $y = 0.05$ mm. (c) Sensitivity pattern in xy plane, $z = 5$ mm. (d) Sensitivity pattern in yz plane, $x = 0.05$ mm.

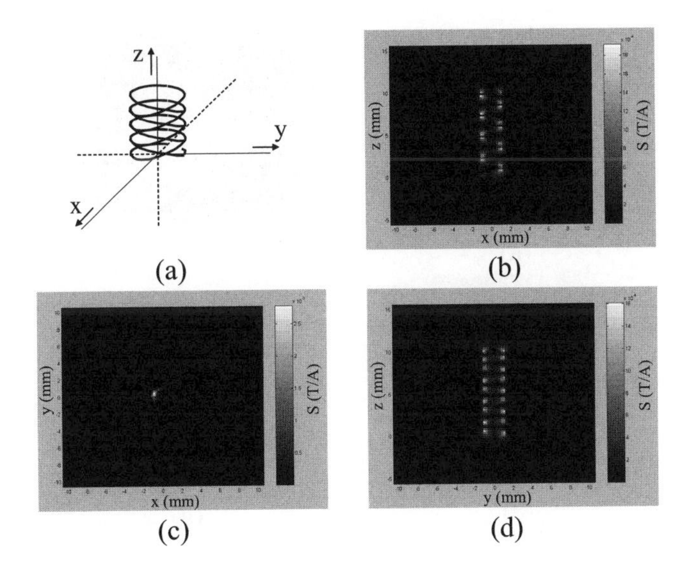

Figure 2.44 Opposed-double-helix antenna: geometry and sensitivity patterns in xz, xy and yz planes. Height 10 mm, radius 1 mm, 4 turns up and 4 turns down, 8 segments per circumference. (a) Antenna geometry. (b) Sensitivity pattern in xz plane, $y = 0.05$ mm. (c) Sensitivity pattern in xy plane, $z = 5$ mm. (d) Sensitivity pattern in yz plane, $x = 0.05$ mm.

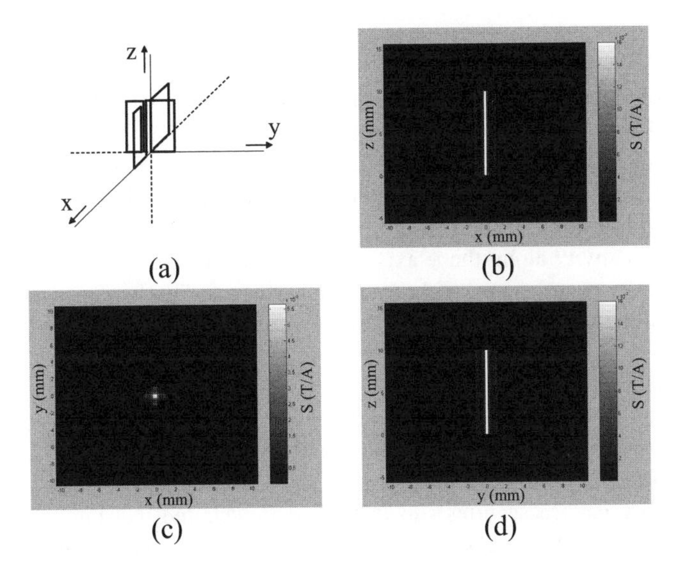

Figure 2.45 Center return antenna: geometry and sensitivity patterns in xz, xy and yz planes. Height 10 mm, radius 1 mm, 4 wires, 8 segments per circumference. (a) Antenna geometry. (b) Sensitivity pattern in xz plane, $y = 0.05$ mm. (c) Sensitivity pattern in xy plane, $z = 5$ mm. (d) Sensitivity pattern in yz plane, $x = 0.05$ mm.

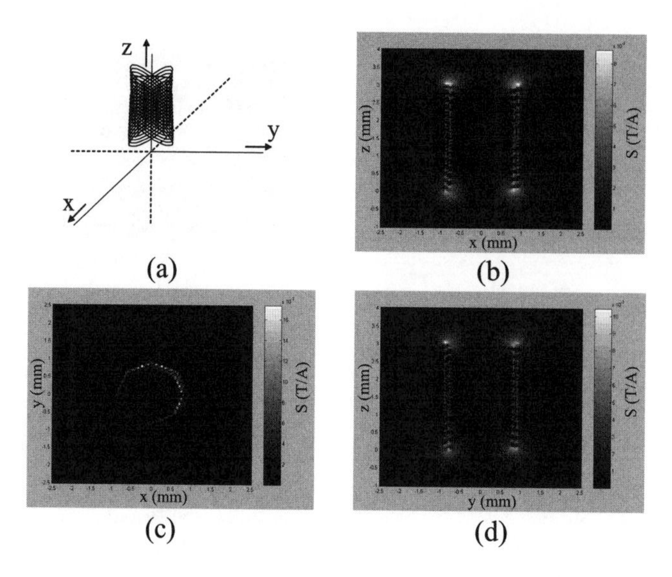

Figure 2.46 Perpendicular-coil antenna: geometry and sensitivity patterns in xz, xy and yz planes. Height 3 mm, radius of inner coil 0.8 mm, radius of outer coil 0.9 mm, 15 turns up and 15 turns down, 8 segments per circumference, coils placed at an angle of $\pi/4$ with respect to the main magnetic-field direction. (a) Antenna geometry. (b) Sensitivity pattern in xz plane, $y = 0.05$ mm. (c) Sensitivity pattern in xy plane, $z = 1.5$ mm. (d) Sensitivity pattern in yz plane, $x = 0.05$ mm.

sensitivity as a function of the distance from the antenna body at antenna half height in the planes $y = 0.05$ mm and $x = 0.05$ mm, respectively.

These figures reveal that the center return antenna shows a highly localized sensitivity on the cylinder axis. The double-helix antenna and dual-opposed helix antenna show – for the chosen z-axis position – good localized sensitivity near the antenna body along the x axis, but poor sensitivity along the y axis. The perpendicular-coils antenna maintains good localized sensitivity near the antenna body along both axes. This becomes more evident from cross sections of the sensitivity patterns when the sensitivity is plotted on a logarithmic scale (Figures 2.49 and 2.50).

Figures 2.49 and 2.50 also show clearly that the sensitivity pattern of the antiparallel-wire antenna is not very localized. Moreover, the sensitivity pattern of this antenna depends strongly on the observation angle in the transverse (xy) plane, which makes this antenna type unsuitable for imaging purposes also.

The sensitivity pattern sections shown have demonstrated that our earlier selection of antenna concepts for active tracking, supplemented with the center return antenna, was a correct one. The quantitative comparison of the antenna concepts reveals that the center return antenna is the best suited for active tracking, judging from magnetic-field considerations only. If we also take manufacturing aspects into account, meaning that we have a preference for an antenna geometry that is situated on the *outside* of a cylindrical body only, the perpendicular-coils antenna is best suited for the job. This antenna combines a localized sensitivity that is

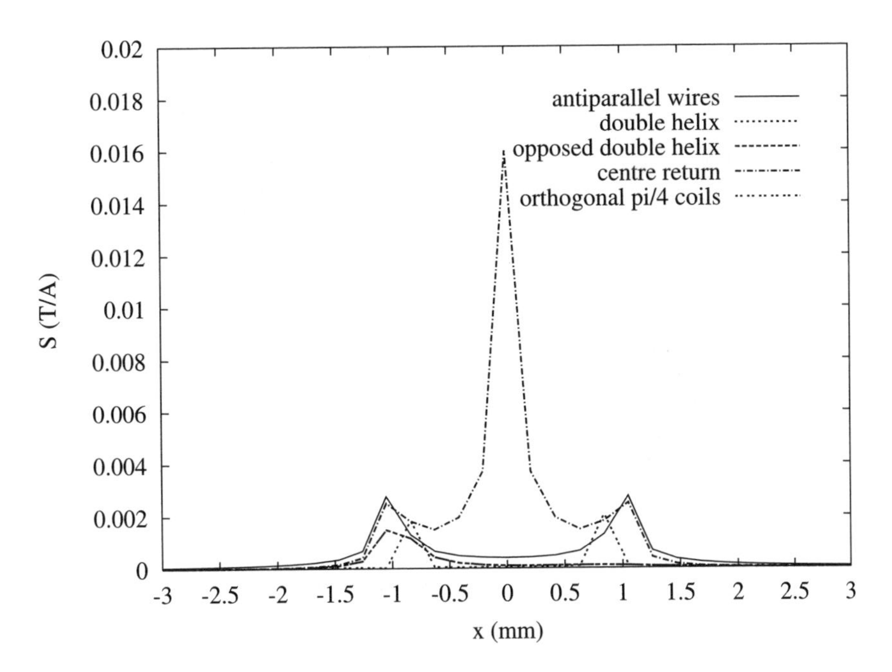

Figure 2.47 Cross section of sensitivity patterns of antennas at antenna half height along the x axis.

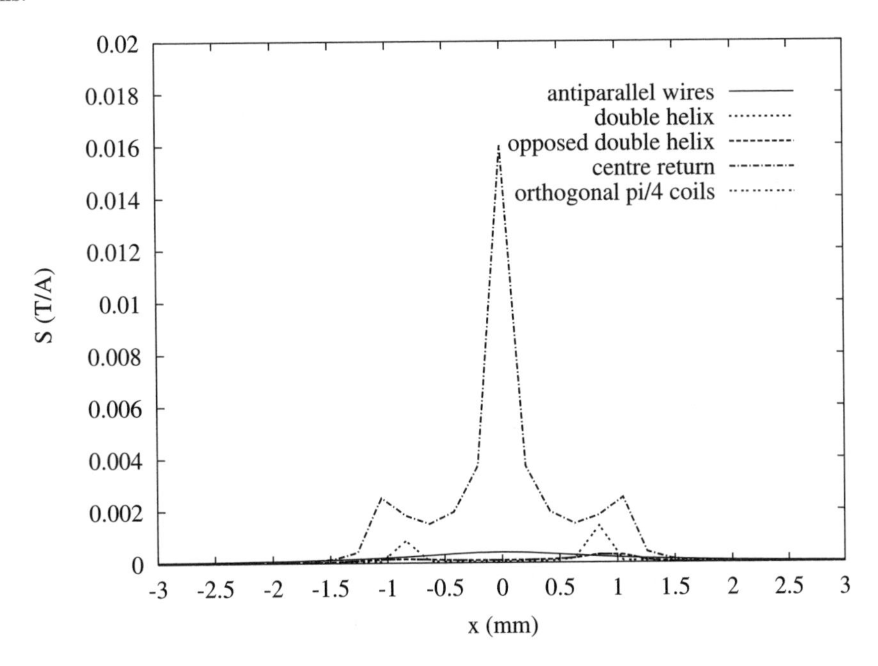

Figure 2.48 Cross section of sensitivity patterns of antennas at antenna half height along the y axis.

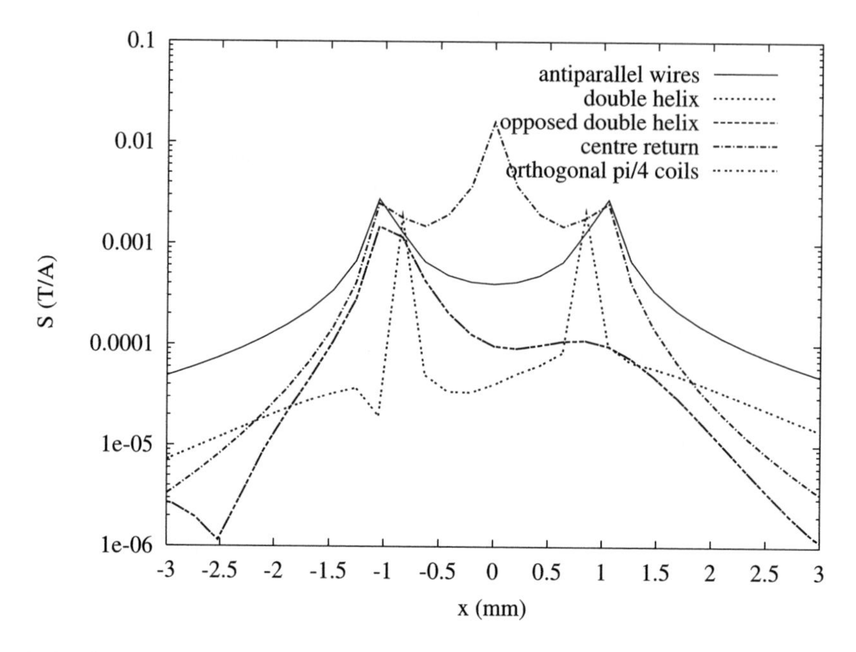

Figure 2.49 Cross section of sensitivity patterns of antennas on a logarithmic scale at antenna half height along the x axis.

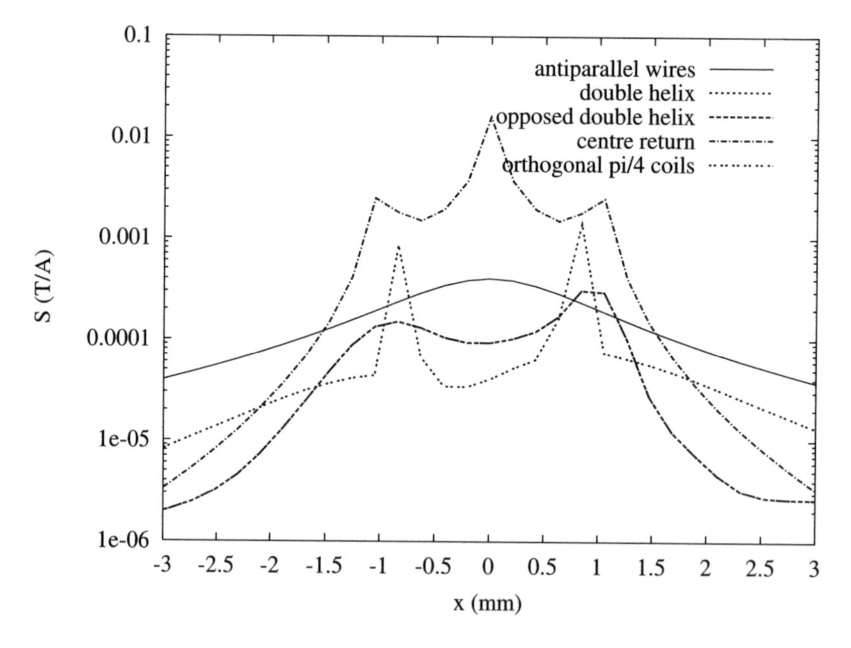

Figure 2.50 Cross section of sensitivity patterns of antennas on a logarithmic scale at antenna half height along the y axis.

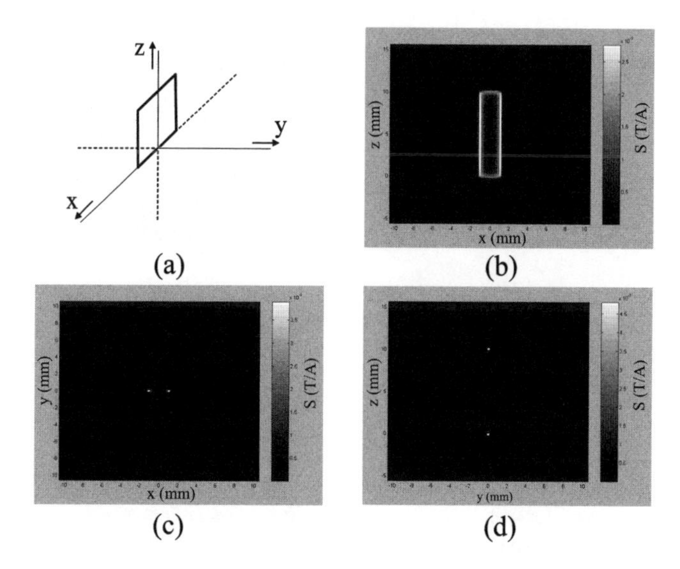

Figure 2.51 Single-loop antenna: geometry and sensitivity patterns in xz, xy and yz planes. Height 10 mm, radius 1 mm. (a) Antenna geometry. (b) Sensitivity pattern in xz plane, $y = 0.05$ mm. (c) Sensitivity pattern in xy plane, $z = 5$ mm. (d) Sensitivity pattern in yz plane, $x = 0.05$ mm.

independent of the transverse observation angle with a geometry that is restricted to the outer surface of the antenna body.

2.5.2 Antennas for Intravascular Imaging

For intravascular imaging, the antenna should show a sensitivity that extends from the antenna body to the vascular wall and is, preferably, homogeneous along the direction of the antenna body. Furthermore, the sensitivity should be independent of the observation angle in the transverse plane. In Figures 2.51–2.56 we show, respectively, the geometry of the single-loop antenna, the double-loop antenna, the triple-loop antenna, the dual-opposed-solenoids antenna, the saddle coil antenna and the birdcage antenna together with, for each antenna, three perpendicular sensitivity profiles. The specifics of the antennas and sensitivity profile positions are specified in the figure captions.

All of the antennas, with the exception of the dual-opposed-solenoids antenna, show a homogeneous sensitivity along the antenna body. It should be noted that the behavior of the single-, double- and triple-loop antennas is very similar. This is not directly visible from Figures 2.51–2.53, owing to the fact that the sensitivity pattern sections parallel to the x axis are taken close to, distant from and close to loops, respectively, in the structure. Although the homogeneity of the sensitivity along the antenna body for the dual-opposed-solenoids antenna is less than that for the other antennas, Figure 2.54 reveals that the sensitivity in the radial direction exceeds those for the other antennas.

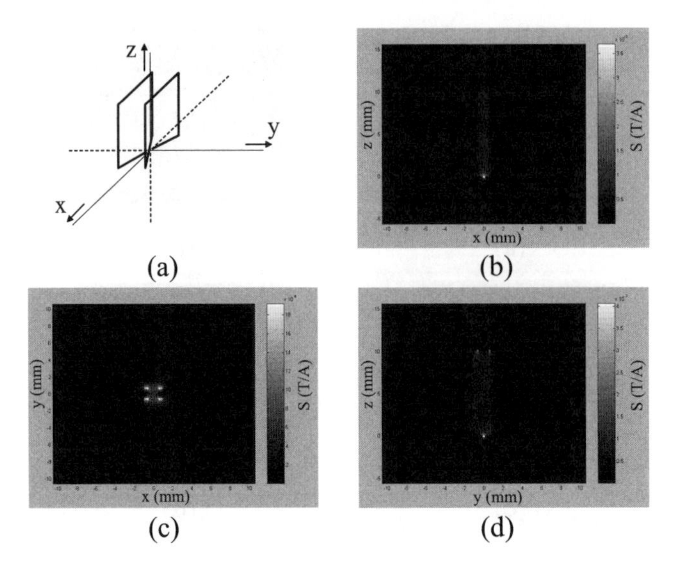

Figure 2.52 Double-loop antenna: geometry and sensitivity patterns in *xz*, *xy* and *yz* planes. Height 10 mm, radius 1 mm, distance between adjacent loops 1.32 mm. (a) Antenna geometry. (b) Sensitivity pattern in *xz* plane, $y = 0.05$ mm. (c) Sensitivity pattern in *xy* plane, $z = 5$ mm. (d) Sensitivity pattern in *yz* plane, $x = 0.05$ mm.

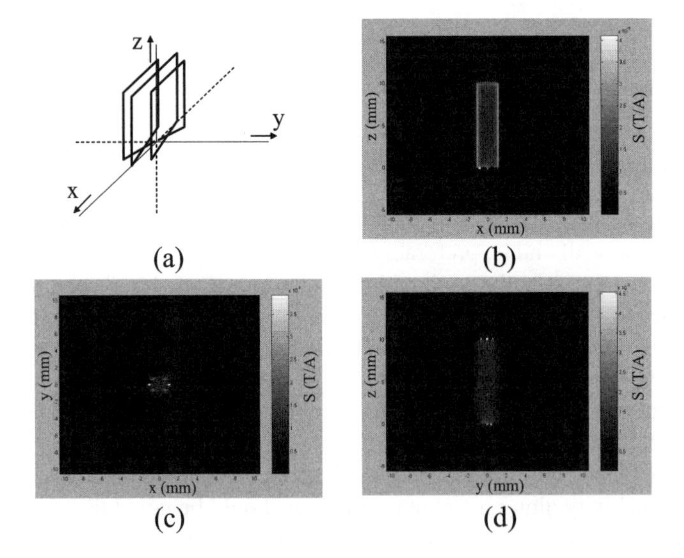

Figure 2.53 Triple-loop antenna: geometry and sensitivity patterns in *xz*, *xy* and *yz* planes. Height 10 mm, radius 1 mm, distance between adjacent loops 0.66 mm. (a) Antenna geometry. (b) Sensitivity pattern in *xz* plane, $y = 0.05$ mm. (c) Sensitivity pattern in *xy* plane, $z = 5$ mm. (d) Sensitivity pattern in *yz* plane, $x = 0.05$ mm.

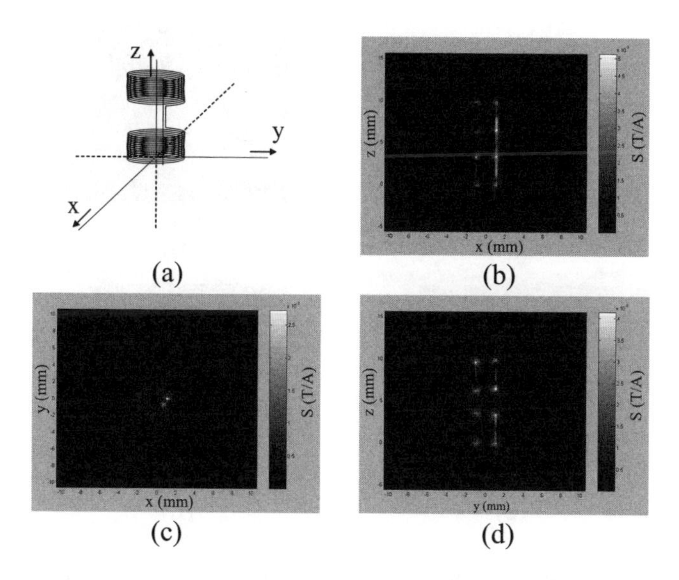

Figure 2.54 Dual-opposed-solenoids antenna: geometry and sensitivity patterns in xz, xy and yz planes. Height per coil 3 mm, gap between coils 3 mm, 15 turns per coil, radius 1 mm, 8 segments per circumference. (a) Antenna geometry. (b) Sensitivity pattern in xz plane, $y = 0.05$ mm. (c) Sensitivity pattern in xy plane, $z = 4.5$ mm. (d) Sensitivity pattern in yz plane, $x = 0.05$ mm.

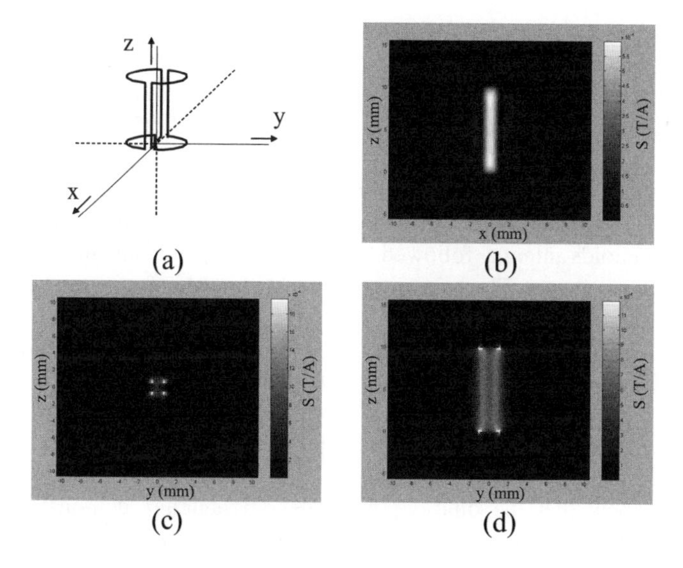

Figure 2.55 Saddle coil antenna: geometry and sensitivity patterns in xz, xy and yz planes. Height 10 mm, radius 1 mm, distance between the two parts 0.8 mm, 8 segments per circumference. (a) Antenna geometry. (b) Sensitivity pattern in xz plane, $y = 0.05$ mm. (c) Sensitivity pattern in xy plane, $z = 5$ mm. (d) Sensitivity pattern in yz plane, $x = 0.05$ mm.

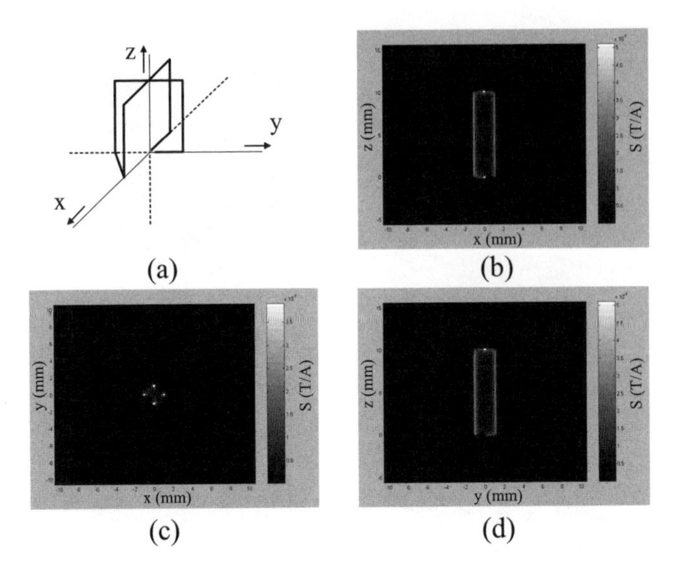

Figure 2.56 Birdcage antenna: geometry and sensitivity patterns in *xz*, *xy* and *yz* planes. Height 10 mm, radius 1 mm, 4 wires. (a) Antenna geometry. (b) Sensitivity pattern in *xz* plane, $y = 0.05$ mm. (c) Sensitivity pattern in *xy* plane, $z = 5$ mm. (d) Sensitivity pattern in *yz* plane, $x = 0.05$ mm.

For a better comparison of the various antenna concepts, we shall look again at the behavior of the sensitivity as a function of the perpendicular distance from the cylindrical antenna body. Figures 2.57 and 2.58 show the sensitivity as a function of the distance from the antenna body at antenna half height in the planes $y = 0.05$ mm and $x = 0.05$ mm, respectively.

For the distances of interest, i.e. where we may expect the vessel wall (2–3 mm for the large arteries [18]), we see that the best antenna, judging from the magnetic field only, is the dual-opposed-solenoids antenna, followed by the triple-loop antenna, the saddle coil antenna and the double-loop antenna. The behavior of the last two antennas is nearly identical for the number of wires and loops chosen.

To complete the quantitative comparison of the antenna concepts, we have to look at the homogeneity of the sensitivity in the *transverse* plane. To that end, we have calculated the sensitivity at half the height of the antenna body as a function of the observation angle. The results are shown in Figures 2.59 and 2.60 for a radial distance from the antenna body axis of, respectively, 2 mm and 4 mm.

The figures show that the dual-opposed-solenoids antenna demonstrates the highest sensitivity levels, but that the triple-loop antenna outperforms the dual-opposed-solenoids antenna with respect to sensitivity homogeneity at distances closer to the antenna body. Further away from the antenna body, the sensitivity becomes comparable for all antennas. The manufacturing of both the dual-opposed-solenoid antenna and the triple-loop antenna is expected to be equal in complexity. Therefore, both antenna types are regarded as suitable for imaging purposes.

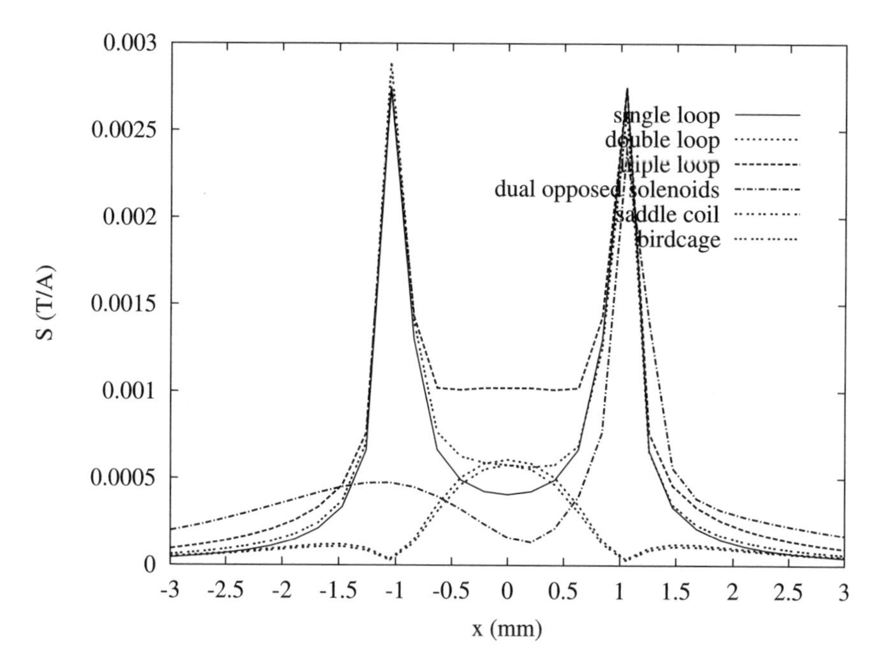

Figure 2.57 Cross section of antenna sensitivity patterns at antenna half height along the x axis.

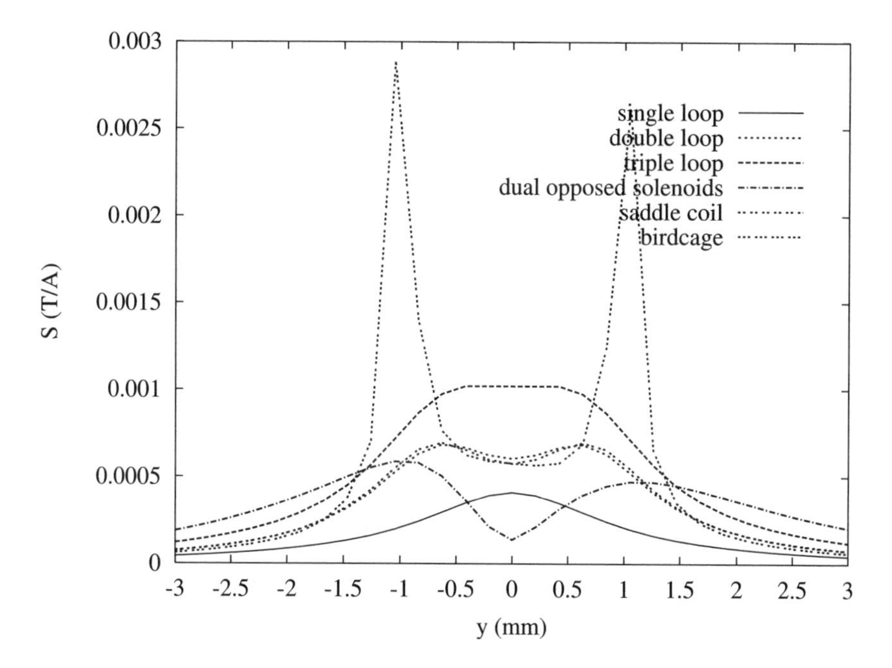

Figure 2.58 Cross section of antenna sensitivity patterns at antenna half height along the y axis.

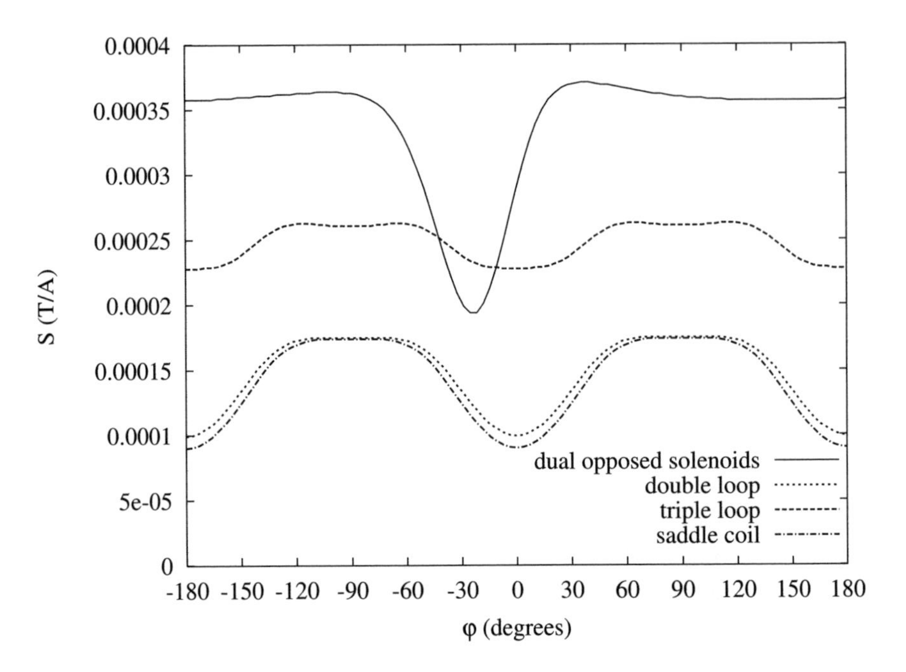

Figure 2.59 Antenna sensitivity in the half-height plane as a function of the observation angle at distance $R = 2$ mm from the antenna body axis.

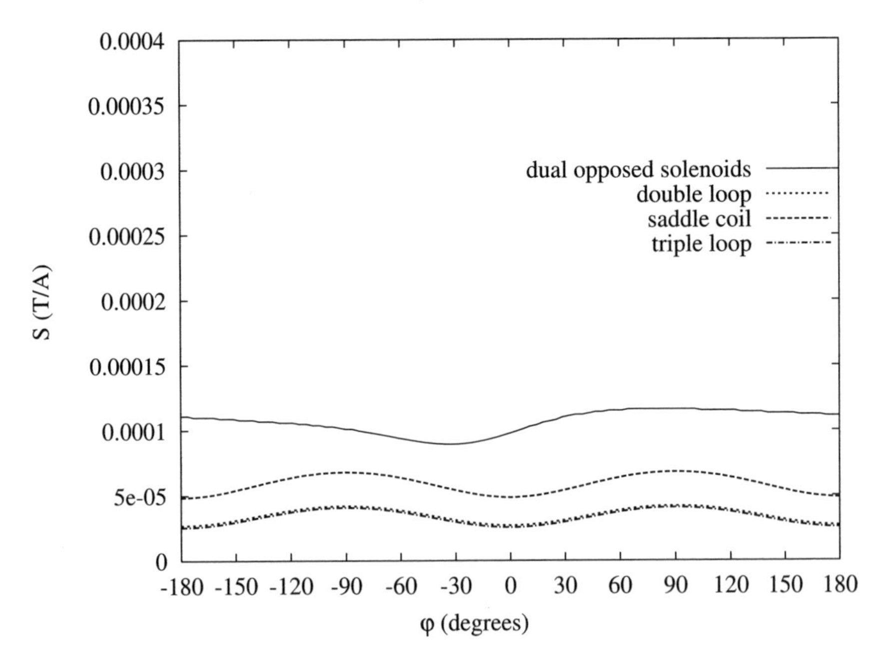

Figure 2.60 Antenna sensitivity in the half-height plane as a function of the observation angle at distance $R = 4$ mm from the antenna body axis.

2.5.3 Antenna Rotation

Although the human vascular system is mainly 'head-to-toe' directed, parts of the system will have different directions. The MR main magnetic field is also 'head-to-toe' directed. For the active tracking of catheters through arteries that are not lined up with the main magnetic field or for imaging the walls of these arteries, it is important to compare the different antenna concepts with respect to antenna rotation.

For all simulations described thus far, the antenna body directed along the direction of the main magnetic field and the sensitivity was calculated for a magnetic-field distortion transverse with respect to the antenna body direction. To analyze the effects of antenna rotation, we make use of the rotation transformations $[R_x(\alpha)]$ for a rotation angle α around the x axis, $[R_y(\beta)]$ for a rotation angle β around the y axis and $[R_z(\gamma)]$ for a rotation angle γ around the z axis [40]:

$$[R_x(\alpha)] = \begin{bmatrix} 1 & 0 & 0 \\ 0 & \cos(\alpha) & \sin(\alpha) \\ 0 & -\sin(\alpha) & \cos(\alpha) \end{bmatrix}, \tag{2.64}$$

$$[R_y(\beta)] = \begin{bmatrix} \cos(\beta) & 0 & -\sin(\beta) \\ 0 & 1 & 0 \\ \sin(\beta) & 0 & \cos(\beta) \end{bmatrix}, \tag{2.65}$$

$$[R_z(\gamma)] = \begin{bmatrix} \cos(\gamma) & \sin(\gamma) & 0 \\ -\sin(\gamma) & \cos(\gamma) & 0 \\ 0 & 0 & 1 \end{bmatrix}. \tag{2.66}$$

For an antenna rotated by the angles α, β and γ, we find the magnetic-flux-density components (B'_x, B'_y, B'_z) from the 'unrotated' components (B_x, B_y, B_z) using

$$\begin{bmatrix} B'_x \\ B'_y \\ B'_z \end{bmatrix} = \begin{bmatrix} R_{xx} & R_{xy} & R_{xz} \\ R_{yx} & R_{yy} & R_{yz} \\ R_{zx} & R_{zy} & R_{zz} \end{bmatrix} \begin{bmatrix} B_x \\ B_y \\ B_z \end{bmatrix}, \tag{2.67}$$

where the transformation matrix is obtained by multiplication of $[R_x(\alpha)]$, $[R_y(\beta)]$ and $[R_z(\gamma)]$. This is equivalent to first rotating by γ, then rotating by β and finally rotating by α.[16] The matrix elements R_{ij}, $i, j = x, y, z$, are then found to be

$$R_{xx} = \cos(\beta)\cos(\gamma), \tag{2.68}$$

$$R_{xy} = \cos(\beta)\sin(\gamma), \tag{2.69}$$

$$R_{xz} = -\sin(\beta), \tag{2.70}$$

$$R_{yx} = -\cos(\alpha)\sin(\gamma) + \sin(\alpha)\sin(\beta)\cos(\gamma), \tag{2.71}$$

[16]These rotation operations are commutative and associative.

$$R_{yy} = \cos(\alpha)\cos(\gamma) + \sin(\alpha)\sin(\beta)\sin(\gamma), \quad (2.72)$$
$$R_{yz} = \sin(\alpha)\cos(\beta), \quad (2.73)$$
$$R_{zx} = \sin(\alpha)\sin(\gamma) + \cos(\alpha)\sin(\beta)\cos(\gamma), \quad (2.74)$$
$$R_{zy} = -\sin(\alpha)\cos(\gamma) + \cos(\alpha)\sin(\beta)\sin(\gamma), \quad (2.75)$$
$$R_{zz} = \cos(\alpha)\cos(\beta). \quad (2.76)$$

The sensitivity of the rotated antenna is given by

$$S = \frac{1}{I}\sqrt{B_x'^2 + B_y'^2}. \quad (2.77)$$

To demonstrate the effects of antenna rotation, we shall rotate the antennas that we found best for tracking and the ones that we found best for imaging around the x axis. The sensitivity, expressed in terms of the unrotated magnetic-flux-density components, is then given by

$$S = \frac{1}{I}\sqrt{B_x^2 + (B_y\cos(\alpha) + B_z\sin(\alpha))^2}, \quad (2.78)$$

where α is the rotation angle.

2.5.3.1 Rotation of Antennas for Active Tracking In section 2.5.1, we found that the center return antenna and the perpendicular-coils antenna were best suited for active-tracking purposes. In Figures 2.61 and 2.62, we show the sensitivity patterns of these antennas in the xy plane at half height for rotation angles of 0° (no rotation), 45°, 60° and 90°. The scaling of the sensitivity in both figures has been adjusted to maximize the visibility of the effects of rotation on the sensitivity pattern sections.

These figures show that for both antennas, up to large angles, the sensitivity patterns remain homogeneous. The perpendicular-coils antenna maintains a homogeneous pattern even up to 90° rotation. This behavior, added to the ease of manufacturing, makes this antenna concept stand out for active-tracking purposes.[17]

2.5.3.2 Rotation of Antennas for Imaging In section 2.5.2, we found that the triple-loop antenna and the dual-opposed-solenoids antenna were the most promising for imaging purposes. In Figures 2.63 and 2.64, we show the sensitivity patterns of these antennas in the xy plane at half height for rotation angles of 0° (no rotation), 45°, 60° and 90°. The scaling of the sensitivity has again been adjusted to maximize the visibility of the effects of rotation on the sensitivity pattern sections.

These figures show that for angles from 45° upwards, the radial sensitivity rapidly loses homogeneity. For a rotation angle of 90°, the radial sensitivity of the triple-loop antenna has formed four distinct lobes and that of the dual-opposed-solenoids antenna has assumed the form of a two-lobe pattern. The increase in radial inhomogeneity with rotation angle seems to be more severe for the triple-loop antenna. Both antennas appear to be suitable for imaging up to rotation angles of 45°.

[17]This antenna is positioned completely on the outside of the antenna body, as opposed to the center return antenna, which has an additional wire segment passing through the axis of the cylindrical antenna body.

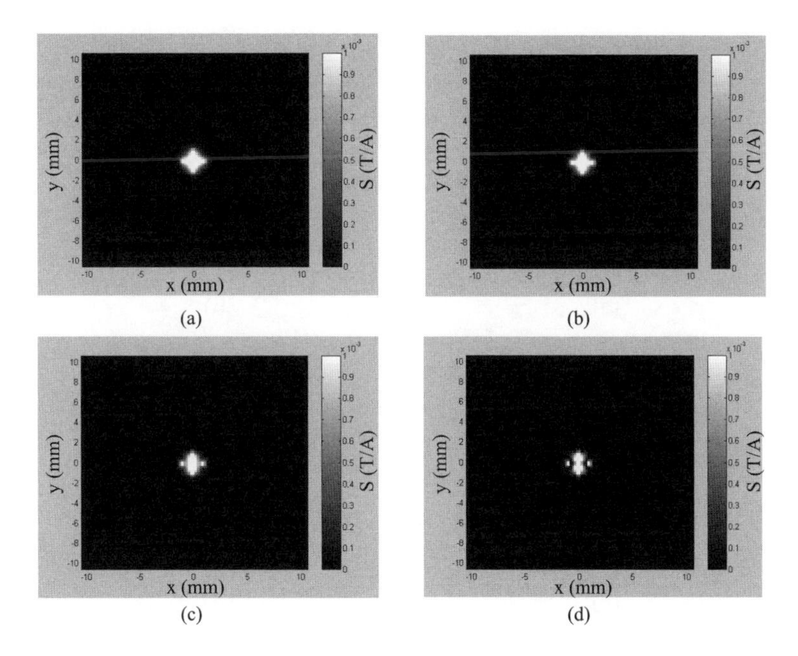

Figure 2.61 Center return antenna: sensitivity in the xy plane at half height for different rotation angles. (a) $\alpha = 0°$, (b) $\alpha = 45°$, (c) $\alpha = 60°$, (d) $\alpha = 90°$.

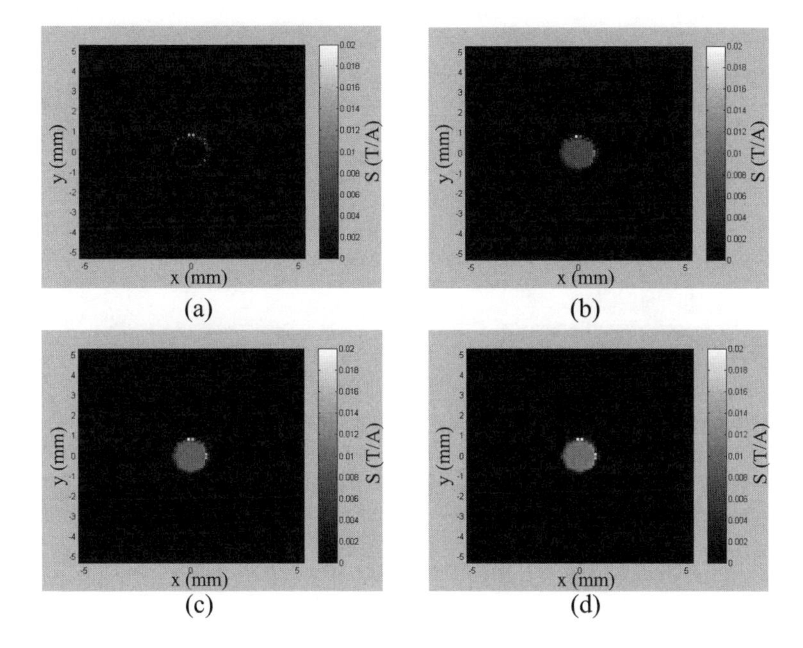

Figure 2.62 Perpendicular-coils antenna: sensitivity in the xy plane at half height for different rotation angles. (a) $\alpha = 0°$, (b) $\alpha = 45°$, (c) $\alpha = 60°$, (d) $\alpha = 90°$.

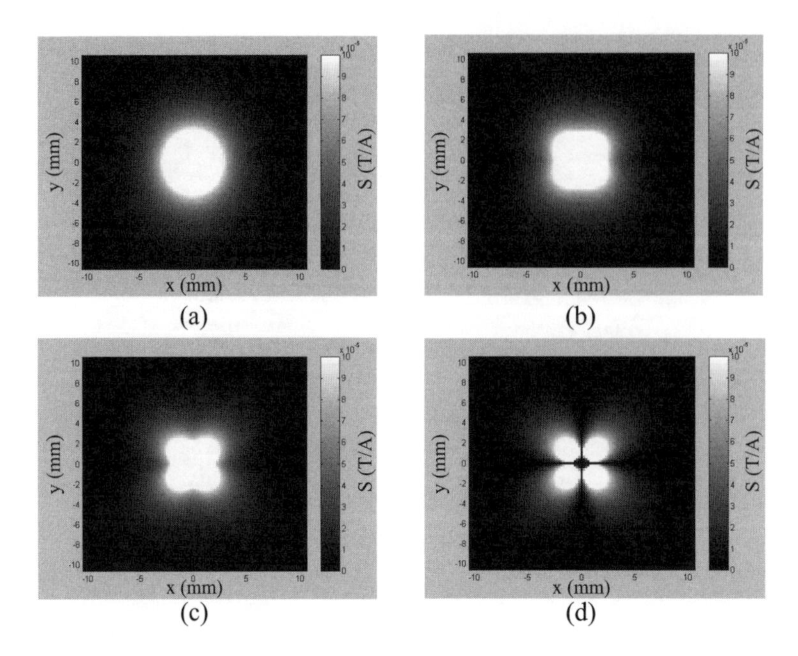

Figure 2.63 Triple-loop antenna: sensitivity in the xy plane at half height for different rotation angles. (a) $\alpha = 0°$, (b) $\alpha = 45°$, (c) $\alpha = 60°$, (d) $\alpha = 90°$.

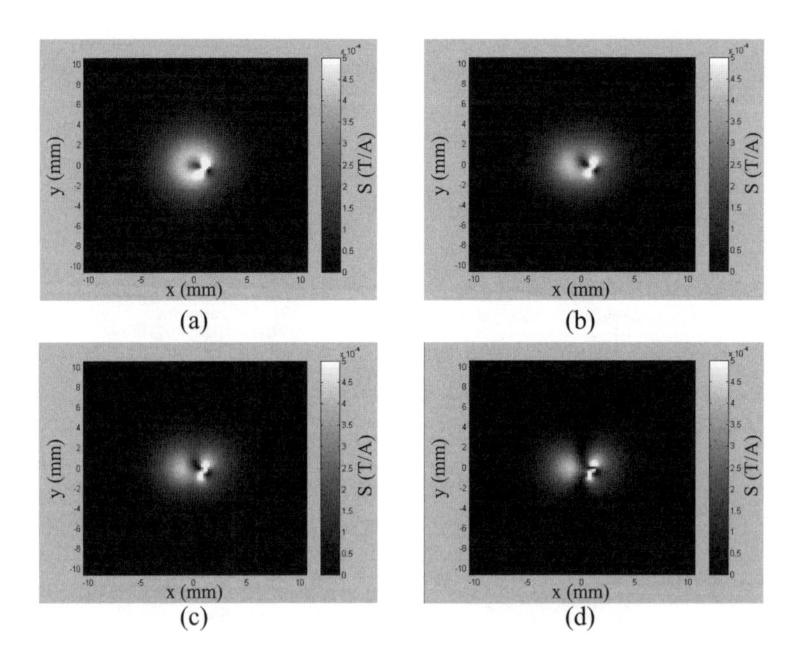

Figure 2.64 Dual-opposed-solenoids antenna: sensitivity in the xy plane at half height for different rotation angles. (a) $\alpha = 0°$, (b) $\alpha = 45°$, (c) $\alpha = 60°$, (d) $\alpha = 90°$.

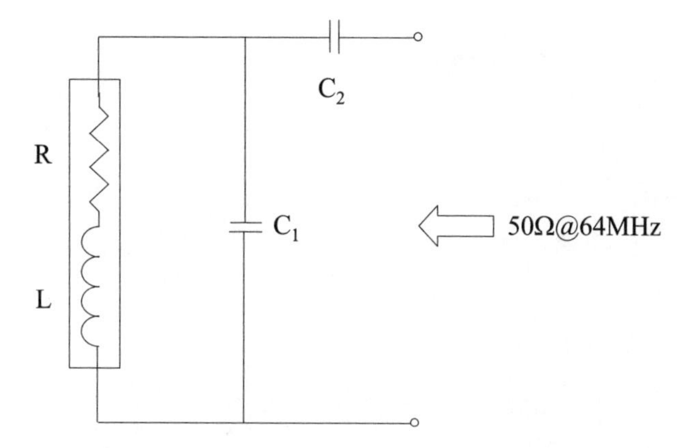

Figure 2.65 Intravascular coil, depicted as a series circuit of an inductor and a resistor with tuning and impedance-matching circuit.

2.6 IN VITRO TESTING

The final validation of the analytical model developed was delivered by comparing the calculated sensitivity patterns for the different antennas with images created by an MR system where the intravascular antennas were used for active tracking. To ensure maximum power transfer between an intravascular antenna and the external circuitry, the antenna needs to be resonant at 64 MHz and impedance-matched to the transmission line that connects the antenna to the external circuitry.

2.6.1 Sensitivity Pattern

A number of antennas were constructed. The complex input impedances of these antennas were measured while the antennas were immersed in tap water, which served for this purpose as a blood-mimicking fluid.[18] Next, a parallel and a series capacitor were added (Figure 2.65) to accomplish tuning at 64 MHz and impedance matching to a 50 Ω coaxial transmission line.

Owing to inaccuracies in the measurements, the unavailability of a well-defined blood-mimicking fluid, the availability of only a limited set of discrete-valued capacitors and, most of all, the fact that the antennas were hand-made, poorly reproducible products (Figure 2.66), the tuning and matching was not optimal for most antennas. Therefore the SNR realized for the antennas left room for improvement.

The first set of prototype antennas were constructed with materials available at that moment and therefore did not have sizes suitable for a clinical application. A center return antenna consisting of four wires, equally spaced around the circumference of a cylindrical body, was constructed (Figure 2.66). The radius of the antenna body was 4 mm, and the height of the antenna was 16 mm. A triple-loop antenna was made on a cylindrical body

[18] As we did not have the possibility to create a saline solution at the time of measurement, the use of tap water ($\sigma \approx 0.01$ S m^{-1} [41]) was preferred over the use of distilled water ($\sigma \approx 0.0001$ S m^{-1} [41]).

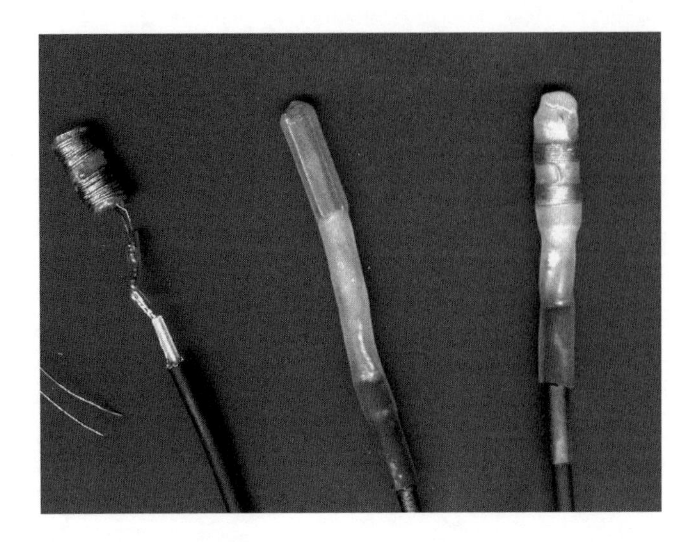

Figure 2.66 Realized intravascular-antenna prototypes. Reproduced by permission on Nicole Op en Camp.

with a radius of 2.5 mm and height of 16 mm. The separation between two adjacent coils was 2 mm. Finally, an opposed-solenoids antenna was constructed (see Figure 2.66 again), consisting of two coils with a height of 3 mm and nine turns, separated by 3 mm and having a total height of 16 mm and a radius of 4 mm.

In Figures 2.67–2.69, we show, for, the center return antenna, the triple-loop antenna and the dual-opposed-solenoids antenna, respectively, the calculated sensitivity patterns and the MR images obtained for various antenna rotation angles [9]. Although the MR images are directly related to the sensitivity, the exact values have been lost in the signal processing. The calculated sensitivity profiles have been scaled to achieve a visual match with the MR images.

The measurements were performed with the antennas in the setting of a phantom made of Perspex. The surrounding medium was a blood-mimicking fluid created by dissolving 2 mg of manganese chloride ($MnCl_2$) per liter of water [42]. By choosing a thicker slice for the measurement, a higher SNR may be achieved. The slice thickness was chosen per measurement. The slice thickness for the 0° rotation angle in the measurement of the center return antenna was 7 mm. The thicknesses for the 45° and 90° rotation angles were 15 mm. The slice thicknesses for all measurements of the triple-loop antenna were 7 mm. The slice thickness for the 0° rotation angle in the measurement of the dual-opposed-solenoids antenna was 30 mm. The thicknesses for the 45° and 90° rotation angles were 7 mm.

To demonstrate the influence of the slice thickness, the MR image for a 0° rotation angle and the dual-opposed-solenoids antenna is shown in Figure 2.70 for two slice thicknesses [9].

Taking the inaccuracies mentioned earlier into account and noting that in the construction of the center return antenna, the lumen of the antenna was filled with a contrast agent and that in the construction of the triple-loop antenna and the dual-opposed-solenoids antenna, the lumen of the antenna was filled with a silicone gel [9], not accounted for by the analytical model, the calculations and measurements show good agreement. So, again, the validity of

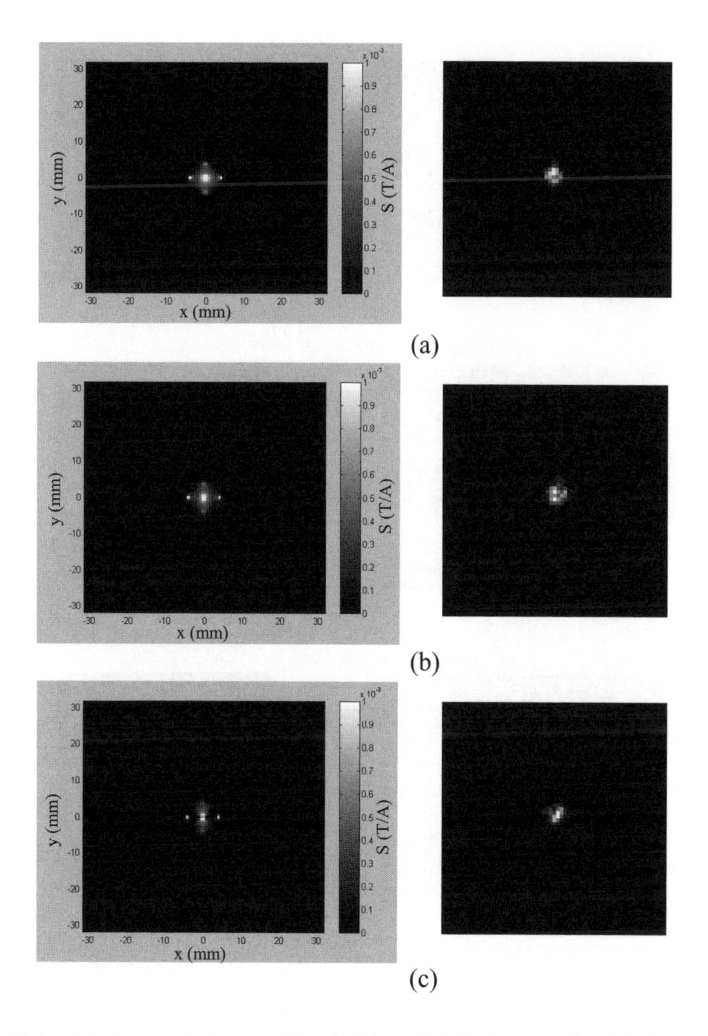

Figure 2.67 Calculated sensitivity profiles (left) and MR images (right) for the prototype center return antenna. (a) Rotation angle 0°. (b) Rotation angle 45°. (c) Rotation angle 90°.

the model developed has been proved. This leaves us with the task of investigating whether tracking works in practice.

2.6.2 Tracking

For the tracking experiments, a new prototype antenna was constructed based on the perpendicular-coils antenna. The antenna consisted of only one coil, of height 3 mm, at 45° from the antenna body axis, consisting of 15 turns wound around the tip of a 1.67 mm diameter catheter with 0.09 mm diameter insulated copper wire [12]. The reason for having only a single coil instead of two was the ease of realizing this one-coil antenna. The copper

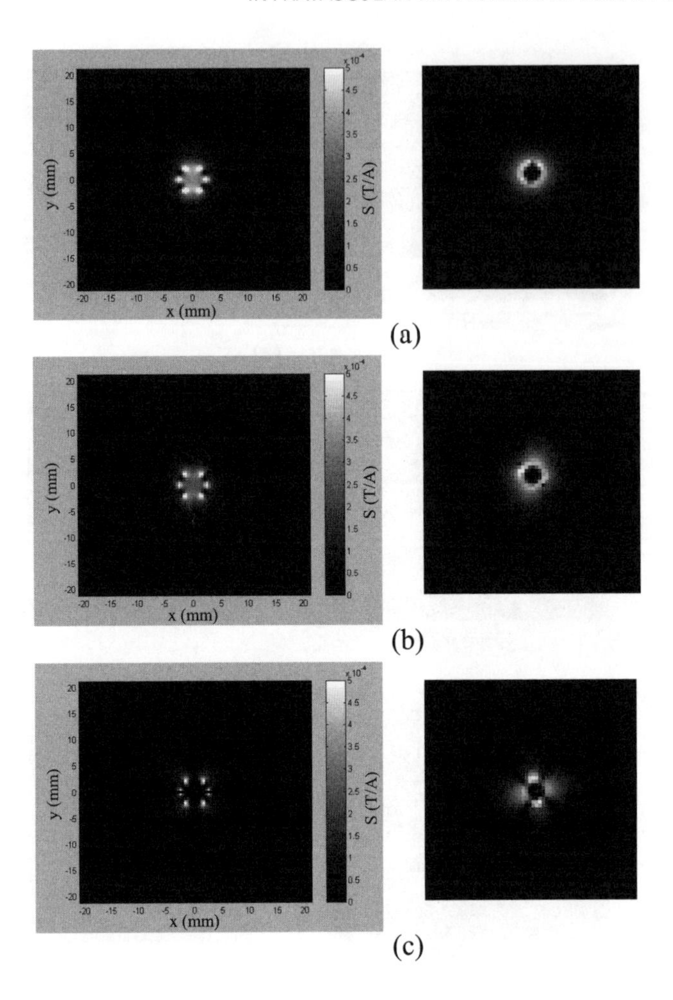

Figure 2.68 Calculated sensitivity profiles (left) and MR images (right) for the prototype triple-loop antenna. (a) Rotation angle 0°. (b) Rotation angle 45°. (c) Rotation angle 90°.

wire leads were twisted over the length of the catheter, to minimize their influence on the magnetic field, and were connected to a coaxial transmission line at the end of the catheter [12].

With the catheter immersed in a phantom filled with a blood-mimicking fluid and the lumen of the catheter also filled with the same fluid, interactive MR scans with active tracking were performed for various rotation angles of the antenna. Measurements were taken over a period of one minute for every rotation angle. The measured antenna positions (indicated with dots) are shown in Figure 2.71 for this '45° coil antenna' at the pixel level, where the squares in the figure represent the pixels of the underlying MR image [12]. The pixel size is 1.3722 mm × 1.3722 mm.

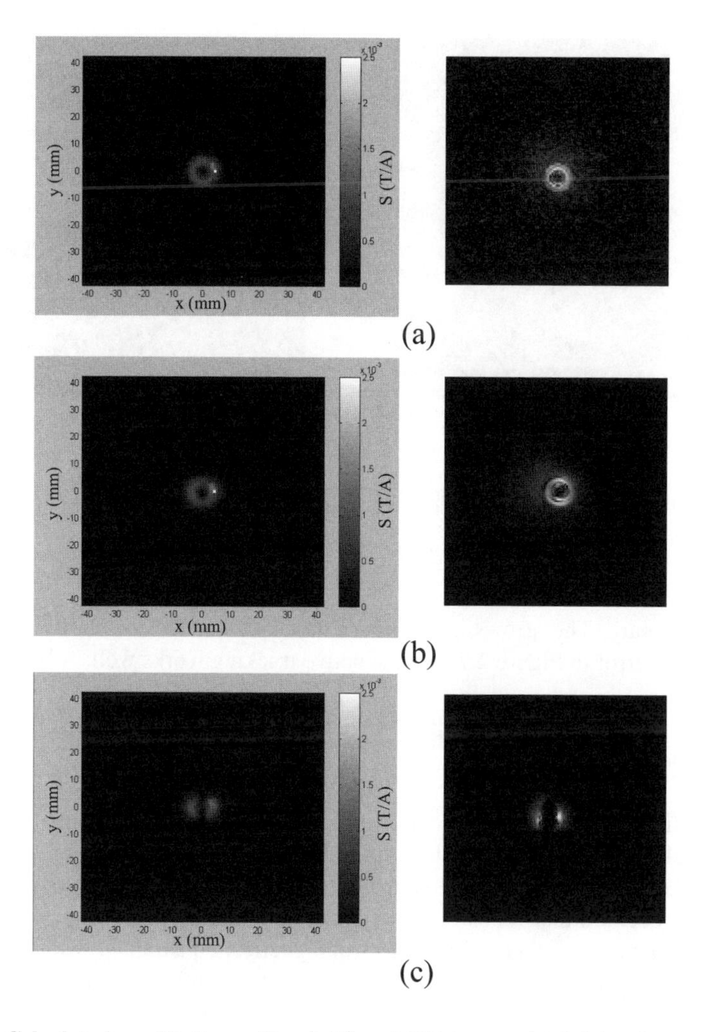

Figure 2.69 Calculated sensitivity profiles (left) and MR images (right) for the prototype dual-opposed-solenoids antenna. (a) Rotation angle 0°. (b) Rotation angle 45°. (c) Rotation angle 90°.

For reference, the same exercise was repeated with an 'ordinary coil antenna', the results of which are shown in Figure 2.72. The figures indicate the superiority of the 45° coil antenna over the ordinary coil antenna. Even better results may be expected from employing a perpendicular-coils antenna, since the sensitivity pattern of this antenna will be more concentrated, as explained in section 2.3.1.

Finally, the 45° coil antenna was inserted into a human-abdomen vascular phantom. The catheter carrying the antenna was inserted via a guide wire into the phantom and then guided through the vessels. The catheter tip positions were measured during this movement, and snapshots of this process are shown in Figure 2.73 [12]. The catheter tip position is indicated

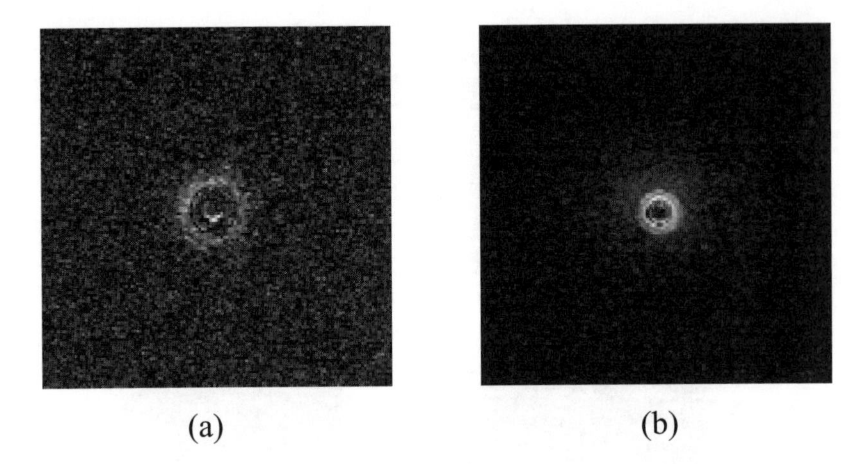

Figure 2.70 MR images for nonrotated prototype dual-opposed-solenoids antenna. (a) Slice thickness 7 mm. (b) Slice thickness 30 mm.

by a white '+' mark. The arrows in the figure have been added for clarity. Apart from the single horizontal error in Figure 2.73(b), the active tracking works well.

2.7 ANTENNA SYNTHESIS

With the availability of an analytical model that – when implemented in software – generates reliable results in a very short time, the possibility has been created to generate or synthesize antenna designs automatically within a reasonable time frame. With an optimization procedure that relies on function evaluations only, a design may be realized, subject to user-defined mechanical and electromagnetical constraints, within a few minutes on standard office computing equipment. Two examples of optimization procedures that need function evaluations only are *simulated annealing* [29] and *genetic algorithms* [43,44]. Here we have opted specifically for the latter method, owing to its 'natural' appeal and its ease of software implementation.

2.7.1 Genetic-Algorithm Optimization

Genetic algorithms (GAs) are optimization methods based upon the principles of natural selection and evolution. The concepts used in the optimization process are genes, chromosomes, generations, populations, parents, children and fitness. A gene is a coded version of one of the parameters of the problem. A possible coding is a binary coding, making the gene a string of zeros and ones. A chromosome is a series of genes and is thus a solution of the problem. A gene is also known as an 'individual'. A population is a set of individuals. A generation is a population iteratively formed from the previous one. A parent is an individual from the previous generation, and a child is an individual from the current generation. The fitness is a number assigned to an individual and is a measure of 'how good' this individual is.

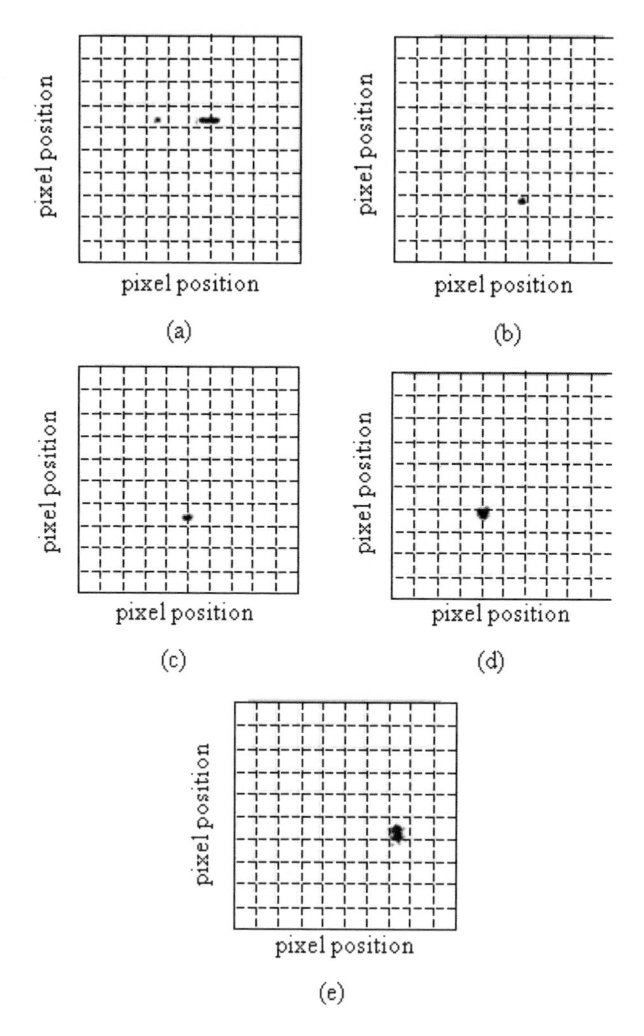

Figure 2.71 Position measurements made with a '45° coil antenna'. (a) Rotation angle 0°. (b) Rotation angle 30°. (c) Rotation angle 45°. (d) Rotation angle 60°. (e) Rotation angle 90°.

In a typical GA optimization problem, a starting population is created randomly, or intelligently if the general direction of the solution is known. In our intravascular-antenna problem, we want to find the number of coils, the number of turns per coil, the coil heights and the turn directions that give the highest sensitivity at the antenna body surface or at a distance where we may expect the artery wall to be present. The population thus consists of sets of numbers, heights and directions. A fitness is assigned to every individual from this population. Here, this fitness could be the amplitude of the sensitivity parameter. Next, parents are selected from the population (several different selection processes exist) and, by means of crossover and mutation, children of a new generation are created. In the crossover process, the parameters of two antenna configurations are intermixed. In the mutation process, one or

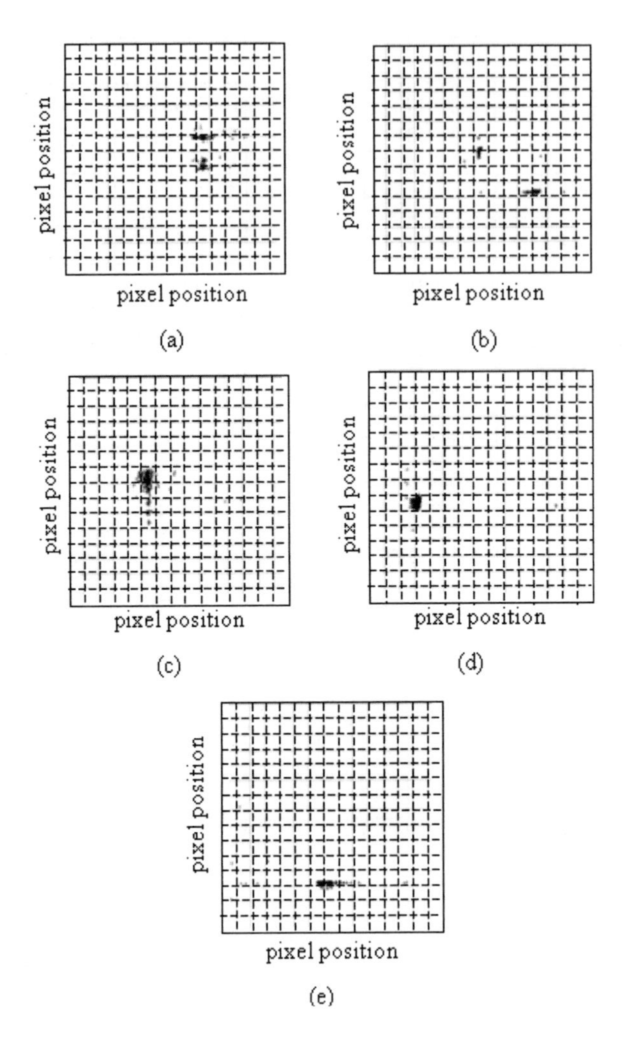

Figure 2.72 Position measurement made with an 'ordinary coil antenna'. (a) Rotation angle 0°. (b) Rotation angle 30°. (c) Rotation angle 45°. (d) Rotation angle 60°. (e) Rotation angle 90°.

a few of the parameters change randomly. The process is depicted in Figure 2.74 and, more specifically, in Figure 2.75 specifically for a five-parameter problem [43, 44].

As an example, we shall look at an intravascular antenna consisting of a discrete number of coils wound around a cylindrical antenna body. The maximum number of coils was three, the height of every coil was allowed to vary between 0.1 mm and 4 mm, the gap between two adjacent coils could vary between 0.1 mm and 3 mm, the number of turns per coil could vary between one and 15, and every coil could be wound clockwise or counterclockwise. The radius of the antenna was 1 mm and one circumference was approximated by 12 straight-line segments. We generated designs for tracking and for imaging.

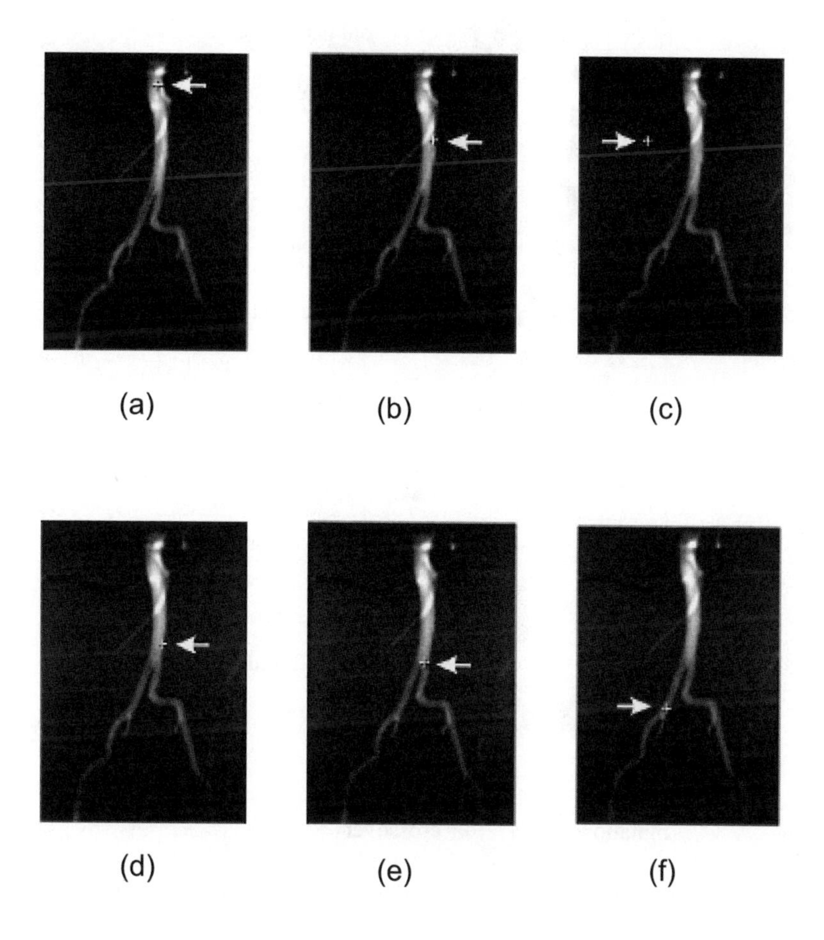

Figure 2.73 MR images obtained during active tracking using the 45° coil antenna in a human-abdomen phantom.

First, we generated an antenna for tracking purposes. As the fitness parameter, we used the minimum value of the sensitivity parameter, sampled on the axis of the antenna between 3 mm and 6 mm in height. The optimization process is a maximisation process, and so, by selecting the fitness parameter in this way, we demanded a high sensitivity on the axis between 3 and 6 mm measured from the base of the antenna.

The optimization process generated (within a few minutes) a design consisting of three coils. The first coil started at 2.64 mm from the antenna base. The gaps between the first and second and between the second and third coil were, respectively, 2.91 mm and 2.68 mm. The heights of the coils, from bottom to top, were 3.76 mm, 3.75 mm and 3.13 mm, and the

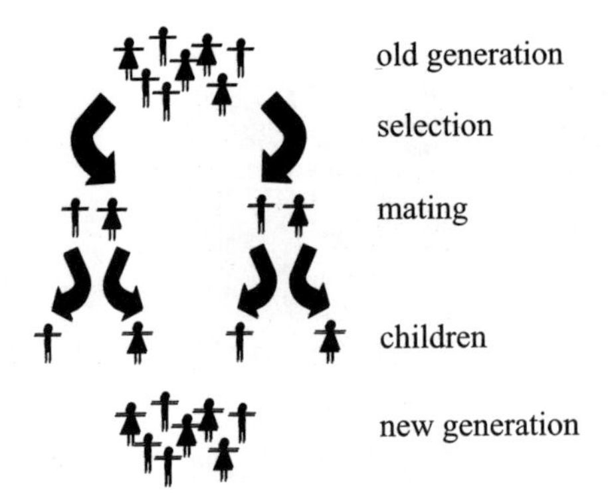

Figure 2.74 Genetic-algorithm process.

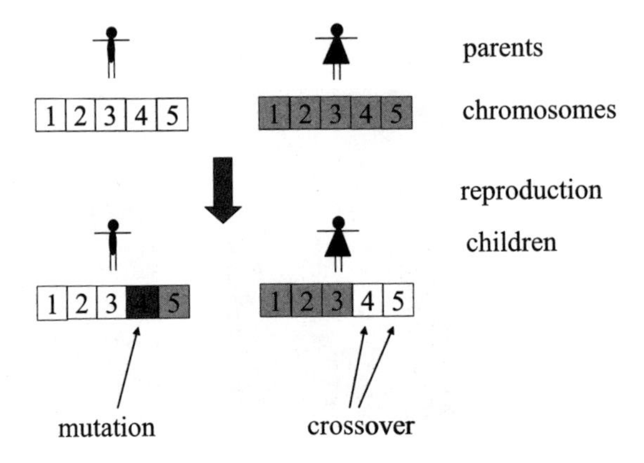

Figure 2.75 Genetic-algorithm iteration for a five-parameter problem.

numbers of turns were 13, 5 and 1. The first and third coils were wound counterclockwise, and the middle coil was wound clockwise.

Sensitivity profiles in the *xz*, *yz* and *xy* planes were calculated and are shown in Figure 2.76. This figure shows that, between 3 mm and 6 mm from the antenna base, an increased sensitivity is indeed present on the axis of the antenna body. Of course, the sensitivity is still below the sensitivity obtained at the surface of the antenna body. So the figure shows that the optimization procedure works, but also that care must be taken in formulating the fitness parameter. Alternatively, for a tracking antenna, one could aim at an increased sensitivity on the antenna body surface only.

As a second example, we generated an antenna for imaging purposes. As the fitness parameter, we used the minimum value of the sensitivity parameter, sampled over the outer surface of a cylinder with a radius of 2.5 mm encapsulating the antenna, between 3 mm and 6 mm in height.

The optimization process generated (within a few minutes) a design consisting of three coils. The first coil started at 1.73 mm from the antenna base. The gaps between the first

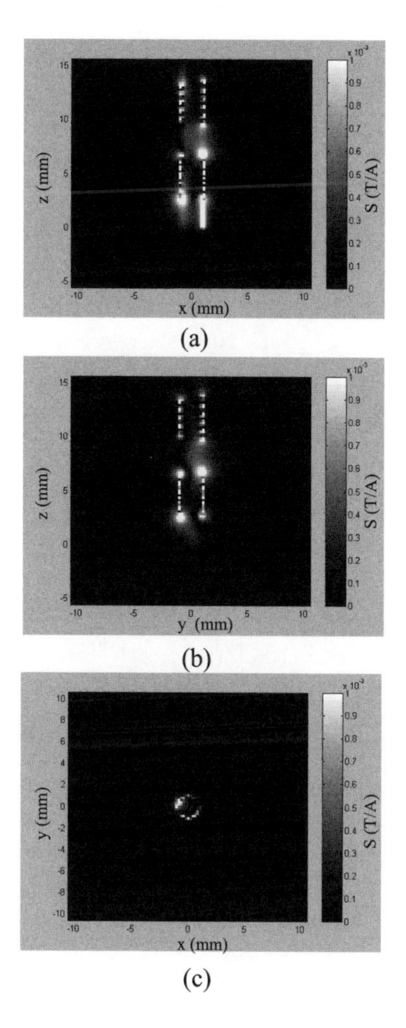

Figure 2.76 Sensitivity patterns in xz, xy and yz planes for a multiple-coil antenna optimized for tracking. (a) Sensitivity pattern in xz plane, $y = 0.05$ mm. (b) Sensitivity pattern in yz plane, $x = 0.05$ mm. (c) Sensitivity pattern in xy plane, $z = 4.5$ mm.

and second and between the second and third coil were, respectively, 0.28 mm and 0.19 mm. The heights of the coils, from bottom to top, were 0.41 mm, 0.12 mm and 3.20 mm, and the numbers of turns were 12, 8 and 11. All coils were wound counterclockwise.

Sensitivity profiles in the xz, yz and xy planes were calculated and are shown in Figure 2.77. A strong, rotationally homogeneous sensitivity is visible in the radial and axial range specified.

The values of the heights and gap widths, however, show that a more accurate and reproducible construction method is needed than the one that has been used up to now. The handwork used for the construction of these prototype antennas thus far will no longer be

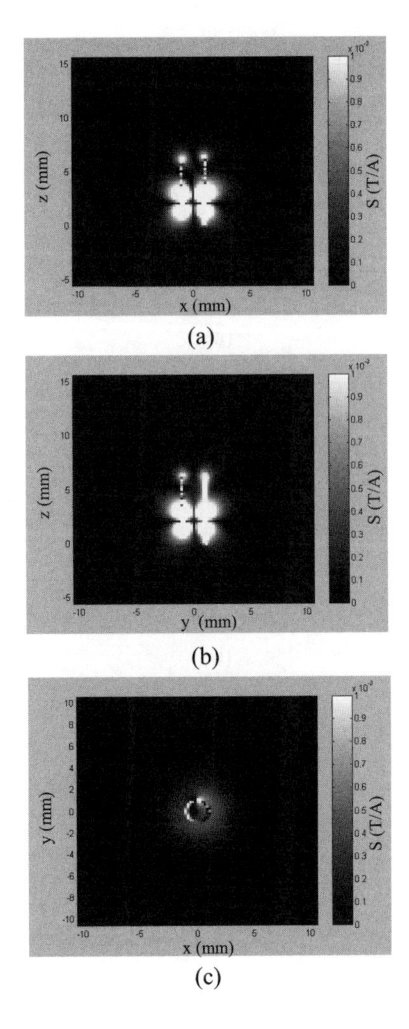

Figure 2.77 Sensitivity patterns in xz, xy and yz planes for a multiple-coil antenna optimized for imaging. (a) Sensitivity pattern in xz plane, $y = 0.05$ mm. (b) Sensitivity pattern in yz plane, $x = 0.05$ mm. (c) Sensitivity pattern in xy plane, $z = 4.5$ mm.

sufficient. The construction accuracy needs to be improved up to the level of the modeling accuracy. For the present, there is no reason to improve the modeling accuracy. The tracking antennas perform as expected, and the imaging antennas first need to undergo tests in an MR environment.

2.8 SAFETY ASPECTS

Thus far, we have been looking at intravascular antennas from a modeling or a constructional point of view, disregarding safety aspects. Since the ultimate goal is to employ these antennas in living persons, we need to address these aspects as well. We have briefly mentioned the risk of heating of the antenna leads. For a complete treatment, however, we need to address all intrinsic potential sources of hazard in an MR environment [12, 45]: static magnetic fields and spatial gradients, pulsed gradient magnetic fields and, finally, pulsed RF fields and the associated heating problem. For a properly operating MR system, the hazards associated with direct interactions of these fields with the body are negligible. It is the interaction of these fields with medical devices placed within them that create concerns for safety [45]. Before we discuss the potential sources of hazard, we first need to define what we mean by the term 'safe'. According to [45], the term 'MR safe' indicates that the device, when used in an MR environment, has been demonstrated to present no additional risk to the patient, but it may affect the quality of the diagnostic information.

Closely connected with the definition of 'MR safe' is the definition of 'MR compatible'. The term 'MR compatible' indicates that a device, when used in an MR environment, is 'MR safe' and has also been demonstrated neither to significantly affect the quality of the diagnostic information nor to have its operations affected by the MR device.

Understanding now what is meant by safety, we may proceed with the potential sources of hazard.

2.8.1 Static Magnetic Fields and Spatial Gradients

A static magnetic field in the range of 0.2 T to 2.0 T, and possibly extending to 4 T or 5 T, is always present in an MR scanner, even when the scanner is not imaging [45]. This strong magnetic field decreases rapidly, on moving away from the magnet, producing a large spatial gradient. This large gradient may cause magnetizable objects to be accelerated, thus possibly causing injuries to patients and/or medical staff.[19]

In addition to the potential hazard of acceleration of magnetizable objects *outside* the patient, magnetizable objects *inside* the patient may undergo torque and displacement forces when brought into the MR main magnetic field, possibly resulting in the tearing of soft tissue.[20] Furthermore, certain cardiac pacemakers are known to function erratically even in relatively weak magnetic fields.[21]

[19] A pair of scissors was pulled out of a nurse's hand as she entered a magnet room. The scissors hit a patient, causing a cut on the patients head (8/2/93). A patient was struck by an oxygen bottle while being placed in a magnet bore. The patient received injuries requiring sutures (6/2/91). Two steel tines (parts of a fork lift truck) weighing 80 pounds each were accelerated by a magnet, striking a technician and knocking him a distance of over 15 feet, resulting in serious injury (6/5/86) [45].

[20] A patient with an implanted intracranial aneurysm clip died as a result of an attempt to scan her. The clip reportedly shifted when exposed to the magnetic field. The staff had apparently obtained information indicating that the material in this clip could be scanned safely (11/11/92). Dislodgement of an iron filing in a patient's eye during MR imaging resulted in vision loss in that eye (1/8/85). A patient complained of double vision after an MR examination. The MR examination, as well as an X-ray, revealed the presence of metal near the patient's eye. The patient was sedated at the time of the examination and was not able to inform anyone of this condition (12/15/93) [45].

[21] A patient with an implanted cardiac pacemaker died during an MR examination (12/2/92). A patient with an implanted cardiac pacemaker died during or shortly after an MR examination. The coroner determined that the death was due to interruption of the pacemaker by the MR system (9/18/89) [45].

2.8.2 Pulsed Gradient Magnetic Fields

A pulsed gradient magnetic field is used for signal localization. During the rise time of the magnetic field, currents are induced in electrical conductors. In most MRI systems, the amplitudes of these currents, however, are about three orders of magnitude smaller than those induced by the pulsed RF field [45]. Therefore, thermal injuries due to pulsed gradient magnetic fields are not of great concern. More important are the biological effects due to pulsed gradient magnetic fields. One of these effects is the electrical stimulation of nerves and the generation of light flashes (magnetophosphenes), which may result from a slight torque exerted on the retinal cones [45]. Current limits on $\partial \mathbf{B}/\partial t$ prevent painful peripheral-nerve stimulation.

2.8.3 Pulsed RF Fields and Heating

Concerning pulsed RF fields, one needs to be aware of the production of heat in tissue and the production of heat by electrical currents induced in metal implants and medical equipment. The rate at which RF energy is deposited in tissue is measured by the specific absorption rate (SAR). The SAR is measured in watts per kilogram and is limited for whole-body exposure to avoid heating problems[22] [45].

As we have already mentioned, one needs to be aware of the length of electrical leads. If this length is equal to or greater than half a wavelength (in the surrounding medium), standing (current) waves may be induced in the leads. Radiation will take place at the tips of the leads, causing an increase in temperature by dissipation in the surrounding medium, which may become harmful for the patient[23] [12, 45]. In all intravascular-antenna designs presented thus far, we have not paid attention to the length of the electrical leads. Our main concern was the development and validation of an antenna model. However, practical solutions for the problem of leads becoming too long have been reported in the open literature [46, 47].

To avoid the leads becoming resonant, quarter-wavelength chokes or traps may be inserted into the cable. The drawback of these countermeasures is that the chokes or traps need to be designed for the correct resonance frequency, and they may give rise to local energy dissipation [48]. A better solution to the heating problem seems to be to divide the cable into sections that are too short to become resonant. This technique was employed in [48], where compact, inductive transformers were used to interconnect the cable sections, which ensured that there was a signal path without the risk of the electrical leads becoming resonant.

[22] A patient received small blistered burns to the left thumb and left thigh. Reportedly, the operator input an inaccurate patient weight, resulting in an incorrect SAR value (2/10/93) [45].

[23] An electrically conductive lead was looped and placed against bare skin, causing a burn on the patient's upper arm (5/19/95). A child received a burn to the right hand from an ECG cable while the patient was anesthetized. A skin graft was required to treat the affected area (1/26/95). A patient received a 1.5 inch $\times$ 4 inch blistered burn to the left side of the back near the pelvis from an ECG gating cable (9/23/91). A patient received blistered burns on a finger where a pulse oximeter was attached during MR scanning. A skin graft was required to treat the affected area (2/27/95) [45].

2.9 CONCLUSIONS

The formation of MR images is accomplished by trading off SNR, imaging speed and spatial resolution. For temporally efficient MRI, local receiver coils are being developed to improve the SNR without compromising imaging speed and spatial resolution. During intravascular interventions, passive methods may be employed to visualize catheter positions and orientations. However, these passive methods suffer from a severe time inefficiency, limiting their feasibility for intravascular, interventional MRI purposes. The visualization of catheter position and orientation is therefore expected to be accomplished best by employing active, intravascular devices (i.e. antennas).

Taking the concept of local receiver coils one step further, intravascular imaging is expected to be feasible too, by employing intravascular receiver coils or antennas that will improve the SNR beyond levels feasible by employing local receiver coils outside the body. Although various intravascular-antenna concepts have been described in the literature and have been evaluated by means of MR imaging, a quantitative comparison of the various concepts has not been conducted until recently [49]. For the purpose of such a quantitative comparison, a fast approximate model, based on the static magnetic field induced by a direct current in a straight wire segment, has been employed. This model was originally developed for the design of surface coils, for which it is now regarded as unsuitable, since the magnetic field at the positions of interest is not expected to behave as a static magnetic field, nor is the current in a surface coil expected to behave as a direct current.

For intravascular antennas, though, positions in or near the artery wall are expected to be in the radiating near field of the antenna, where the fields are locally inversely proportional to the square of the distance. The static magnetic field induced by a direct current is also inversely proportional to the square of the distance. The current in a small intravascular antenna is expected to be well approximated by a uniform current, and therefore an approximation of the dynamic fields by static ones in and near the artery wall should yield acceptable results.

To assess the validity of a static model, comparisons were made between the static model and the small-loop approximation for a loop antenna immersed in blood. Before this assessment was performed, the small-loop uniform current approximation was validated. It turns out that we may consider a bare loop antenna, immersed in blood and subject to a 1.5 T main MR magnetic field, to carry a uniform current for radii up to 1.7 mm.

Having thus put a practical limit on the radius of our reference, we have verified the static model. For several different loop orientations, we compared the 'static' sensitivity with the 'dynamic' sensitivity, where the sensitivity S is defined by $S = (1/I)\sqrt{B_x^2 + B_y^2}$. On the axis of the loop, the dynamic sensitivity is approximated to within 13% for small loop antennas (radius 0.5 mm) within the region of interest, i.e. a circular cylinder with a radius between 2 mm and 3 mm. This cylinder corresponds to a large artery. Moreover, the behavior of the sensitivity as a function of the distance from the loop center is similar for both the static and the dynamic model, which means that the static model may be employed for comparison of loop antenna designs.

For bare wire antennas larger than a single loop and contained in a small volume, the static approach will fail owing to coupling effects. However, when a thin insulation layer is used a uniform current is maintained when the antenna is immersed in a highly conducting medium such as blood. For multiturn loop antennas where the turns are not too closely spaced, up to about 35 turns with a radius of the order of 0.5 mm may be employed without compromising the model.

A comparison of the sensitivity patterns of a number of intravascular antennas described in literature for tracking and imaging purposes, not keeping too strictly to the limits defined earlier, confirms the results obtained from our qualitative comparisons. The center return antenna is best suited for active tracking, judging from magnetic-field considerations only. If we also take manufacturing aspects into account, the perpendicular-coils antenna may be better suited for the job, combining a localized sensitivity that is independent of the transverse angle of observation with a geometry that is restricted to the outer surface of the antenna body. Furthermore, the perpendicular-coils antenna performs better when it is rotated with respect to the MR main magnetic field.

For imaging purposes, both the dual-opposed-solenoids antenna and the triple-loop antenna are considered favorites. They exhibit comparable sensitivity profiles and the manufacturing of both antennas is expected to be equally complex. Neither antenna should be used for rotation angles with respect to the MR main magnetic field in excess of 45°.

The calculated sensitivity profiles compare well with images created with an MR system for a number of realized prototype antennas, even though the geometrical limits were not observed too strictly and the use of contrast fluid in the antenna body lumen was not taken into account in the model.

Having established the availability of a fast analytical model that is of practical use in the analysis of intravascular antennas, we have incorporated the model into a genetic-algorithm optimization environment. It has been demonstrated that antenna designs may be generated – subject to user-defined mechanical and electromagnetic constraints – within minutes, employing standard office computing equipment. For the realization of the antenna designs thus generated, more precise manufacturing methods are required than the handwork used for the construction of the prototype antennas so far. A preliminary investigation of creating copper strip patterns on a cylindrical dielectric body, by applying laser patterning, reveals that precise manufacturing is feasible.

To prevent heating of the antenna leads, dissecting the transmission line connecting the antenna to the MR hardware into sections that are too short to become resonant at the Larmor frequency is recommended. A technique involving inductive coupling from transmission line section to transmission line section, as described in [48], could be employed for transferring signals between the antenna and the MR hardware.

APPENDIX 2.A. BIOT–SAVART LAW FOR QUASI-STATIC SITUATION

To derive the Biot–Savart law for the quasi-static situation, we start with Maxwell's equations for a homogeneous, lossless, isotropic medium,

$$\nabla \times \boldsymbol{E} = -\mathrm{j}\omega\mu\boldsymbol{H}, \tag{2.A.1}$$

$$\nabla \times \boldsymbol{H} = \boldsymbol{J} + \mathrm{j}\omega\varepsilon\boldsymbol{E}, \tag{2.A.2}$$

where $\boldsymbol{E}$ is the electric field, $\boldsymbol{H}$ is the magnetic field, $\boldsymbol{J}$ is the current density, ω is the angular frequency, μ is the permeability of the medium and ε is the permittivity of the medium. Next, we need the continuity equation, which is given by

$$\nabla \cdot \boldsymbol{J} = -\mathrm{j}\omega\rho, \tag{2.A.3}$$

where ρ is the charge density, and Gauss's laws, which are given by

$$\nabla \cdot \boldsymbol{H} = 0, \tag{2.A.4}$$

$$\nabla \cdot \boldsymbol{E} = \frac{\rho}{\varepsilon}. \tag{2.A.5}$$

Now, we assume that $\mathrm{j}\omega\varepsilon\boldsymbol{E}$ is negligible compared with $\boldsymbol{J}$, so that equation (2.A.2) may be approximated by

$$\nabla \times \boldsymbol{H} = \boldsymbol{J}. \tag{2.A.6}$$

Since $\nabla \cdot \nabla \times \boldsymbol{H} = 0$, equation (2.A.6) results in $\nabla \cdot \boldsymbol{J} = 0$, and this, when substituted in equation (2.A.3), means that $\rho = 0$ in equation (2.A.3). This is known as the *quasi-static* approach, where $\boldsymbol{J}$ is assumed to be 'almost stationary'. A static charge in equation (2.A.5) remains possible.

Next, we introduce the magnetic vector potential $\boldsymbol{A}$ through

$$\boldsymbol{H} = \nabla \times \boldsymbol{A}. \tag{2.A.7}$$

Then, from equation (2.A.6),

$$\nabla \times \nabla \times \boldsymbol{A} = \nabla(\nabla \cdot \boldsymbol{A}) - \nabla^2\boldsymbol{A} = \boldsymbol{J}. \tag{2.A.8}$$

Assuming the Coulomb gauge $\nabla \cdot \boldsymbol{A} = 0$ then results in

$$\nabla^2\boldsymbol{A} = -\boldsymbol{J}, \tag{2.A.9}$$

and [30]

$$\boldsymbol{A} = \frac{1}{4\pi}\iiint_{V_{\text{source}}} \frac{\boldsymbol{J}(\boldsymbol{r}')}{|\boldsymbol{r} - \boldsymbol{r}'|}\, dv'. \tag{2.A.10}$$

In the above, primed coordinates are associated with the source volume, and unprimed coordinates refer to the observation point. The magnetic field may be written as

$$\boldsymbol{H} = \iiint_{V_{\text{source}}} \nabla \times d\boldsymbol{A}, \tag{2.A.11}$$

where

$$d\boldsymbol{A} = \frac{\boldsymbol{J}(\boldsymbol{r}')\, dv'}{4\pi\boldsymbol{R}}. \tag{2.A.12}$$

Here $R = |\boldsymbol{r} - \boldsymbol{r}'|$.

For a current-carrying wire, the product $\boldsymbol{J}(\boldsymbol{r}')\, dv'$ may be written (Figure 2.A.1) as

$$\boldsymbol{J}(\boldsymbol{r}')\, dv' = J(\boldsymbol{r}')\, d\boldsymbol{S}'\, d\ell' = I(\boldsymbol{r}')\, d\ell', \tag{2.A.13}$$

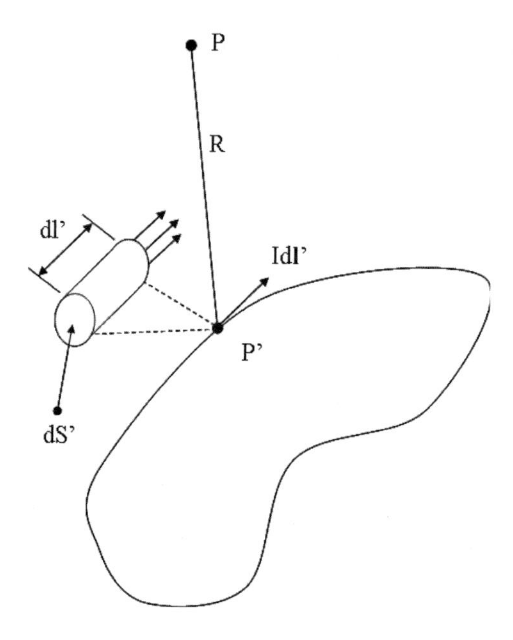

Figure 2.A.1 Current-carrying wire.

so that

$$\nabla \times d\boldsymbol{A} = \frac{I(\boldsymbol{r}')}{4\pi}\left[\nabla\left(\frac{1}{R}\right) \times d\boldsymbol{\ell}' + \frac{1}{R}\nabla \times d\boldsymbol{\ell}'\right]. \tag{2.A.14}$$

Since the nabla operator works on the observation point coordinates, the second term in the above equation equals zero, and since

$$\nabla\left(\frac{1}{R}\right) = -\frac{\boldsymbol{R}/R}{R^2}, \tag{2.A.15}$$

equation (2.A.11) may be written as

$$\boldsymbol{H} = \int_C \frac{I(\boldsymbol{r}')\, d\boldsymbol{\ell}' \times \boldsymbol{R}}{4\pi R^3}, \tag{2.A.16}$$

which is the Biot–Savart law stated in equation (2.36). C is the contour in Figure 2.A.1, carrying the current $I(\boldsymbol{r}')$.

REFERENCES

1. B. Blümich, *NMR Imaging of Materials*, Clarendon Press, Oxford, 2000.

2. H. Hricak, S. White, D. Vigneron, J. Kurhanewickz, A. Cosko, D. Levin, J. Weiss, P. Narayan and P.R. Carroll, 'Carcinoma of the prostate gland: MR imaging with pelvic

phased-array coils versus integrated endorectal-pelvic phased-array coils', *Radiology*, Vol. 193, No. 3, pp. 703–709, December 1994.

3. N.M. deSouza, R. Dina, G.A. McIndoe and W.P. Soutter, 'Cervical cancer: Value of an endovaginal coil magnetic resonance imaging technique in detecting small volume disease and assessing parametrial extension', *Gynecologic Oncology*, Vol. 102, pp. 80–85, 2006.
4. J.P. Hornbak, *The Basics of MRI*, available at www.cis.rit.edu/htbooks/mri, 1996–2007.
5. A. Glowinski, J. Kürsch, G. Adam, A. Bücker, T.G. Noll and R.W. Günther, 'Device visualization for interventional MRI using local magnetic fields: Basic theory and its application to catheter visualization', *IEEE Transactions on Medical Imaging*, Vol. 17, No. 5, pp. 786–793, October 1998.
6. H.H. Quick, M.E. Ladd, G.G. Zimmermann-Paul, P. Erhart, E. Hofmann, G.K. von Schulthess and J.F. Debatin, 'Single-loop coil concepts for intravascular magnetic resonance imaging', *Magnetic Resonance in Medicine*, Vol. 41, pp. 751–758, 1999.
7. P.A. Rivas, K.S. Nayak, G.C. Scott, M.V. McConnell, A.B. Kerr, D.G. Nishimura, J.M. Pauly and B.S. Hu, 'In vivo real-time intravascular MRI', *Journal of Cardiovascular Magnetic Resonance*, Vol. 4, No. 2, pp. 223–232, 2002.
8. H.H. Quick, J.-M. Serfaty, H.K. Pannu, R. Genadry, C.J. Yeung and E. Atalar, 'Endourethral MRI', *Magnetic Resonance in Medicine*, Vol. 45, pp. 138–146, 2001.
9. N.A.A. Op Den Kamp, *Analysis and Design of Intravascular MR Antennas*, MSc thesis, Eindhoven University of Technology, Faculty of Electrical Engineering, February 2003.
10. L.W. Bartels, 'MRI voor het Geleiden en Evalueren van Behandelingen van het Bloedvatstelsel', *NVS Nieuws*, pp. 18–21, December 2001.
11. O. Ocali and E. Atalar, 'Intravascular magnetic resonance imaging using a loopless catheter antenna', *Magnetic Resonance in Medicine*, Vol. 37, pp. 112–118, 1997.
12. M.J.H. Aben, *Aspects of Active Tracking in MRI*, MSc thesis, Eindhoven University of Technology, Faculty of Electrical Engineering, June 2004.
13. S. Weiss, T. Kuehne and M. Zenge, 'Switchable resonant fiducial marker for safe instrument localisation at all marker orientations', *Proceedings of the 10th ISMRM Scientific Meeting and Exhibition*, p. 2245, 2002.
14. M. Mohammad-Zadeh, H. Soltanian-Zadeh, M. Shah-Adabi and A. Tavakkoli, 'New double-turn loop probe for intravascular MRI', *Proceedings of the 26th Annual International Conference of the IEEE EMBS*, pp. 1151–1154, September 2004.
15. P.J. Cassidy, K. Clarke and D.J. Edwards, 'Validation of the transmission-line modelling method for the electromagnetic characterization of magnetic resonance imaging radio-frequency coils', *Proceedings of the IEE Seminar on Validation of Computational Electromagnetics*, pp. 37–41, March 2004.

16. J.-M. Jin, 'Electromagnetics in magnetic resonance imaging', *IEEE Antennas and Propagation Magazine*, Vol. 40, No. 6, pp. 7–22, December 1998.

17. S.M. Michaelson and J.C. Lin, *Biological Effects and Health Implications of Radiofrequency Radiation*, Plenum Press, New York, p. 120, 1987.

18. J.D. Bronzino (ed.), *The Biomedical Engineering Handbook*, second edition, Vol. 1, CRC Press, Boca Raton, pp. 89-4–89-5, 2000.

19. R.J. Johnson, *Antenna Engineering Handbook*, third edition, McGraw-Hill, New York, 1993.

20. S. Ramo, J.R. Whinnery and T. Van Duzer, *Fields and Waves in Communication Electronics*, second edition, John Wiley & Sons, New York, 1984.

21. C.A. Balanis, *Antenna Theory Analysis and Design*, second edition, John Wiley & Sons, New York, 1997.

22. H. Werner, 'An exact integration procedure for vector potentials of thin circular loop antennas', *IEEE Transactions on Antennas and Propagation*, Vol. 44, No. 2, pp. 157–165, February 1996.

23. E. Lepelaars, *Transient Electromagnetic Excitation of Biological Media by Circular Loop Antennas*, PhD thesis, Eindhoven University of Technology, 1997.

24. R.W.P. King, in R.E. Collin and F.J. Zucker (eds.), *Antenna Theory*, Chapter 11, Part I, McGraw-Hill, New York, pp. 458–482, 1969.

25. R.W.P. King, C.W. Harrison and D.G. Tingley, 'The admittance of bare circular loop antennas in a dissipative medium', *IEEE Transactions on Antennas and Propagation*, pp. 434–438, July 1965.

26. R.W.P. King, C.W. Harrison and D.G. Tingley, 'The current in bare circular loop antennas in a dissipative medium', *IEEE Transactions on Antennas and Propagation*, pp. 529–531, July 1965.

27. J.D. Kraus, *Antennas*, McGraw-Hill, New York, 1950.

28. C.-L. Chen and R.W.P. King, 'The small bare loop antenna immersed in a dissipative medium', *IEEE Transactions on Antennas and Propagation*, pp. 266–269, May 1963.

29. W.H. Press, B.P. Flannery, S.A. Teukolsky and W.T. Vetterling, *Numerical Recipes: The Art of Scientific Computing*, Cambridge University Press, 1988.

30. C.T.A. Johnk, *Engineering Electromagnetic Fields and Waves*, John Wiley & Sons, New York, 1975.

31. J.H. Letcher, 'Computer-assisted design of surface coils used in magnetic resonance imaging. I. The calculation of the magnetic field', *Magnetic Resonance Imaging*, Vol. 7, pp. 581–583, 1989.

32. R.K. Moore, 'Effects of a surrounding conducting medium on antenna analysis', *IEEE Transactions on Antennas and Propagation*, pp. 216–225, May 1963.

33. K. Ilzuka, R.W.P. King and C.W. Harrison Jr., 'Self- and mutual admittances of two identical circular loop antennas in a conducting medium and in air', *IEEE Transactions on Antennas and Propagation*, Vol. AP-14, No. 4, pp. 440–450, July 1966.

34. C.D. Taylor and C.W. Harrison Jr., 'On thin-wire multiturn loop antennas', *IEEE Transactions on Antennas and Propagation*, Vol. 22, No. 3, pp. 407–413, May 1974.

35. S.-G. Pan, T. Becks, D. Heberling, P. Nevermann, H. Rosmann and I. Wolff, 'Design of loop antennas and matching networks for low-noise RF receivers: Analytic formula approach', *IEE Proceedings*, Part H, Vol. 144, No. 4, pp. 274–280, August 1997.

36. R.C. Hansen, 'Radiation and reception with buried and submerged antennas', *IEEE Transactions on Antennas and Propagation*, pp. 207–215, May 1963.

37. J. Galeijs, 'Admittance of insulated loop antennas in a disipative medium', *IEEE Transactions on Antennas and Propagation*, pp. 229–235, March 1965.

38. J.A. Stratton, *Electromagnetic Theory*, McGraw-Hill, New York, 1941.

39. M. Abramowitz and I.A. Stegun, *Handbook of Mathematical Functions*, Dover Publications, New York, 1965.

40. H. Reichardt (ed.), *Kleine Enzyklopädie Mathematik*, VEB Bibliographisches Institut Leipzig, 1986.

41. C.A. Balanis, *Advanced Engineering Electromagnetics*, John Wiley & Sons, New York, 1989.

42. S.E. Langerak, P.K. Kunz, H.W. Vliegen, J.W. Jukema, A.H. Zwinderman, P. Steendijk, H.J. Lamb, E.E. van der Wall and A. Roos, 'MR flow mapping in coronary artery bypass grafts: A validation study with Doppler flow measurements', *Radiology*, Vol. 122, No. 1, pp. 127–135, January 2002.

43. J.M. Johnson and Y. Rahmatt-Samii, 'Genetic algorithms in engineering electromagnetics', *IEEE Antennas and Propagation Magazine*, Vol. 39, No. 4, pp. 7–11, August 1997.

44. Y. Rahmatt-Samii and E. Michielsen, *Electromagnetic Optimization by Genetic Algorithms*, John Wiley & Sons, New York, 1999.

45. R.A. Philips and M. Skopec, *A Primer on Medical Device Interactions with Magnetic Resonance Imaging Systems*, draft document, US Department of Health and Human Services, Food and Drug Administration, Center for Devices and Radiological Health, February 7, 1997.

46. E. Atalar, 'Safe coaxial cables for MRI', *Proceedings of ISMRM Annual Meeting*, p. 1006, 1999.

47. M.E. Ladd and H.H. Quick, 'Reduction of resonant RF heating in intravascular catheters using coaxial chokes', *Magnetic Resonance in Medicine*, Vol. 43, pp. 615–619, 2000.

48. P. Vernickel, V. Schulz, S.N. Weiss and B. Gleich, 'A safe transmission line for MRI', *IEEE Transactions on Biomedical Engineering*, Vol. 52, No. 6, pp. 1094–1102, June 2005.

49. N.A.A. Op den Kamp, J.H. Seppenwoolde, H.J. Visser, A.G. Tijhuis and C.J.G. Bakker, 'Intravascular MR antenna designs by simulation of sensitivity profiles', *Proceedings of the International Society of Magnetic Resonance in Medicine*, p. 1187, July 2003.

3

PCB Antennas: Printed Monopoles

The dipole antenna is one of the oldest antennas used in practice. Heinrich Hertz used a half-wave dipole antenna in the first ever radio experiment in 1886. A wire dipole antenna may be easily constructed from a two-wire transmission line by bending the ends of the open transmission line outward by 90°. The half-wave dipole antenna has a 'near-omnidirectional' radiation pattern, i.e. the radiation pattern looks like a torus with a maximum in directions perpendicular to the antenna and 'nulls' in directions along the antenna. A monopole antenna may be derived from a dipole antenna by mounting one arm of the dipole above a ground plane. The monopole and its image in the ground plane then form a dipole antenna. The input impedance of this monopole antenna is equal to half that of the corresponding dipole antenna, and the radiation pattern above the (infinite) ground plane is identical to the upper half of the radiation pattern of the corresponding dipole antenna. Near-omnidirectional ultrawideband (UWB) antennas may be realized, starting from a dipole or monopole antenna, using physical reasoning and 'trial and error' employing full-wave analysis software. These antennas may be realized as planar printed circuit board (PCB) antennas. For less wide-frequency-band applications, the effect of the ground plane on the behavior of printed monopole antennas will play an important role. For this class of antennas, it is worthwhile to develop analytical models to aid in the design process.

3.1 INTRODUCTION

The current trend in miniaturization of handheld mobile wireless devices puts high constraints on the antenna or antennas to be employed. The antenna has to be small and has to possess omnidirectional radiation and sensitivity characteristics. For a single-frequency-band applications, these two requirements make the choice of a monopole antenna natural, limiting

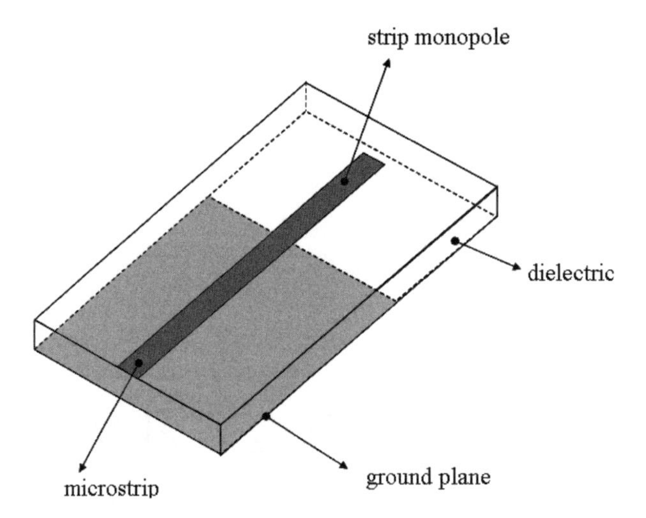

Figure 3.1 Printed (microstrip) monopole antenna integrated into a PCB. The monopole antenna is excited by a microstrip transmission line. The strip monopole starts at the rim of the microstrip ground plane and is a continuation of the top conductor of the microstrip.

the size to about a quarter of a wavelength at the center frequency. For aesthetic reasons, the antenna often needs to be placed inside the device, and the combination of this constraint with the ever-present pressure to reduce production costs leads to the choice of employing a printed monopole antenna. This printed monopole antenna needs to be integrated into the RF printed circuit board (PCB), as shown in Figure 3.1.

The antenna shown in this figure is excited by a microstrip transmission line. Coplanar-waveguide (CPW) excitation is an alternative possible feeding mechanism. The antennas should preferably be realized on standard FR4[1] PCB material, instead of on special microwave laminates. Microwave laminates, although they have a very stable relative permittivity and – in general – low losses, are expensive and difficult to process.

These antennas, as well as other antenna types, may be designed on the basis of physical reasoning and 'educated' trial and error employing a commercially off-the-shelf (COTS) full-wave analysis program. This approach, however, is only recommended for the design of a one-of-a-kind antenna. As soon as it is foreseen that similar but not identical antennas,[2] i.e. a class of antennas, need to be designed, it is worthwhile to invest in the development of an analytical model for this class of antennas. In the end, this will speed up the entire design process.

In the following, we shall demonstrate this by discussing the design of a printed UWB antenna [1] and the development of an analytical model for microstrip-excited monopole

[1] 'FR' here means 'flame-retardant', and '4' means fiber glass epoxy.

[2] These could be antennas for similar applications but now for different frequency bands, on different dielectric substrates or in different environments.

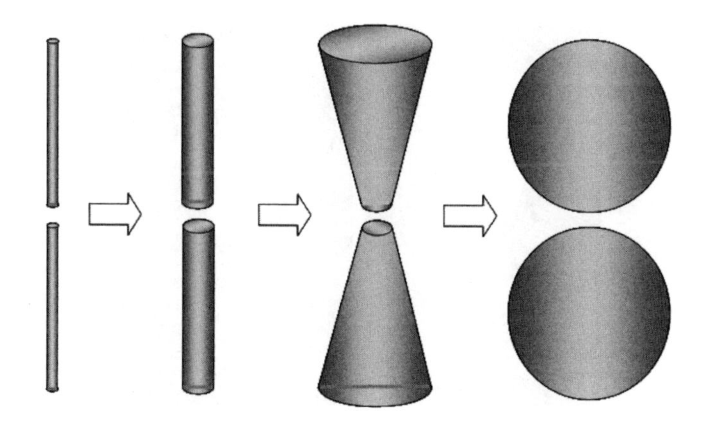

Figure 3.2 Evolution from a narrowband, thin-wire dipole antenna to a broadband, spherical dipole antenna.

antennas of the kind shown in Figure 3.1. The design constraint for all antennas will be an impedance match to 50 Ω.

3.2 PRINTED UWB ANTENNAS

High-data-rate wireless communications need wide bandwidths. In the UWB frequency band from 3.1 GHz to 10.6 GHz, information may be spread over a large bandwidth at low power levels, thus creating the possibility of sharing the spectrum with other users. To prevent interference with existing wireless systems, such as IEEE 802.11a WLAN, stop band characteristics are required from 5 GHz to 6 GHz. In general, a UWB system and thus its antenna should be small and inexpensive. These constraints, added to the low power levels, make the antenna a critical component, a fact often undervalued by electronic designers (even RF designers).

3.2.1 Ultrawideband Antennas

From the 1930s on, antenna engineers have been searching for wideband antenna elements. They soon discovered that, starting from a dipole or monopole antenna, thickening the arms resulted in an increased bandwidth. The reason for this is that for a thick dipole or monopole antenna, the current distribution is – unlike for the thin dipole and monopole – no longer sinusoidal. While this hardly affects the radiation pattern of the antenna, it severely influences the input impedance [2]. This band-widening effect is even more severe if the thick dipole is given the shape of a biconical antenna. A further evolution may be found in dipole and monopole antennas formed from spheres or ellipsoids [3]. Figure 3.2 shows the evolution from a thin-wire dipole antenna to a spherical dipole antenna.

For practical, compact applications however, a planar antenna is preferred. One planar version of the biconical antenna (the third antenna from the left in Figure 3.2) is the bow

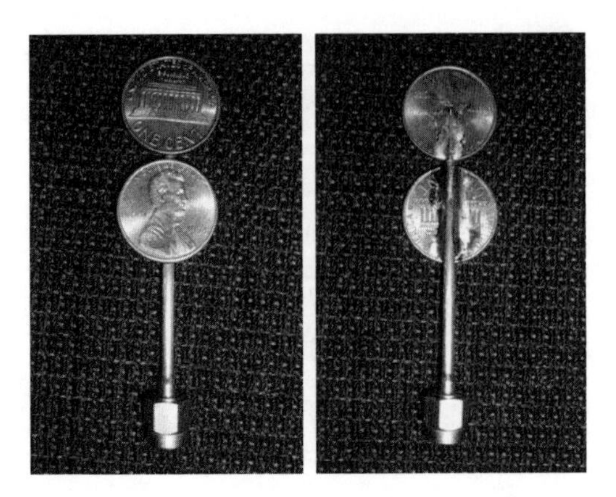

Figure 3.3 Two-penny dipole antenna. Front (left) and back (right).

tie antenna. The angular discontinuities in the bow tie antenna, however, make it difficult to create an impedance match over a large frequency bandwidth [3]. Therefore, a planar antenna structure with a curved outline is preferred. A planar version of the spherical dipole antenna may be found in the 'two-penny dipole antenna'.

3.2.2 Two-Penny Dipole Antenna

A circular planar dipole antenna may be constructed using two US cents ('pennies') and a semirigid coaxial piece of transmission line [3, 4] (Figure 3.3). The measured return loss of this antenna as a function of frequency is presented in Figure 3.4.

A good match to 50 Ω ($S_{11} < -10$ dB) may be observed for the UWB frequency band (3.1 GHz to 10.6 GHz) and beyond. The measurements show the first resonance just above 3 GHz. Considering the size of the antenna – the diameter of a penny is 19 mm – this resonance may be attributed to the dipole[3] (Figure 3.5(a)). For higher frequencies, the current is concentrated in the rims of the pennies and the good impedance match is now due to the fact that the dipole with circular elements has transitioned to a dual-notch horn antenna formed by the rims of the pennies [3] (Figure 3.5(b)).

From the two-penny UWB antenna, it should be a relatively small step towards the design of a compact PCB UWB antenna.

3.2.3 PCB UWB Antenna Design

A PCB microstrip version of the two-penny antenna has been designed and manufactured. Here, the upper circular dipole arm was realized on the upper PCB plane and was connected

[3]The 'dipole' arm length is 19 mm. The total length of the dipole – including a 1 mm gap between the arms – is thus 39 mm, corresponding to half a wavelength at resonance (for a thin dipole). The first resonance frequency thus equals 3.8 GHz. Since we are dealing not with a thin but with a thick dipole, the first resonance frequency should be a little lower than this value. This is demonstrated in Figure 3.4.

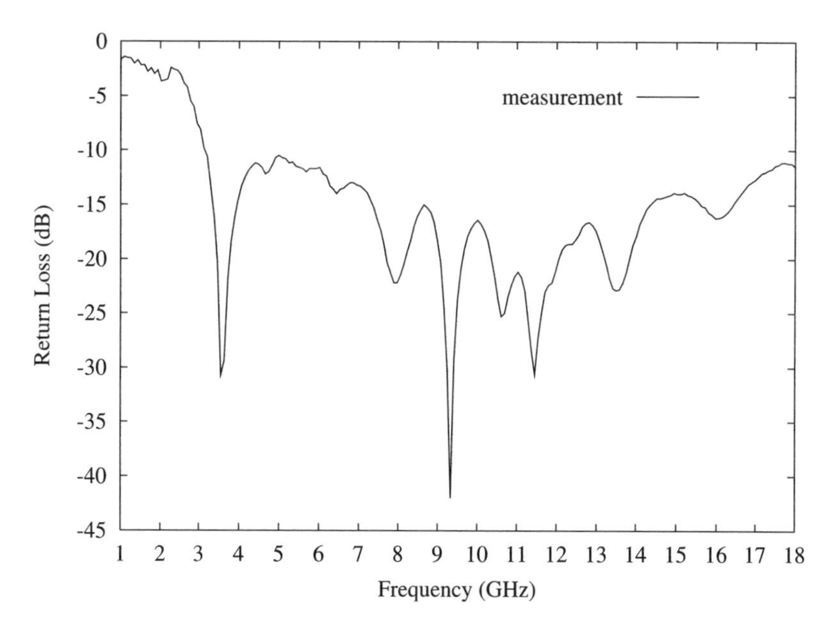

Figure 3.4 Measured return loss as a function of frequency for the antenna shown in Figure 3.3.

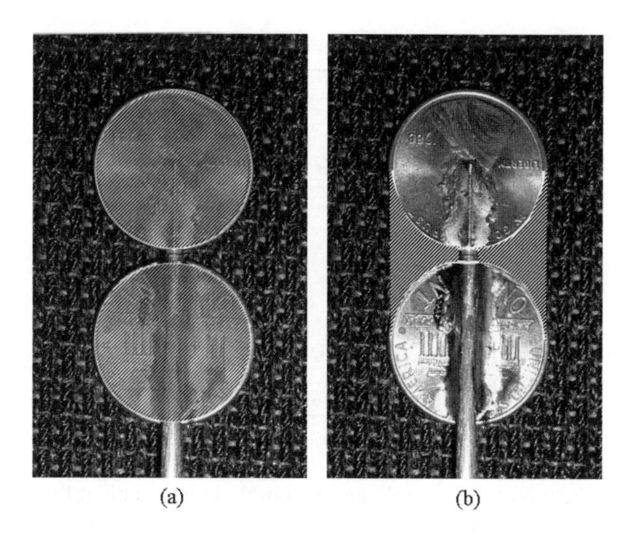

Figure 3.5 Two-penny dipole UWB antenna. (a) Dipole structure around 3 GHz. (b) Dual-notch horn structure for higher frequencies.

to a microstrip transmission line (Figure 3.6). The lower circular dipole arm was integrated with the microstrip ground plane, thus forming a pseudo-monopole [5].

The lower circular arm was integrated into a rectangular ground plane. This does not need to disturb the dual-notch horn antenna behavior seriously, as long as the reflection level

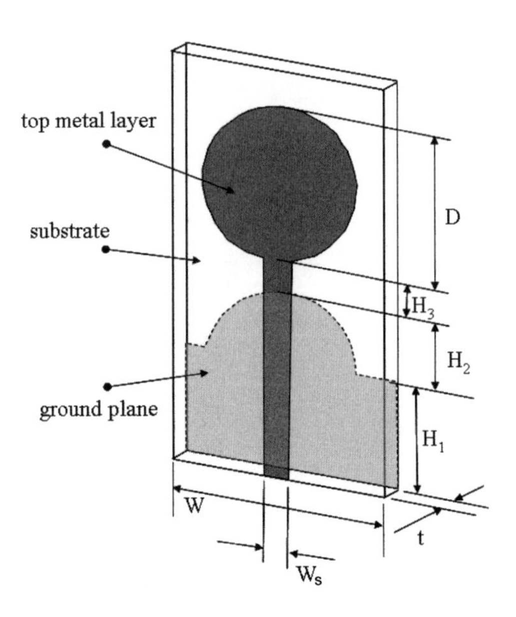

Figure 3.6 Microstrip-excited UWB antenna.

Table 3.1 Dimensions of the microstrip-excited UWB antenna.

Parameter	Value (mm)
W	22
W_S	1.44
t	1.6
th	0.07
H_1	14.38
H_2	4.62
H_3	1.51
D	19
ε_r	4.28
$\tan\delta$	0.016

at the discontinuity formed by the circle and rectangle is low. This reflection condition may be controlled by a height parameter H_2 (Figure 3.6). The antenna may be regarded as an evolution of the stripline version demonstrated in [6]. The microstrip version is less costly in production than the stripline version and easier to integrate into an existing RF PCB design.

The simulated return loss as a function of frequency, after the dimensions indicated in Figure 3.6 had been optimized manually, is shown in Figure 3.7. For the optimization, use was made of the full-wave finite-integration technique (FIT) software package Microwave Studio© from CST [7]. The dimensions used are stated in Table 3.1. The parameters r, $\tan\delta$ and th are the relative permittivity of the PCB substrate, the loss tangent of the substrate and the thickness of the copper layers, respectively.

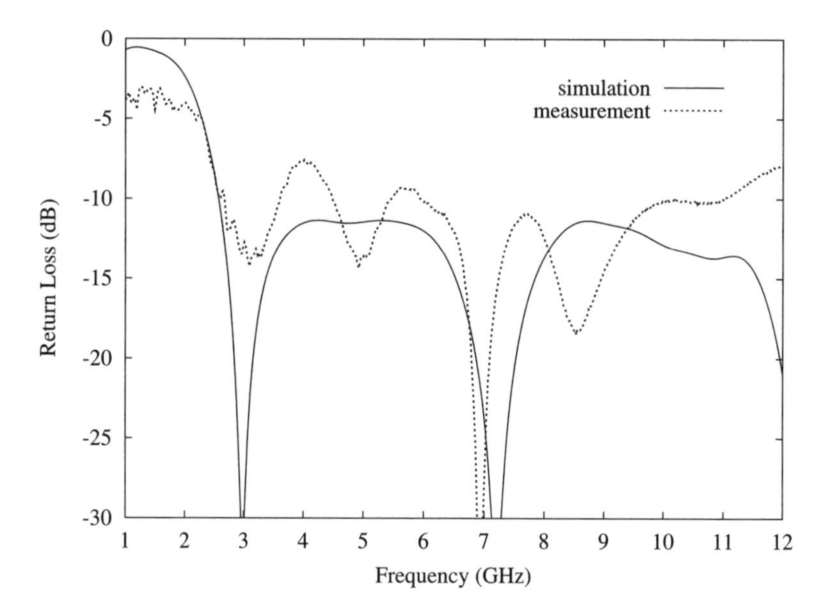

Figure 3.7 Simulated and measured return loss as a function of frequency for the microstrip-excited UWB antenna shown in Figure 3.6.

In Figure 3.7, we also show measurement results for the antenna that was constructed, shown in Figure 3.8. It was observed that cable currents greatly influenced the measurement results. This is the main reason for the differences between the simulation and measurement results visible in Figure 3.7. This phenomenon may also be observed, although it is not always explicitly mentioned, in recently published UWB antenna simulation and measurement results (e.g. [5, 8–10]). Nevertheless, the author is convinced that with proper actions for suppressing these currents [5, 11], close agreement may be reached. One has to bear in mind, though, that the antenna is intended for application in an integrated on-PCB solution.

3.2.3.1 Feed Line The disadvantage of the center-fed dipole is that a transmission line must be brought to the gap between the dipole arms. Since the transmission line will be positioned inside the reactive near field of the antenna, it will be vulnerable to undesired sheet coupling. The radiation pattern of the antenna may be distorted owing to this coupling [3]. In [3], a solution to this possible problem was shown that consists of a strip transmission line feed and a tapered balun. In [6], a 'hidden' stripline feed was used. The stripline was positioned halfway between the two layers of the bottom dipole arm. Both dipole arms consisted of two metal layers on opposite sides of the PCB substrate, electrically connected through metallized vias located on the rims of the circular arms. This latter antenna will be used as a benchmark.

Our pseudo-monopole UWB antenna with a nonhidden microstrip feed, however, does not seem to be prone to the above-mentioned negative effects. This is demonstrated in Figure 3.9, which shows the three-dimensional radiation patterns for 3 GHz and 6 GHz. The gain value

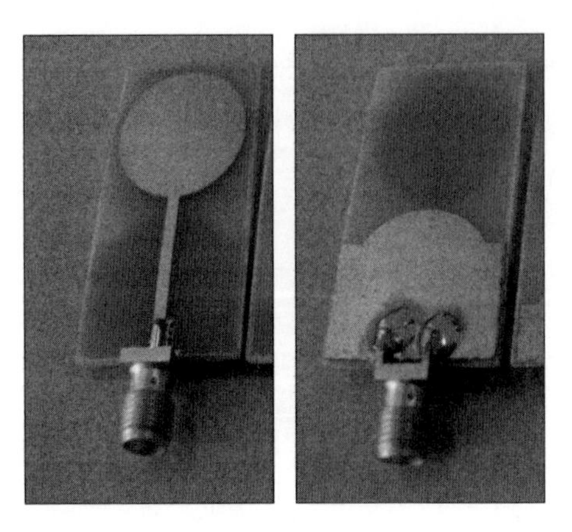

Figure 3.8 The microstrip-excited UWB antenna constructed. Top (left) and bottom (right).

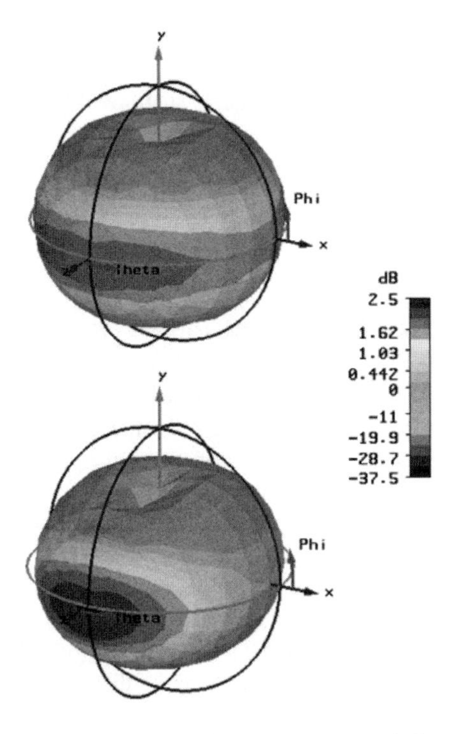

Figure 3.9 Simulated three-dimensional radiation patterns of the pseudo-monopole printed UWB antenna for 3 GHz (top) and 6 GHz (bottom).

is indicated in the figure at the right. The antenna PCB was positioned parallel to and in the xy plane shown in Figure 3.9.

For the stripline dipole antenna discussed in [6], as well as for our microstrip pseudo-monopole antenna, the radiation patterns for frequencies above 6 GHz start to deviate

Table 3.2 Azimuthal gain and maximum variation of the azimuthal gain function for a stripline dipole (SL) and a microstrip pseudo-monopole (MS) antenna.

Frequency (GHz)	Azimuthal gain, SL (dBi)	Azimuthal gain, MS (dBi)	Maximum variation, SL (dBi)	Maximum variation, MS (dBi)
3	2.19	2.46	0.39	0.84
4	2.63	3.02	0.79	1.69
5	3.09	3.63	1.84	2.54
6	3.82	4.02	3.26	3.35
7	3.65	4.24	4.42	5.22
8	2.46	2.16	5.10	8.16
9	–	−3.19	–	9.41
10	–	0.11	–	15.91

seriously from the half-wave dipole patterns shown in Figure 3.9. The origin of this deviation will be explained in section 3.2.3.2. First, we shall take a closer look at the azimuthal (xz-plane) radiation patterns, which demonstrate this deviation. In Table 3.2, the azimuthal gain and the maximum variation of the azimuthal gain function are shown for both antennas for a number of discrete frequencies.

This table demonstrates, together with Figure 3.7, that the two antennas are comparable in behavior. The stripline antenna shows a slightly more uniform radiation pattern, close to that of a half-wave dipole antenna, but the microstrip antenna is easier and thus less costly to manufacture.

To demonstrate how the radiation pattern changes with frequency, simulated azimuthal (xz-plane) radiation patterns of the pseudo-monopole antenna are shown in Figure 3.10 for frequencies from 3 GHz to 6 GHz and in Figure 3.11 for frequencies from 7 GHz to 10 GHz. Zero degrees coincides with $x = 0$.

Figure 3.11 clearly shows how, for frequencies in excess of 6 GHz, the azimuthal pattern deviates seriously from that of a half-wave dipole antenna. Since the antenna behaves around 3 GHz as a half-wave dipole antenna, above 6 GHz the length of the antenna becomes larger than a whole wavelength and 'elevational lobes' will evolve with increasing frequency, which disturb the azimuthal sections of the radiation pattern. The occurrence of elevational lobes is demonstrated in Figure 3.12, which shows the three-dimensional radiation patterns at 7 GHz and 10 GHz. A more half-wave-dipole-like pattern over the whole UWB frequency band can thus be created by shortening the antenna.

3.2.3.2 Antenna Shortening If we shorten the antenna, the generation of elevational lobes will start at a higher frequency. The first resonance, however, will also occur at a higher frequency. Looking at the results for return loss versus frequency (simulation) in Figure 3.7, we observe that we still have some margin if we require a return loss of less than −10 dB over the frequency band from 3.1 GHz to 10.6 GHz.

After, again, a manual optimization using a full-wave-analysis software package, the length $H_1 = 14.38$ mm (Figure 3.6) was replaced by $H_1 = 6.38$ mm. The simulated and measured return loss as a function of frequency are shown in Figure 3.13.

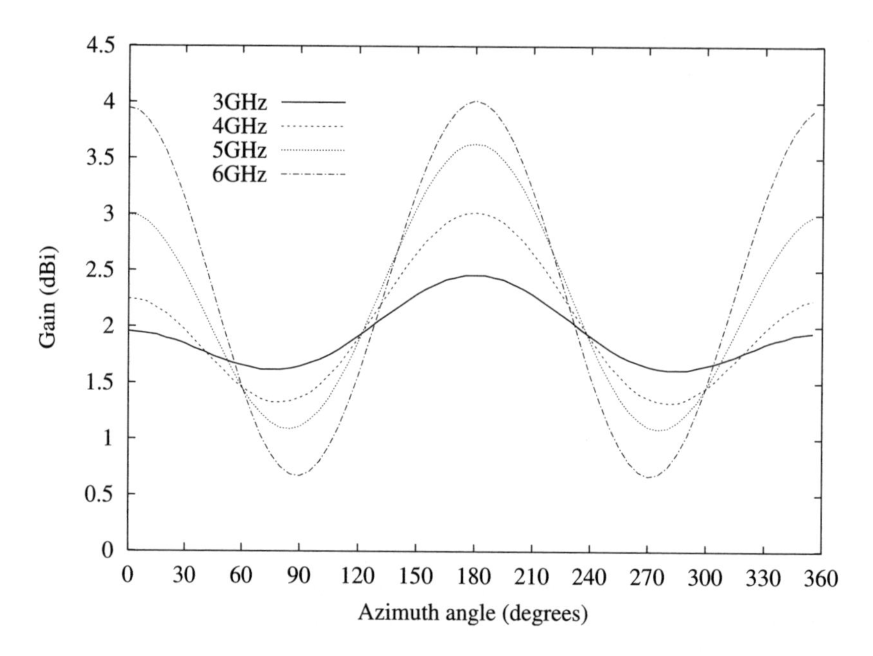

Figure 3.10 Simulated gain of the pseudo-monopole printed UWB antenna as a function of azimuthal angle for frequencies of 3 GHz, 4 GHz, 5 GHz and 6 GHz.

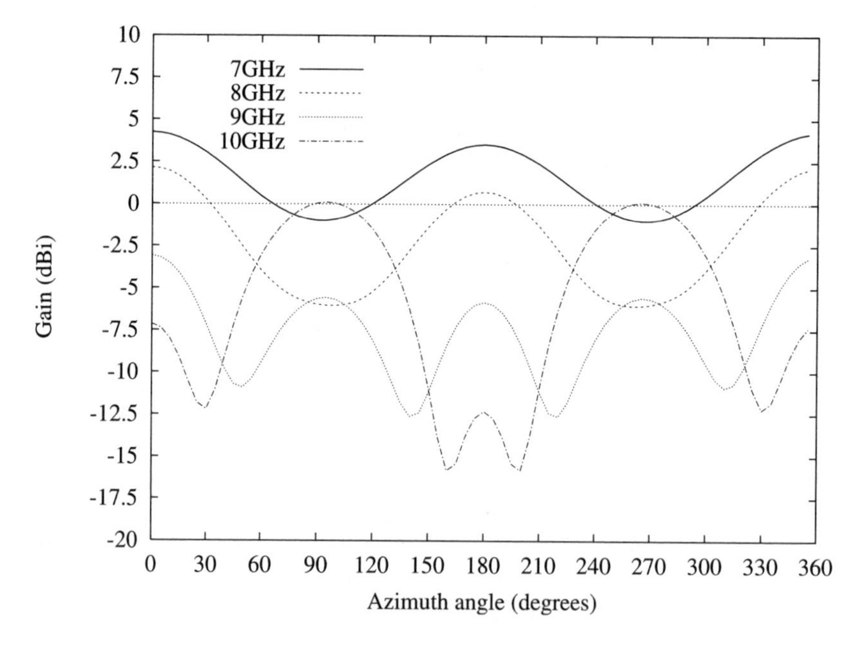

Figure 3.11 Simulated gain of the pseudo-monopole printed UWB antenna as a function of azimuth angle for frequencies of 7 GHz, 8 GHz, 9 GHz and 10 GHz.

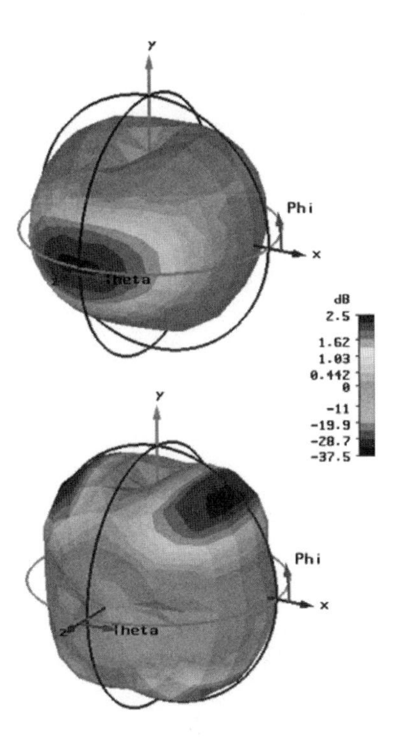

Figure 3.12 Simulated three-dimensional radiation patterns of the pseudo-monopole printed UWB antenna for 7 GHz (top) and 10 GHz (bottom).

The measurements again suffered from cable current effects. These effects are more severe than for the previous antenna, since this antenna is shorter. The main characteristics, i.e. a high return loss for frequencies below 3 GHz and a low return loss for frequencies above 3 GHz, are still present in the measurement results. Comparison of the simulation results with those shown in Figure 3.7 shows that it is possible to shorten the antenna without compromising the return loss characteristics. One has to be careful, though, for the return loss around 3 GHz. In Table 3.3, we compare the azimuthal gain characteristics of the shortened pseudo-monopole antenna with those of the strip line dipole antenna of [6]. The simulated azimuthal (xz-plane) radiation patterns of the shortened pseudo-monopole antenna are shown in Figure 3.14 for frequencies from 3 GHz to 6 GHz and in Figure 3.15 for frequencies from 7 GHz to 10 GHz. Zero degrees coincides with $x = 0$.

Table 3.3 and Figures 3.14 and 3.15 (to be compared with Table 3.2 and Figures 3.10 and 3.11, respectively) show that the azimuthal behavior of the gain has improved with respect to the original microstrip pseudo-monopole antenna. This is demonstrated again in Figure 3.16, which shows the maximum of the gain function and the maximum variation of the gain function in the azimuthal (xz) plane for the original and the shortened pseudo-monopole UWB antenna.

This figure shows that the shortened antenna exhibits a more constant gain over the frequency band. A comparison of Table 3.3 with Table 3.2 reveals further that the shortened

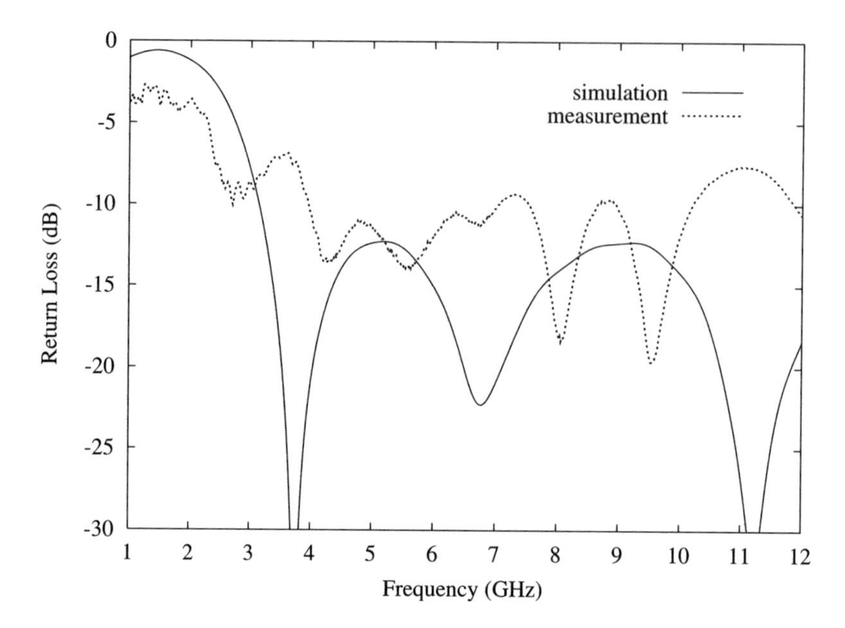

Figure 3.13 Simulated and measured return loss as a function of frequency for the shortened microstrip-excited UWB antenna.

Table 3.3 Azimuthal gain and maximum variation of the azimuthal gain function for a stripline dipole (SL) and a shortened microstrip pseudo-monopole (MS) antenna.

Frequency (GHz)	Azimuthal gain, SL (dBi)	Azimuthal gain, MS (dBi)	Maximum variation, SL (dBi)	Maximum variation, MS (dBi)
3	2.19	2.30	0.39	0.554
4	2.63	2.68	0.79	1.17
5	3.09	3.07	1.84	1.90
6	3.82	3.14	3.26	2.84
7	3.65	2.99	4.42	5.58
8	2.46	2.13	5.10	6.50
9	–	0.62	–	6.45
10	–	1.36	–	3.68

microstrip UWB antenna shows a behavior more similar to that of the stripline UWB antenna. The dimensions, however, are smaller (22 mm × 33 mm × 1.6 mm versus 20.5 mm × 40 mm × 1 mm [6]), and the antenna is easier to produce and does not have a critical feeding and transition region [12].

With this shortened microstrip pseudo-monopole UWB antenna as a basis, we shall now look at measures to suppress signals in the 5 GHz to 6 GHz frequency band.

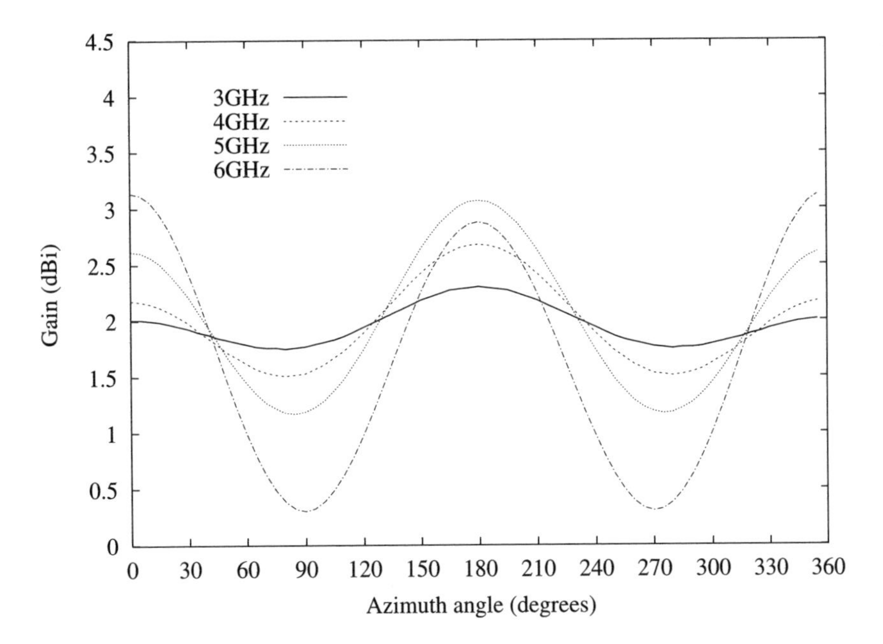

Figure 3.14 Simulated gain of the shortened pseudo-monopole printed UWB antenna as a function of azimuth angle for frequencies of 3 GHz, 4 GHz, 5 GHz and 6 GHz.

Table 3.4 Slot dimensions of the microstrip UWB antenna.

Parameter	Value (mm)
W_{U1}	7.2
W_{U2}	1.2
W_{L}	1.2
L_{U1}	15.1
L_{U2}	5.9

3.2.4 Band-Stop Filter

To create a frequency band notch function, we may either change the current flow in the metal parts of the antenna or insert a filter before or in the feed line of the antenna.

3.2.4.1 Slot in Radiator To influence the current flow (in such a way that destructive interference would occur for frequencies between 5 GHz and 6 GHz), we introduced a slot into the upper arm of our antenna. Since, at higher frequencies, the current will be concentrated at the rims of the two circular arms, the slot has to be positioned in the neighbourhood of the rim of the circle. For ease of drawing, we chose a U-shaped slot as shown in Figure 3.17. The slot dimensions, after manual optimization, were as stated in Table 3.4. All other dimensions were those of the shortened UWB antenna.

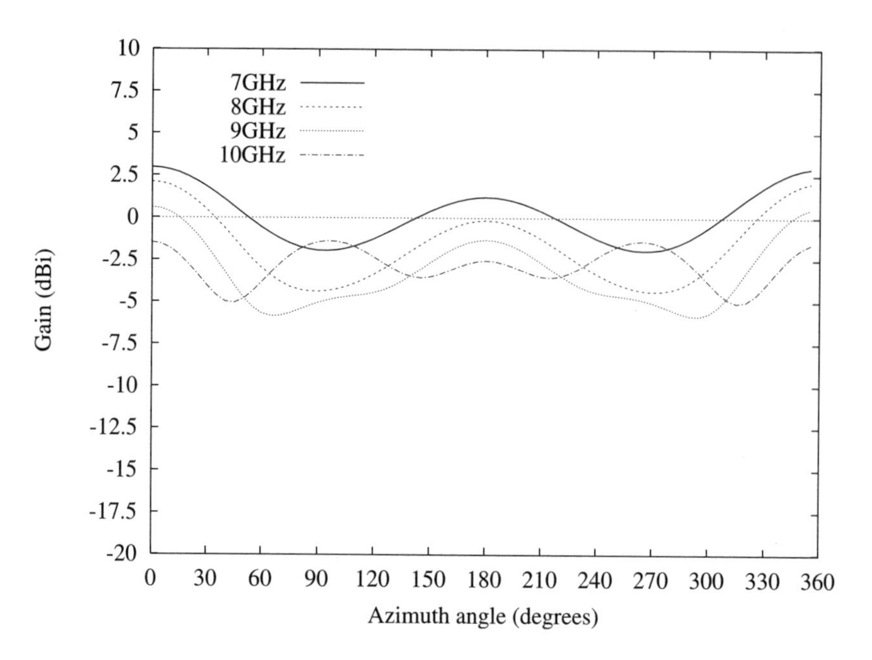

Figure 3.15 Simulated gain of the shortened pseudo-monopole printed UWB antenna as a function of azimuth angle for frequencies 7 GHz, 8 GHz, 9 GHz and 10 GHz.

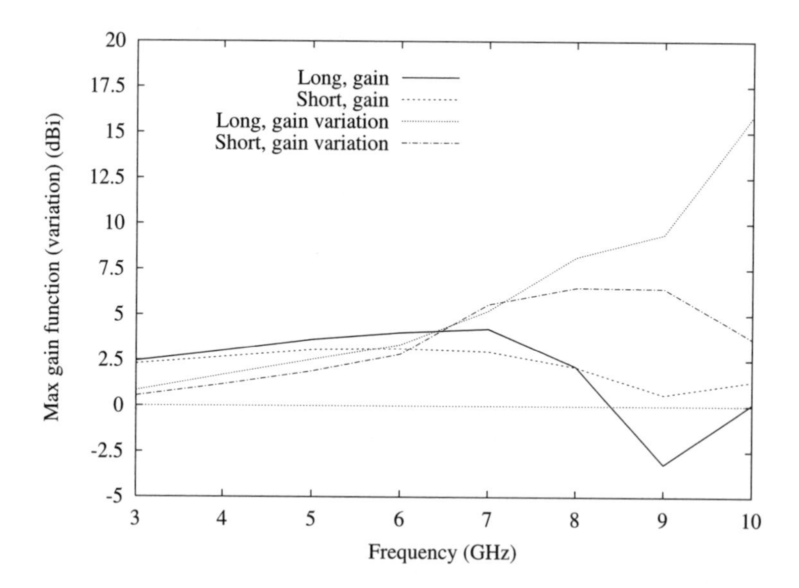

Figure 3.16 Simulated maximum of the gain function and maximum variation of the gain function with azimuthal angle versus frequency for original ('long') and shortened ('short') microstrip UWB antenna.

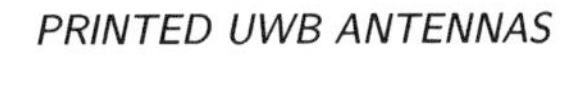

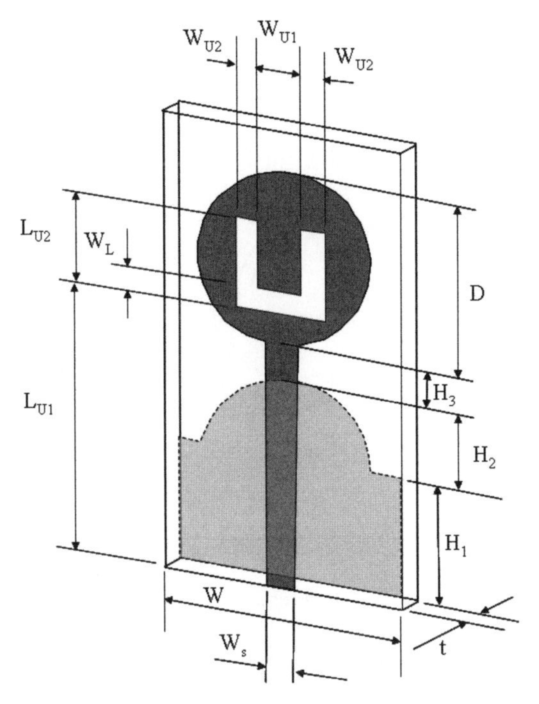

Figure 3.17 Microstrip UWB antenna with U-shaped slot to create a frequency band notch function.

Figure 3.18 Shortened microstrip UWB antenna with U-shaped slot to create a frequency band notch function.

The antenna constructed is shown in Figure 3.18. The simulated and measured return loss as a function of frequency are shown in Figure 3.19.

Although the measurements were still hindered by cable-current effects, we can clearly observe that the U-shaped slot adds the desired frequency band notch functionality.

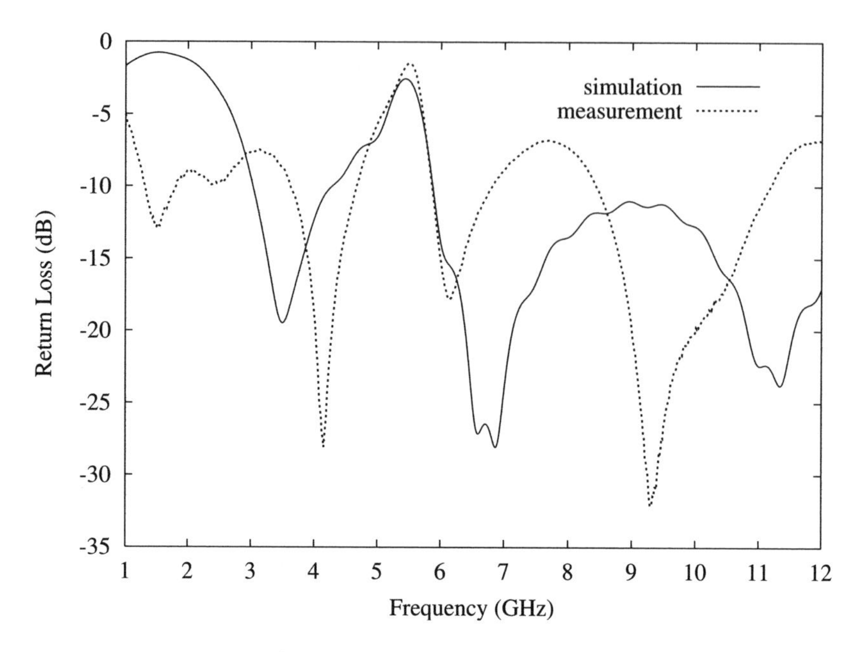

Figure 3.19 Simulated and measured return loss as a function of frequency for a shortened microstrip UWB antenna with a U-shaped slot.

3.2.4.2 Spurline Filter in Microstrip A microstrip spurline filter [13] acts as a band-stop filter. A nice feature of a spurline filter is that the physical structure is completely contained within the boundaries of the microstrip transmission line (Figure 3.20). The length L is equal to a quarter of the wavelength in the transmission line [14].

As an example, a spurline filter was incorporated into the microstrip transmission line of the planar UWB antenna shown in Figure 3.6. The widths S and G were taken to be $S = G = 0.3$ mm and the length L was 7 mm. The spurline filter was positioned symmetrically in the microstrip transmission line at a distance of 1 mm from the edge of the substrate. These values were found after several iterations employing a full-wave analysis program. Figure 3.21 shows the simulated return loss as a function of frequency for the original antenna, i.e. without a filter, and for the antenna incorporating a spurline filter.

The figure clearly shows the stop band behavior between 5 GHz and 6 GHz. The figure also shows that additional optimization is needed to correct the return loss characteristics between 3 GHz and 5 GHz. Since the microstrip transmission line in the UWB antenna is part of the antenna (and the characteristic impedance is not equal to 50 Ω), this optimization may involve many lengthy full-wave iterations. Therefore, it may be advantageous to incorporate the spurline filter into a 50 Ω microstrip transmission line that will be connected to the antenna.

The reflection and transmission coefficients of the spurline filter may be calculated relatively easily using the closed-form equations for the elements of the *ABCD* matrix given in [13]. In these equations, use may be made of the quasi-static even- and odd-mode effective

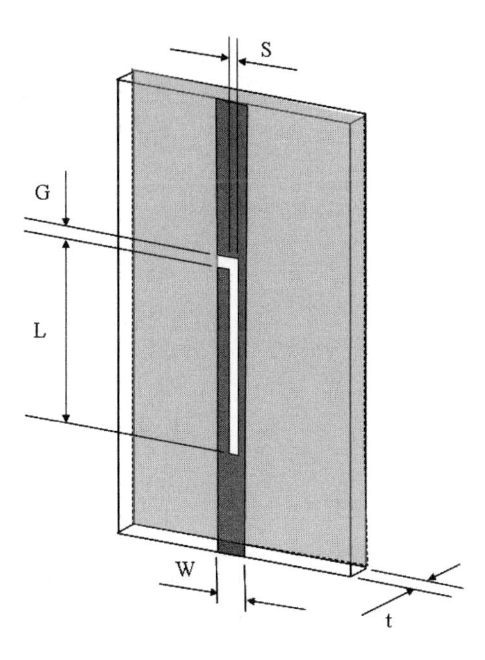

Figure 3.20 Microstrip spurline filter. The filter structure is 'cut out' of the top layer of the microstrip.

permittivities and characteristic impedances for parallel coupled microstrip lines that can be found in [15].

The *ABCD* matrix of the spurline filter shown in Figure 3.20 is given by [13]

$$\begin{bmatrix} A & B \\ C & D \end{bmatrix} = \begin{bmatrix} \cos(\vartheta_e) & j(1/2)[Z_{0e}\sin(\vartheta_e) + Z_{0o}\tan(\vartheta_o)\cos(\vartheta_e)] \\ j(2/Z_{0e})\sin(\vartheta_e) & \cos(\vartheta_e) - (Z_{0o}/Z_{0e})\sin(\vartheta_e)\tan(\vartheta_o) \end{bmatrix}, \tag{3.1}$$

where Z_{0e} and Z_{0o} are the even-mode and odd-mode characteristic impedances, respectively, and ϑ_e and ϑ_o are the even-mode and odd-mode electrical lengths. These are given by

$$\vartheta_e = \beta_e L, \tag{3.2}$$

$$\vartheta_o = \beta_o L, \tag{3.3}$$

where

$$\beta_e = \frac{2\pi\sqrt{\varepsilon_{\mathrm{reff}_e}}}{\lambda_0}, \tag{3.4}$$

$$\beta_o = \frac{2\pi\sqrt{\varepsilon_{\mathrm{reff}_o}}}{\lambda_0}, \tag{3.5}$$

Here, λ_0 is the free-space wavelength and $\varepsilon_{\mathrm{reff}_e}$ and $\varepsilon_{\mathrm{reff}_o}$ are the effective relative permittivities of the even mode and odd mode, respectively.

The *ABCD* matrix of the microstrip spurline filter is derived from the impedance matrix of a section of coupled microstrip transmission lines, applying the correct termination

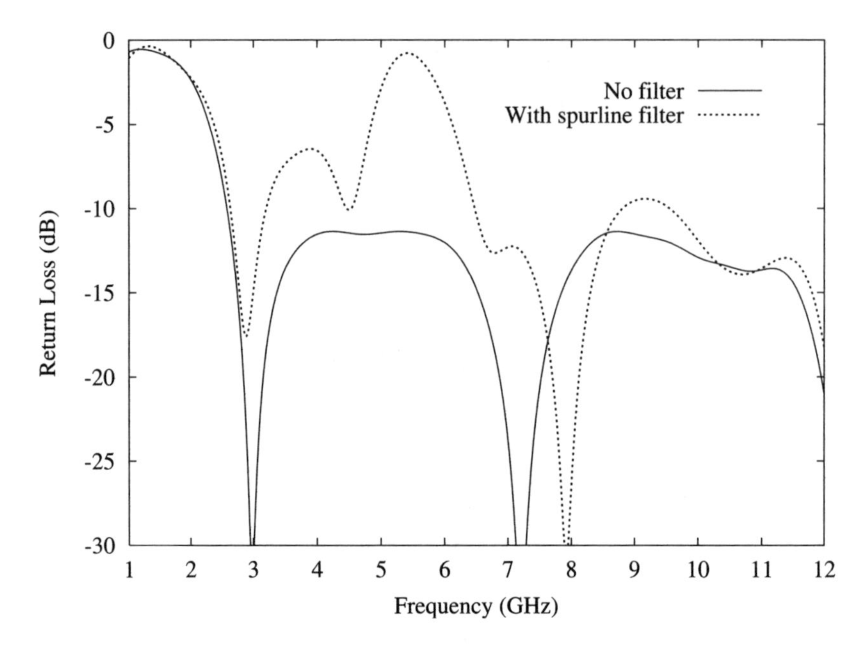

Figure 3.21 Simulated return loss versus frequency for original (nonshortened) pseudo-monopole UWB antenna without and with an integrated spurline filter.

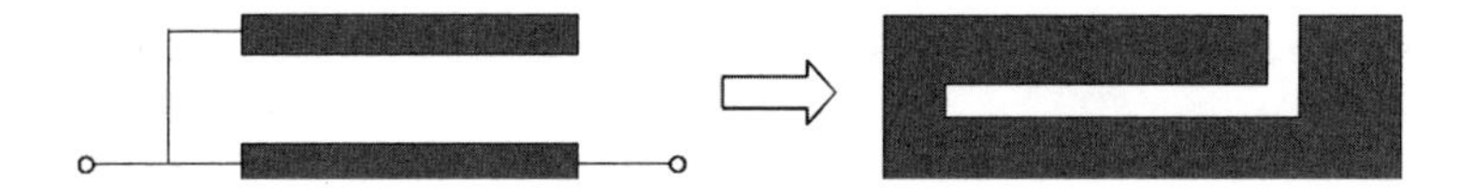

Figure 3.22 Applying termination conditions to coupled transmission lines to create a spurline filter.

conditions, i.e. the two coupled microstrip transmission lines are connected together at one side while one of the transmission lines is left open at the other side (Figure 3.22).

The derivation of the impedance matrix of a section of coupled microstrip transmission lines follows the derivation in [16, 17] for coupled TEM transmission lines, corrected for the non-TEM nature of a microstrip transmission line. This correction consists of employing different phase velocities for the even and odd modes.

The even- and odd-mode characteristic impedances and effective permittivities follow from the treatment in [15], see also Figure 3.23. The normalized strip width and the normalized gap width (with respect to the substrate height) are given by

$$u = \frac{W}{h}, \tag{3.6}$$

$$g = \frac{S}{h}. \tag{3.7}$$

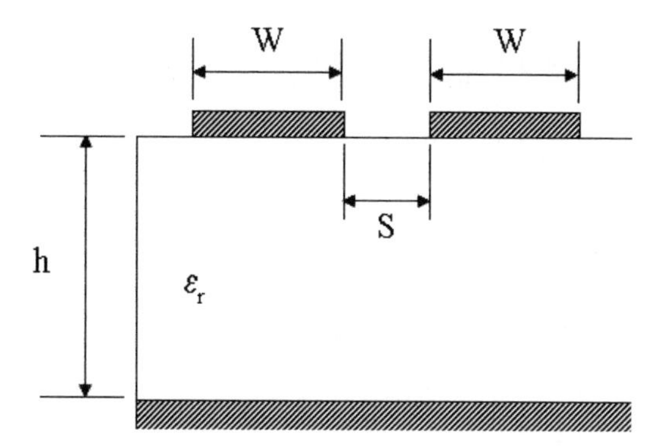

Figure 3.23 Coupled microstrip transmission lines.

The mode characteristic impedances are given by

$$Z_{0m}(u, g) = \frac{Z_0(u)}{1 - Z_0(u)\phi_m(u, g)/\eta}, \tag{3.8}$$

where $m = e$ (even mode) or $m = \mathrm{o}$ (odd mode) and where

$$Z_0(u) = \frac{\eta}{u + 1.98u^{0.172}} \tag{3.9}$$

and

$$\eta = \sqrt{\frac{\mu_0}{\varepsilon_0 \varepsilon_{\mathrm{reff}_m}}}. \tag{3.10}$$

Further,

$$\phi_\mathrm{e}(u, g) = \frac{\varphi(u)}{\Psi(g)\{\alpha(g)u^{m(g)} + [1 - \alpha(g)]u^{-m(g)}\}}, \tag{3.11}$$

$$\varphi(u) = 0.8645u^{0.172}, \tag{3.12}$$

$$\Psi(g) = 1 + \frac{g}{1.45} + \frac{g^{2.09}}{3.95}, \tag{3.13}$$

$$\alpha(g) = 0.5\mathrm{e}^{-g}, \tag{3.14}$$

$$m(g) = 0.2175 + \left[4.113 + \left(\frac{20.36}{g}\right)^6\right]^{-0.251} + \frac{1}{323}\ln\left[\frac{g^{10}}{1 + (g/13.8)^{10}}\right], \tag{3.15}$$

$$\phi_\mathrm{o}(u, g) = \phi_\mathrm{e}(u, g) - \frac{\theta(g)}{\Psi(g)}\mathrm{e}^{[\beta(g)u^{-n(g)}\ln(u)]}, \tag{3.16}$$

$$\theta(g) = 1.729 + 1.175\ln\left[1 + \frac{0.627}{g + 0.327g^{2.17}}\right], \tag{3.17}$$

$$\beta(g) = 0.2306 + \frac{1}{301.8}\ln\left[\frac{g^{10}}{1+(g/3.73)^{10}}\right] + \frac{1}{5.3}\ln[1+0.646g^{1.175}], \tag{3.18}$$

$$n(g) = \left\{\frac{1}{17.7} + e^{[-6.424-0.76\ln(g)-(g/0.23)^5]}\right\} \times \ln\left[\frac{10+68.3g^2}{1+32.5g^{3.093}}\right], \tag{3.19}$$

$$\varepsilon_{\mathrm{reff}_m}(u, g, \varepsilon_r) = \frac{\varepsilon_r+1}{2} + \frac{\varepsilon_r-1}{2}F_m(u, g, \varepsilon_r), \tag{3.20}$$

$$F_e(u, g, \varepsilon_r) = \left[1 + \frac{10}{\mu(u, g)}\right]^{-a(u)b(\varepsilon_r)}, \tag{3.21}$$

$$\mu(u, g) = ge^{-g} + u\frac{20+g^2}{10+g^2}, \tag{3.22}$$

$$a(u) = 1 + \frac{1}{49}\ln\left[\frac{u^4+(u/52)^2}{u^4+0.432}\right] + \frac{1}{18.7}\ln\left[1+\left(\frac{u}{18.1}\right)^3\right], \tag{3.23}$$

$$b(\varepsilon_r) = 0.564\left(\frac{\varepsilon_r-0.9}{\varepsilon_r+3}\right)^{0.053}, \tag{3.24}$$

$$F_o(u, g, \varepsilon_r) = f_o(u, g, \varepsilon_r)\left[1+\frac{10}{u}\right]^{-a(u)b(\varepsilon_r)}, \tag{3.25}$$

$$f_o(u, g, \varepsilon_r) = f_{o1}(g, \varepsilon_r)e^{[p(g)\ln(u)+q(g)\sin(\pi\ln(u)/\ln(10))]}, \tag{3.26}$$

$$p(g) = \frac{e^{-0.745g^{0.295}}}{\cosh(g^{0.68})}, \tag{3.27}$$

$$q(g) = e^{-1.366-g}, \tag{3.28}$$

$$f_{o1}(g, \varepsilon_r) = 1 - e^{[-0.179g^{0.15}-0.328g^{r(g,\varepsilon_r)}/\ln[e+(g/7)^{2.8}]]}, \tag{3.29}$$

$$r(g, \varepsilon_r) = 1 + 0.15\left[1 - \frac{e^{1-(\varepsilon_r-1)^2/8.2}}{1+g^{-6}}\right]. \tag{3.30}$$

As an example, in Figure 3.24, the transmission coefficient is shown as a function of frequency for a spurline filter with $W = 3.3$ mm, $t = 1.6$ mm, $\varepsilon_r = 4.28$, $th = 0.07$ mm, $\tan\delta = 0.016$ (50 Ω characteristic impedance), $S = G = 0.3$ mm and $L = 7$ mm. The transmission coefficient was calculated from the closed-form expressions and is compared here method-of-moments simulation results.

The figure shows that the calculations based on the quasi-static, closed-form expressions result in transmission characteristics very close to those calculated with a full-wave method. The differences still present must be attributed to the fact that the gap (G; see Figure 3.20) is not accounted for in the quasi-static calculations. This gap may be accounted for by employing an effective spurline filter length. In Figure 3.24, employing an effective length $L_{\mathrm{eff}} = 1.065L$ results in closer agreement between the quasi-static and full-wave simulation results. As will be shown in Chapter 5, the concept of an equivalent length may be employed

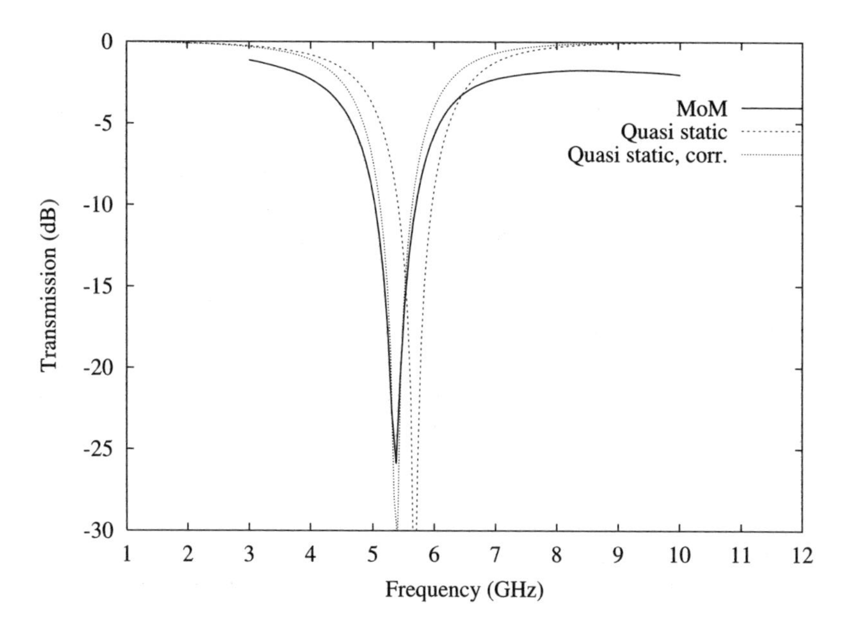

Figure 3.24 Simulated transmission coefficient as a function of frequency for a spurline filter.

to correct for resonance. A general length extension equation needs to be derived, but this is beyond the scope of this chapter. Nevertheless, it has been demonstrated how analytical models may help in speeding up the design process. The time needed to generate the quasi-static results is orders of magnitude smaller than the time needed by full-wave methods. An analytical model for the antenna itself would therefore help considerably in speeding up the design process.

3.3 PRINTED STRIP MONOPOLE ANTENNAS

Most printed UWB antennas reported in the literature may be considered as pseudo-monopole antennas, acting as half-wave dipole antennas around 3 GHz and as two tapered-slot antennas for higher frequencies. Since printed pseudo-monopole antennas (Figure 3.1), may also be of interest for non-UWB applications, owing to their small size and easy integration into a PCB, the availability of a model that is fast when implemented in software but is also accurate would be advantageous. The modeling of a monopole antenna at the edge of an infinite sheet has been performed by several authors employing the dyadic Green's function for a perfectly conducting wedge [18, 19]. Modeling of monopole antennas on the edge of finite half-sheets has been conducted in [20], amongst others. These models, however, still rely heavily on numerical methods and are not considered fit for our purpose. We seek an analytical model that is relatively easy to implement in software and generates results quickly. A model relying heavily on numerical methods could be employed to generate a database of analysis results for various configurations, after which, through interpolation and extrapolation, an antenna design could be generated in a relatively short time. The additional advantage of employing an analytical model, next to the overall speed benefit, however, is that it provides insight by showing how different parameters are related to the analysis results. The price that we are prepared to pay is a limited but still acceptable accuracy that will allow us to use the model for generating initial designs that will, eventually, need to be fine-tuned, employing slower but more accurate methods. We have found such an analytical model.

This analytical model is based on the 'three-term model' for a cylindrical dipole antenna, where an imperfect conductor is modeled by means of a distributed impedance [21, 22]. We specifically chose an analytical dipole model in favor of a numerical model for the reasons mentioned above. By using a distributed *inductance* [23] for the distributed impedance, it becomes possible to model a cylindrical dipole antenna that has a dielectric or magnetic coating. Next, a strip dipole antenna on a dielectric slab is modeled as an equivalent magnetically coated cylindrical dipole antenna [24]. The input impedance of a strip monopole antenna is then found as half that of the corresponding strip dipole antenna; the radiation pattern above an infinite, perpendicular ground plane is identical to the upper half of the radiation pattern of the corresponding strip dipole antenna. Next, the (finite) ground plane is placed parallel to the strip monopole antenna as shown in Figure 3.1. In the following, we shall briefly discuss the model of an imperfectly conducting dipole antenna with a circular cross section, the introduction of a distributed inductance representing a magnetic coating and the use of a generalization of the concept of the equivalent radius to convert a strip dipole antenna to a magnetically coated wire dipole antenna with a circular cross section.

3.3.1 Model of an Imperfectly conducting Dipole Antenna

The admittance Y of a circularly cylindrical, imperfectly conducting dipole antenna with half-length h and cylinder radius a, excited centrally by a delta-gap voltage generator V (Figure 3.25), is given by [25]

$$Y = \mathrm{j}\frac{2\pi k_0}{\xi_0 k \psi_{\mathrm{dR}} \cos(kh)}\left[\sin(kh) + T_{\mathrm{U}}\{1-\cos(kh)\} + T_{\mathrm{D}}\left\{1-\cos\left(\frac{1}{2}k_0 h\right)\right\}\right]. \tag{3.31}$$

As will be shown, the distributed impedance is included in the wave number k. In the above equation,

$$\xi_0 = \sqrt{\frac{\mu_0}{\varepsilon_0}} \tag{3.32}$$

is the characteristic impedance of free space; $\mu_0 = 4\pi \times 10^{-7}$ H m^{-1} is the permeability of free space and $\varepsilon_0 \approx 8.854 \times 10^{-12}$ F m^{-1} is the permittivity of free space. Also, in equation (3.31),

$$k_0 = \omega\sqrt{\varepsilon_0 \mu_0} \tag{3.33}$$

is the free-space wave number, where $\omega = 2\pi f$, f being the frequency.

The wave number k in equation (3.31) is defined by [25]

$$k^2 = k_0^2\left[1 - \mathrm{j}\frac{4\pi z^{\mathrm{i}}}{k_0 \xi_0 k \psi_{\mathrm{dR}}}\right], \tag{3.34}$$

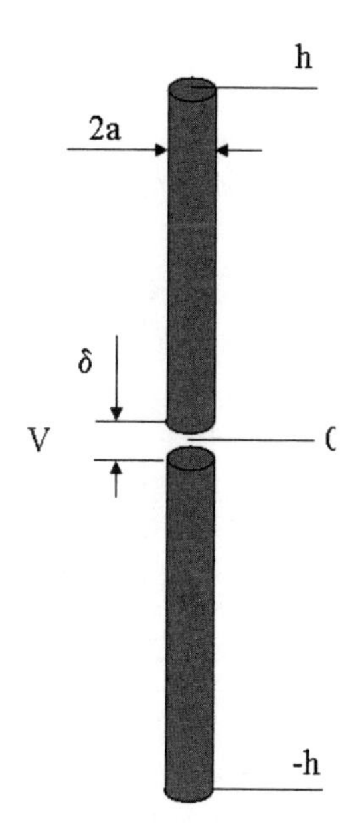

Figure 3.25 Circularly cylindrical dipole antenna of half-length h.

where z^{i} is the distributed impedance.

The expansion parameter ψ_{dR} is defined by

$$\psi_{\mathrm{dR}} = \begin{cases} \psi_{\mathrm{dR}}(0) & \text{if } k_0 h \leq \pi/2 \\ \psi_{\mathrm{dR}}(h - \lambda/2) & \text{if } \pi/2 \leq k_0 h \leq 3\pi/2, \end{cases} \tag{3.35}$$

where λ is the wavelength and

$$\psi_{\mathrm{dR}}(z) = \csc(k[h - |z|]) \int_{z'=-h}^{h} \sin(k[h - |z'|]) \left[\frac{\cos(k_0 r)}{r} - \frac{\cos(k_0 r_h)}{r_h} \right] dz', \tag{3.36}$$

where

$$r = \sqrt{(z - z')^2 + a^2} \tag{3.37}$$

and

$$r_h = \sqrt{(h - z')^2 + a^2}. \tag{3.38}$$

Further,

$$T_{\mathrm{U}} = \frac{C_{\mathrm{V}} E_{\mathrm{D}} - C_{\mathrm{D}} E_{\mathrm{V}}}{C_{\mathrm{U}} E_{\mathrm{D}} - C_{\mathrm{D}} E_{\mathrm{U}}} \tag{3.39}$$

and

$$T_{\mathrm{D}} = \frac{C_{\mathrm{U}} E_{\mathrm{V}} - C_{\mathrm{V}} E_{\mathrm{U}}}{C_{\mathrm{U}} E_{\mathrm{D}} - C_{\mathrm{D}} E_{\mathrm{U}}}, \tag{3.40}$$

where

$$C_{\mathrm{U}} = \left(1 - \frac{k^2}{k_0^2}\right)(\psi_{\mathrm{dUR}} - \psi_{\mathrm{dR}})(1 - \cos[kh]) - \psi_{\mathrm{dUR}} \cos[kh] + \mathrm{j}\psi_{\mathrm{dUI}}\left(\frac{3}{4} - \cos\left[\frac{1}{2}k_0 h\right]\right) + \psi_{\mathrm{U}}(h), \tag{3.41}$$

$$C_{\mathrm{D}} = \psi_{\mathrm{dD}}\left(\frac{3}{4} - \cos\left[\frac{1}{2}k_0 h\right]\right) - \left(1 - \frac{k^2}{k_0^2}\right)\psi_{\mathrm{dR}}\left(1 - \cos\left[\frac{1}{2}k_0 h\right]\right) + \psi_{\mathrm{D}}(h), \tag{3.42}$$

$$C_{\mathrm{V}} = -\left\{\mathrm{j}\psi_{\mathrm{dI}}\left(\frac{3}{4} - \cos\left[\frac{1}{2}k_0 h\right]\right) + \psi_{\mathrm{V}}(h)\right\}, \tag{3.43}$$

$$E_{\mathrm{U}} = -\frac{k^2}{k_0^2}\psi_{\mathrm{dUR}} \cos[kh] - \mathrm{j}\frac{1}{4}\psi_{\mathrm{dUI}} \cos\left[\frac{1}{2}k_0 h\right] + \psi_{\mathrm{U}}(h), \tag{3.44}$$

$$E_{\mathrm{D}} = -\frac{1}{4}\psi_{\mathrm{dD}} \cos\left[\frac{1}{2}k_0 h\right] + \psi_{\mathrm{D}}(h), \tag{3.45}$$

$$E_{\mathrm{V}} = \mathrm{j}\frac{1}{4}\psi_{\mathrm{dI}} \cos\left[\frac{1}{2}k_0 h\right] - \psi_{\mathrm{V}}(h). \tag{3.46}$$

In equations (3.41)–(3.46),

$$\psi_{\mathrm{V}}(h) = \int_{z'=-h}^{h} \sin[k(h - |z'|)] \frac{\mathrm{e}^{-\mathrm{j}k_0 r_h}}{r_h}\, dz', \tag{3.47}$$

$$\psi_{\mathrm{U}}(h) = \int_{z'=-h}^{h} \{\cos[kz'] - \cos[kh]\} \frac{\mathrm{e}^{-\mathrm{j}k_0 r_h}}{r_h}\, dz', \tag{3.48}$$

$$\psi_{\mathrm{D}}(h) = \int_{z'=-h}^{h} \left\{\cos\left[\frac{1}{2}k_0 z'\right] - \cos\left[\frac{1}{2}k_0 h\right]\right\} \frac{\mathrm{e}^{-\mathrm{j}k_0 r_h}}{r_h}\, dz', \tag{3.49}$$

$$\psi_{\mathrm{dUR}} = \{1 - \cos[kh]\}^{-1} \times \int_{z'=-h}^{h} \{\cos[kz'] - \cos[kh]\} \left\{\frac{\cos[k_0 r_0]}{r_0} - \frac{\cos[k_0 r_h]}{r_h}\right\} dz', \tag{3.50}$$

$$\psi_{\mathrm{dD}} = \{1 - \cos[kh]\}^{-1} \times \int_{z'=-h}^{h} \{\cos[k_0 z'] - \cos[k_0 h]\} \left\{\frac{\mathrm{e}^{-\mathrm{j}k_0 r_0}}{r_0} - \frac{\mathrm{e}^{-\mathrm{j}k_0 r_h}}{r_h}\right\} dz', \tag{3.51}$$

$$\psi_{\mathrm{dI}} = -\left\{1 - \cos\left[\frac{1}{2}k_0 h\right]\right\}^{-1} \times \int_{z'=-h}^{h} \sin[k(h - |z'|)] \left\{\frac{\sin[k_0 r_0]}{r_0} - \frac{\sin[k_0 r_h]}{r_h}\right\} dz', \tag{3.52}$$

$$\psi_{\mathrm{dUI}} = -\left\{1 - \cos\left[\frac{1}{2}k_0 h\right]\right\}^{-1} \times \int_{z'=-h}^{h} \{\cos[kz'] - \cos[kh]\}\left\{\frac{\sin[k_0 r_0]}{r_0} - \frac{\sin[k_0 r_h]}{r_h}\right\} dz'. \tag{3.53}$$

Equations (3.34), (3.35) and (3.36) are implicit, meaning that the expansion parameter ψ_{dR} is needed for the calculation of the wave number k and, equally, the wave number k is needed for the calculation of the expansion parameter ψ_{dR}.

To obtain a solution, an iterative method was used. First, the expansion parameter ψ_{dR} was calculated with k_0 (the free-space wave number) substituted for k in equation (3.36). The expansion parameter thus found was then used to obtain a better solution for k by substituting the value found into equation (3.34). With the newly found value for k, a better solution for ψ_{dR} was calculated, after which the whole procedure was repeated. Since the expansion parameter is relatively insensitive to the value of the wave number [25, 26], a stable solution was obtained, in general, after one or two iterations.

3.3.2 Dipole Antenna with Magnetic Coating

In this section, we shall briefly explain how any (analytical) expression for the current in a wire antenna where we have the facility to impose an arbitrary impedance per unit length may be used to obtain results for the same configuration when the wire has a magnetic coating. To this end, we start with the electric-field integro-differential equation for the unknown current $I(\ell)$ in the inner wire of a wire configuration where the wires are coated with a material having a relative permeability $\mu_{\mathrm{r}}(\ell)$ and a relative permittivity $\varepsilon_{\mathrm{r}}(\ell)$ as a function of the position ℓ along the wire. The core is assumed to have a radius $a(\ell)$ as a function of the position ℓ along the wire, and the radius of the cylindrical coating is $b(\ell)$. The expression is given by [23]

$$\begin{aligned}&\frac{\mathrm{j}\omega\mu_0}{4\pi}\int_{\text{wires}}\left\{\mu_{\mathrm{r}}(\ell')\hat{\ell}\cdot\hat{\ell}' + \frac{1}{\varepsilon_{\mathrm{r}}(\ell')}\frac{1}{k^2}\frac{dI(\ell')}{d\ell'}\frac{\partial}{\partial\ell}\right\}G_a(\ell,\ell')\,d\ell' \\ &\quad - \frac{\mathrm{j}\omega\mu_0}{4\pi}\int_{\text{wires}}\left\{[\mu_{\mathrm{r}}(\ell') - 1]\hat{\ell}\cdot\hat{\ell}' I(\ell')\right. \\ &\quad \left. + \left[\frac{1}{\varepsilon_{\mathrm{r}}(\ell')} - 1\right]\frac{1}{k^2}\frac{dI(\ell')}{d\ell'}\frac{\partial}{\partial\ell}\right\}G_b(\ell,\ell')\,d\ell' \\ &= \hat{\ell}\cdot\boldsymbol{E}^{\mathrm{i}}(\ell) - Z_{\mathrm{i}}(\ell)I(\ell),\end{aligned} \tag{3.54}$$

where $\boldsymbol{E}^{\mathrm{i}}$ is the externally impressed electric field, $Z_{\mathrm{i}}(\ell)$ is the intrinsic impedance per unit length of the inner conductor and $\hat{\ell}$ and $\hat{\ell}'$ are unit vectors parallel to the wire at positions ℓ and ℓ', respectively. $G_a(\ell, \ell')$ and $G_b(\ell, \ell')$ are the Green's functions for the inner and outer radii, respectively. Using $G_{ab}(\ell, \ell') \equiv G_a(\ell, \ell') - G_b(\ell, \ell')$, the above equation may be rewritten as

$$\begin{aligned}
&\frac{j\omega\mu_0}{4\pi}\int_{\text{wires}}\left\{\hat{\ell}\cdot\hat{\ell}' I(\ell') + \frac{1}{k^2}\frac{dI(\ell')}{d\ell'}\frac{\partial}{\partial\ell}\right\}G_a(\ell,\ell')\,d\ell' \\
&\quad + \frac{j\omega\mu_0}{4\pi}\int_{\text{wires}}[\mu_r(\ell') - 1]\hat{\ell}\cdot\hat{\ell}' I(\ell')G_{ab}(\ell,\ell')\,d\ell' \\
&\quad + \frac{j\omega\mu_0}{4\pi}\int_{\text{wires}}\left[\frac{1}{\varepsilon_r(\ell')} - 1\right]\frac{1}{k^2}\frac{dI(\ell')}{d\ell'}\frac{\partial}{\partial\ell}G_{ab}(\ell,\ell')\,d\ell' \\
&= \hat{\ell}\cdot \boldsymbol{E}^{i}(\ell) - Z_i(\ell)I(\ell).
\end{aligned} \tag{3.55}$$

The above equation shows separate contributions for the bare wire (first term), the magnetic effect of the coating (second term) and the dielectric effect of the coating (third term). The equation shows that a magnetic coating is easier to handle than a dielectric coating. Therefore, in section 3.3.4 we shall discus how we can transform an equivalent dipole with a dielectric coating into another equivalent dipole with a magnetic coating.

For a magnetic coating ($\varepsilon_r(\ell') = 1$), the third term in the above equation vanishes. With

$$\begin{aligned}
G_p(\ell,\ell') &= \frac{e^{-jkR_p(\ell,\ell')}}{R_p(\ell,\ell')}, \\
R_p(\ell,\ell') &= \sqrt{p^2 + |\boldsymbol{r}(\ell) - \boldsymbol{r}(\ell')|^2}, \\
p &= a, b,
\end{aligned} \tag{3.56}$$

where $\boldsymbol{r}(\ell)$ is the position vector of the point ℓ on the wire structure and the source point at ℓ' taken on the axis of the wire, $G_{ab}(\ell,\ell')$ shows a contribution concentrated at and around $\ell = \ell'$, so that the second term may be approximated by [23]

$$\frac{j\omega\mu_0}{4\pi}I(\ell)[\mu_r(\ell) - 1]\int_{\text{wires}}G_{ab}(\ell,\ell')\,d\ell' \simeq \frac{j\omega\mu_0}{2\pi}I(\ell)[\mu_r(\ell) - 1]\ln\left[\frac{b(\ell)}{a(\ell)}\right]. \tag{3.57}$$

Then, finally, the integral equation may be written as

$$\frac{j\omega\mu_0}{4\pi}\int_{\text{wires}}\left\{\hat{\ell}\cdot\hat{\ell}' I(\ell') + \frac{1}{k^2}\frac{dI(\ell')}{d\ell'}\frac{\partial}{\partial\ell}\right\}G_a(\ell,\ell')\,d\ell' = \hat{\ell}\cdot\boldsymbol{E}^{i}(\ell) - [Z_i(\ell) + Z_m(\ell)]I(\ell), \tag{3.58}$$

where

$$Z_m(\ell) = \frac{j\omega\mu_0}{2\pi}[\mu_r(\ell) - 1]\ln\left[\frac{b(\ell)}{a(\ell)}\right] \equiv j\omega L(\ell). \tag{3.59}$$

3.3.3 Generalization of the Concept of Equivalent Radius

The equivalent-radius theory of Hallén is based essentially on a two-dimensional electrostatic approximation [24]. We first determine the capacitance per unit length of a two-dimensional conductor with a cross section the same as that of the antenna with respect to some parallel reference conductor at a certain distance. Then we demand that the capacitance per unit length of the equivalent radius conductor with respect to the same reference conductor, at

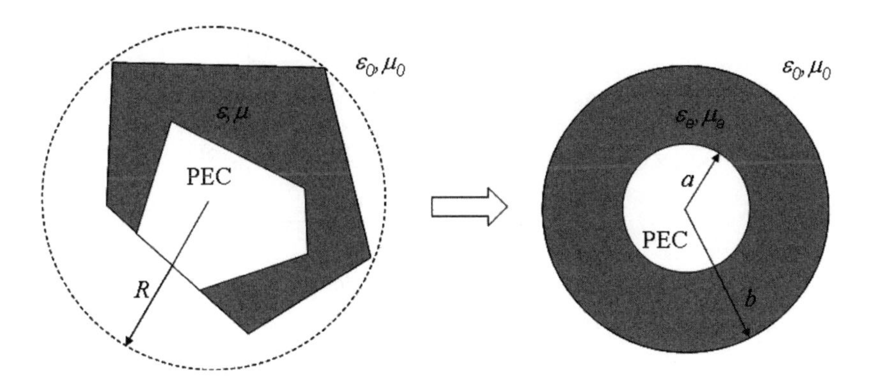

Figure 3.26 Conversion from a cylindrical antenna in the presence of a substrate to an equivalent circular-cross-section antenna (PEC, perfect electric conductor).

the same distance, is equal to this value. This concept may be generalized to the case of a thin cylindrical antenna in the presence of arbitrarily shaped dielectric and/or magnetic materials (Figure 3.26).

If we consider the metal part of the antenna shown on the left of Figure 3.26 as a perfect conductor, the longitudinal component of the electric field on the metallic surface will be zero. Also, the normal component of the magnetic-field vector will be zero, and – owing to the assumed cylindrical shape – the longitudinal component of the magnetic field will be zero as well. Close to the antenna, the two field vectors may therefore considered as quasi-static in nature, and to be due to a current flowing along and a charge on an infinitely long cylinder. For the antenna shown on the right of Figure 3.26 to be equivalent to the one shown on the left, the charge per unit length Q' and the current I should be identical [24]. Furthermore, the electric and magnetic fields at large distances from the antennas should also be identical, whereas the fields near the antennas will in general differ greatly.

Next, we take a two-conductor system where both conductors are of the form shown on the left of Figure 3.26 and a two-conductor system where both conductors are of the form shown on the right of the same figure. We demand that the electrical energies per unit length W'_e for the two systems corresponding to equal and opposite charges Q' and $-Q'$ on the conductors be equal [24]:

$$W'_{\mathrm{e}_\mathrm{left}} = \frac{Q'^2}{2C'_\mathrm{left}} = W'_{\mathrm{e}_\mathrm{right}} = \frac{Q'^2}{2C'_\mathrm{right}}. \tag{3.60}$$

We also demand that the magnetic energies per unit length W'_m for the two systems corresponding to equal and opposite currents I and $-I$ in the conductors be equal [24]:

$$W'_{\mathrm{m}_\mathrm{left}} = \frac{L'_\mathrm{left} I^2}{2} = W'_{\mathrm{m}_\mathrm{right}} = \frac{L'_\mathrm{right} I^2}{2}. \tag{3.61}$$

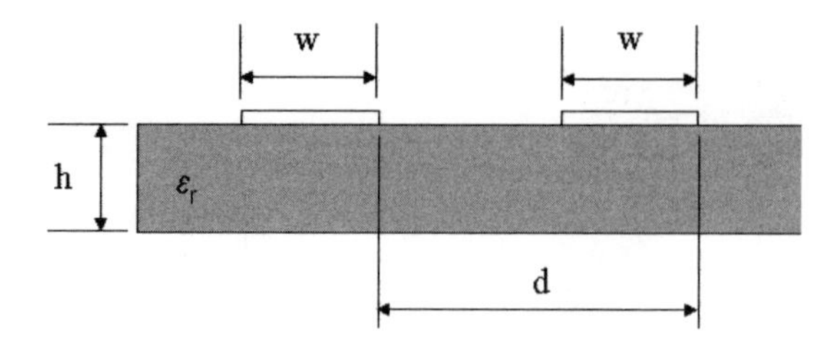

Figure 3.27 Geometry for the calculation of C'_{left}.

With

$$C'_{\text{right}} = \frac{\pi\varepsilon_0}{\ln(d/a) - (1 - 1/\varepsilon_{\text{er}})\ln(b/a)}, \tag{3.62}$$

$$L'_{\text{right}} = \frac{\mu_0}{\pi}\ln\left(\frac{d}{a}\right) + \frac{\mu_{\text{e}} - \mu_0}{\pi}\ln\left(\frac{b}{a}\right), \tag{3.63}$$

where $d \gg b$ is the distance between the two conductors, we find that

$$\left(1 - \frac{1}{\varepsilon_{\text{er}}}\right)\ln\left(\frac{b}{a}\right) = \ln\left(\frac{d}{a}\right) - \frac{\pi\varepsilon_0}{C'_{\text{left}}}, \tag{3.64}$$

$$(\mu_{\text{er}} - 1)\ln\left(\frac{b}{a}\right) = \frac{\pi}{\mu_0}L'_{\text{left}} - \ln\left(\frac{d}{a}\right). \tag{3.65}$$

We now have two equations and four unknowns (a, b, ε_{er} and μ_{er}), so two unknowns may be chosen for convenience.

For a metallic strip on a dielectric slab, we choose $\varepsilon_{\text{er}} = \varepsilon_{\text{r}}$ and $\mu_{\text{er}} = 1$. Then, equation (3.64) leads to

$$\left(\frac{b}{a}\right) = \exp\left\{\frac{\varepsilon_{\text{r}}}{\varepsilon_{\text{r}} - 1}\left[\ln\left(\frac{d}{a}\right) - \frac{\pi\varepsilon_0}{C'_{\text{left}}}\right]\right\}. \tag{3.66}$$

This choice of unknowns, when substituted into equation (3.65), leads to a formulation that applies to an antenna without a dielectric or magnetic covering. This implies that the equivalent radius a should be equal to one-fourth of the strip width, i.e. $a = w/4$.

The capacitance C'_{left} is the capacitance between two identical electrically conducting strips on the dielectric slab, displaced relative to each other a distance d, as shown in Figure 3.27. The capacitance value is calculated as [27]

$$\frac{1}{C'_{\text{left}}} = \frac{1}{\pi\varepsilon_{\text{r}}\varepsilon_0}\int_{x=0}^{\infty}\frac{16}{x^3w^2}\sin^2\left(\frac{xw}{2}\right)\sin^2\left(\frac{xd}{2}\right)\frac{\varepsilon_{\text{r}}^2\cosh(xh) + \varepsilon_{\text{r}}\sinh(xh)}{(\varepsilon_{\text{r}}^2 + 1)\sinh(xh) + 2\varepsilon_{\text{r}}\cosh(xh)}\,dx. \tag{3.67}$$

It is not possible to let the distance d go to infinity, since that would lead to L'_{left}, L'_{right}, C'^{-1}_{left} and C'^{-1}_{right} becoming infinite. A value of 20 to 200 times the radius R in Figure 3.26 is advised in [24].

3.3.4 Equivalent Dipole with Magnetic Coating

To analyze a strip dipole or monopole antenna on a dielectric slab, we prefer to transform the equivalent dielectrically coated wire antenna into an equivalent wire antenna with a purely magnetic coating. The reason for performing this extra transformation lies in the fact that this will lead to a thinner coating and the theory is more accurate for thinner coatings [24]. For this transformation, we replace the parameters a, b, ε_{r} and μ_{r} by a', b', $\varepsilon_{\mathrm{r}}'$ and μ_{r}', respectively, where [23, 24]

$$b' = b, \tag{3.68}$$

$$\mu_{\mathrm{r}}' = \mu_{\mathrm{r}}\varepsilon_{\mathrm{r}}, \tag{3.69}$$

$$a' = b\left(\frac{a}{b}\right)^{1/\varepsilon_{\mathrm{r}}}. \tag{3.70}$$

Substituting these transformed parameters into equation (3.59) gives

$$Z^{e} = \frac{\mathrm{j}\omega\mu_0}{2\pi}\left[\frac{\varepsilon_{\mathrm{r}}-1}{\varepsilon_{\mathrm{r}}}\right]\ln\left(\frac{b}{a}\right). \tag{3.71}$$

Then, with equation (3.66) substituted into equation (3.71) and the latter equation substituted into equation (3.34), we may calculate the input impedance of a strip dipole on a dielectric slab.

3.3.5 Validation

To validate the computer code based on the analysis techniques described above, a strip monopole antenna on a dielectric slab, placed perpendicularly on an infinite ground plane, was analyzed. The configuration and its dimensions are shown in Figure 3.28. The real and imaginary parts of the input admittance ($Y_{\mathrm{in}} = G + \mathrm{j}B$) were calculated as a function of frequency and are shown, together with measured results from [24], in Figure 3.29. The admittance was calculated as twice the admittance of the corresponding dipole antenna. In the same figure, we also show the calculated results for a bare strip antenna, analyzed as an equivalent antenna of circular cross section.[4]

The figure shows, first of all, that the effects of the dielectric need to be included in the analysis. Furthermore, the agreement between the calculated and measured input admittance results is fair to good over the frequency band shown: the difference between the calculated and measured values of G relative to the measured maximum of G remains below 10% and the difference between the calculated and measured values of B relative to the measured maximum of B remains below 16%. Around resonance, these numbers are much lower. By replacing the numerical evaluation of a coated wire antenna by an analytical evaluation, we have simplified the total analysis without severely compromising the accuracy, as demonstrated in [24]. Next, we adapt the analysis technique to analyze planar printed monopoles of the kind shown in Figure 3.1.

[4] In the calculation of the capacitance, the distance between the identical strips was varied between 10 and 200 times the height of the dielectric. No significant difference was observed between the calculated admittance results.

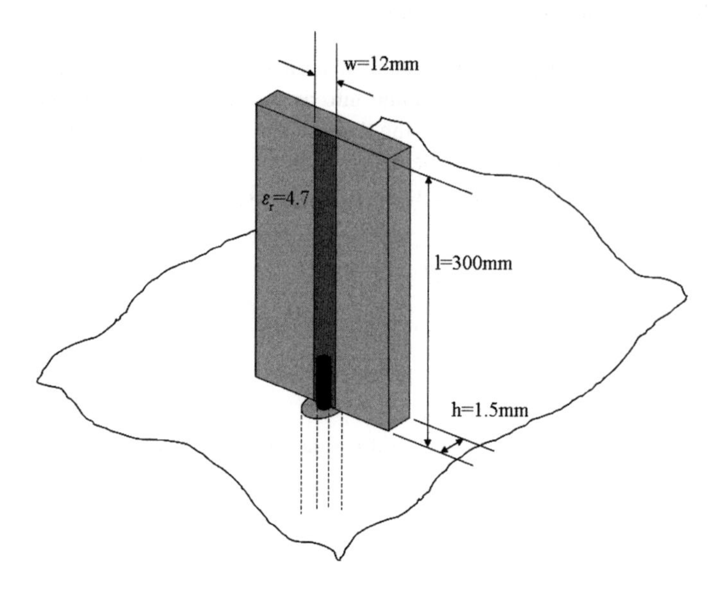

Figure 3.28 Strip monopole antenna on a dielectric slab, placed perpendicularly on an infinite ground plane. The width of the dielectric slab was 51 mm.

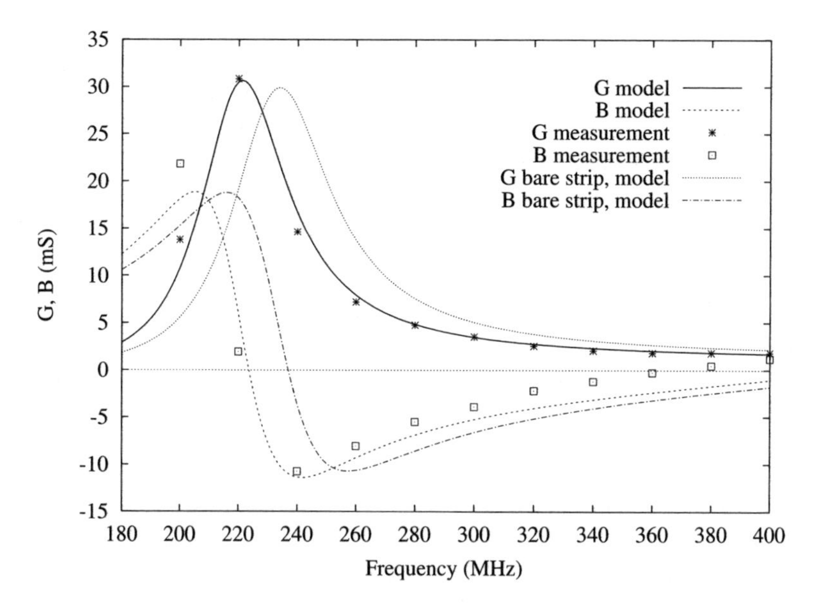

Figure 3.29 Calculated and measured (from [24]) input admittance as a function of frequency for the strip monopole antenna shown in Figure 3.28, and calculated input admittance as a function of frequency for a bare strip monopole antenna.

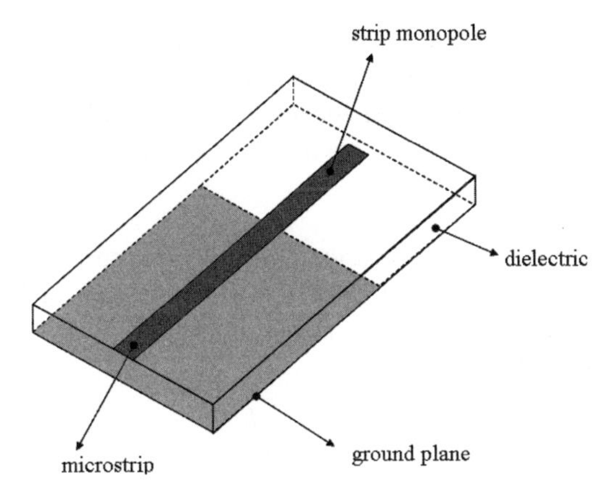

Figure 3.30 Microstrip-excited planar strip monopole antenna.

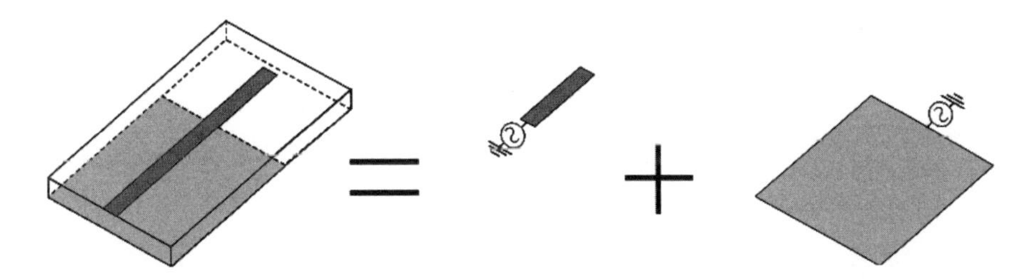

Figure 3.31 Microstrip-excited planar strip monopole antenna considered as an asymmetrically driven strip dipole antenna that may be separated into two strip monopole antennas.

3.3.6 Microstrip-Excited Planar Strip Monopole Antenna

To analyze microstrip-excited planar monopole antennas of the kind shown in Figure 3.30, we make use of an approximate expression for the impedance of an asymmetrically driven antenna [28, 29]. To this end, we consider the structure shown in Figure 3.30 as an asymmetrically driven strip dipole antenna and then separate the structure into two grounded monopole antennas as shown in Figure 3.31, where one of the monopole antennas is the strip monopole and the other monopole antenna is formed by the ground plane of the microstrip transmission line.

Both strip monopole antennas can be analyzed using the theory discussed above. We need to incorporate the microstrip transmission line into the analysis to transfer the input admittance, calculated at the junction between the microstrip and the strip, to the beginning of the microstrip transmission line on the underside of the printed circuit board. We employed copper tape, a knife and a ruler for the construction of prototype antennas [30]. Using this technique, we could not realize 50 Ω microstrip lines with high accuracy.[5] Since we could measure the strip dimensions with high accuracy, however, it was necessary to have an accurate, preferably analytical, model for microstrip transmission lines at our disposal to accurately account for the section of microstrip transmission line. We used the model described in [31].

3.3.6.1 Analysis of Microstrip Transmission Line A microstrip transmission line of strip width W, thickness d, positioned on a grounded dielectric slab of height h and relative permittivity ε_r, is characterized by a characteristic impedance Z_c and a phase constant[6] β, which are defined by, respectively,

$$Z_c = \frac{\eta_0}{\sqrt{\varepsilon_{eff}}} \frac{h}{W_{eff}}, \tag{3.72}$$

$$\beta = k_0 \sqrt{\varepsilon_{eff}}, \tag{3.73}$$

where $\eta_0 = \sqrt{\mu_0/\varepsilon_0}$ is the characteristic impedance of free space and $k_0 = \omega\sqrt{\varepsilon_0\mu_0}$ is the free-space wave number.

In equations (3.72) and (3.73), an effective width W_{eff} and an effective relative permittivity ε_{eff} have been used. The effective width is defined by

$$W_{eff}(f) = \frac{W}{3} + (R_w + P_w)^{1/3} - (R_w - P_w)^{1/3}, \tag{3.74}$$

where

$$P_w = \left(\frac{W}{3}\right)^3 + \frac{S_w}{2}\left[W_{eff}(0) - \frac{W}{3}\right], \tag{3.75}$$

$$Q_w = \frac{S_w}{3} - \left(\frac{W}{3}\right)^2, \tag{3.76}$$

$$R_w = (P_w^2 + Q_w^2)^{1/2}, \tag{3.77}$$

$$S_w = \frac{c_0^2}{4f^2[\varepsilon_{eff}(f) - 1]}, \tag{3.78}$$

$$\varepsilon_{eff}(f) = \varepsilon_r - \frac{\varepsilon_r - \varepsilon_{eff}(0)}{1 + P}, \tag{3.79}$$

with

$$P = P_1 P_2 \{(0.1844 + P_3 P_4) f_n\}^{1.5763}, \tag{3.80}$$

$$P_1 = 0.27488 + \left\{0.6315 + \frac{0.525}{(1 + 0.0157 f_n)^{20}}\right\} u - 0.065683 e^{-8.7513u}, \tag{3.81}$$

$$P_2 = 0.33622\{1 - e^{-0.03442\varepsilon_r}\}, \tag{3.82}$$

[5] The accuracy with which we could realize strips with this technique was, depending on the operator, about half a millimeter.

[6] Ignoring losses for the moment.

$$P_3 = 0.0363e^{-4.6u}\{1 - e^{(f_n/38.7)^{4.97}}\}, \tag{3.83}$$

$$P_4 = 1 + 2.751\{1 - e^{(\varepsilon_r/15.916)^8}\}, \tag{3.84}$$

$$f_n = fh \times 10^{-6}, \tag{3.85}$$

$$u = \frac{W + (W' - W)/\varepsilon_r}{h}, \tag{3.86}$$

and where c_0 is the speed of light in free space.

The static ($f = 0$) effective width is defined by

$$W_{\text{eff}}(0) = \frac{2\pi h}{\ln\{hF/W' + \sqrt{1 + (2h/W')}\}}, \tag{3.87}$$

where

$$F = 6 + (2\pi - 6)e^{-(4\pi^2/3)(h/W')^{3/4}} \tag{3.88}$$

and

$$W' = W + \frac{d}{\pi}\left\{1 + \ln\left(\frac{4}{\sqrt{(d/h)^2 + \frac{(1/\pi)^2}{(W/t+1.1)^2}}}\right)\right\}. \tag{3.89}$$

The static relative permittivity is defined by

$$\varepsilon_{\text{eff}}(0) = \frac{1}{2}\{\varepsilon_r + 1 + (\varepsilon_r - 1)G\}, \tag{3.90}$$

where

$$G = \left(1 + \frac{10h}{W}\right)^{-AB} - \frac{\ln(4)}{\pi}\frac{d}{\sqrt{Wh}}, \tag{3.91}$$

$$A = 1 + \frac{1}{49}\ln\left\{\frac{(W/h)^4 + (W/52h)^2}{(W/h)^4 + 0.432}\right\} + \frac{1}{18.7}\ln\left\{1 + \left(\frac{W}{18.1h}\right)^3\right\}, \tag{3.92}$$

and

$$B = 0.564e^{-0.2/(\varepsilon_r+0.3)}. \tag{3.93}$$

With equations (3.74)–(3.93), we can calculate the transformation from the input admittance of the monopole antenna to the connector at the side of the PCB (Figure 3.30) using the well-known transmission line equation

$$Y_{\text{in}} = Y_c\frac{Y_L + Y_c\tanh(\gamma\ell)}{Y_c + Y_L\tanh(\gamma\ell)}. \tag{3.94}$$

In the above equation, Y_{in}, the input admittance, is the admittance at the edge of the PCB. Y_L, the load admittance, is the input admittance of the monopole antenna at the position where the microstrip continues as a strip without a ground plane. $Y_c = 1/Z_c$ is the characteristic admittance of the microstrip transmission line, ℓ is the length of the transmission line and γ is the propagation constant, which is given by

$$\gamma = \alpha + \mathrm{j}\beta. \tag{3.95}$$

Here α is the attenuation coefficient, which we have ignored so far. If losses cannot be ignored, the attenuation factor is given by

$$\alpha = \alpha_{\mathrm{d}} + \alpha_{\mathrm{cs}} + \alpha_{\mathrm{cg}}, \tag{3.96}$$

where

$$\alpha_{\mathrm{d}} = 0.5\beta \frac{\varepsilon_{\mathrm{r}}}{\varepsilon_{\mathrm{eff}}(f)} \frac{\varepsilon_{\mathrm{eff}}(f) - 1}{\varepsilon_{\mathrm{r}} - 1} \tan(\delta), \tag{3.97}$$

$$\alpha_{\mathrm{cs}} = \alpha_{\mathrm{n}} R_{\mathrm{ss}} F_{\Delta\mathrm{s}} F_{\mathrm{s}}, \tag{3.98}$$

$$\alpha_{\mathrm{cg}} = \alpha_{\mathrm{n}} R_{\mathrm{sg}} F_{\Delta\mathrm{g}}, \tag{3.99}$$

$$R_{\mathrm{ss}} = \sqrt{\frac{\pi f \mu_0}{\sigma_{\mathrm{s}}}}, \tag{3.100}$$

$$R_{\mathrm{sg}} = \sqrt{\frac{\pi f \mu_0}{\sigma_{\mathrm{g}}}}, \tag{3.101}$$

$$\alpha_{\mathrm{n}} = \begin{cases} \dfrac{1}{4\pi h Z_{\mathrm{c}}(0)} \dfrac{32 - (W'/h)^2}{32 + (W'/h)^2} & \text{if } \dfrac{W'}{h} < 1 \\ \dfrac{\sqrt{\varepsilon_{\mathrm{eff}}(0)}}{2\eta_0 W_{\mathrm{eff}}(0)} \left(\dfrac{W'}{h} + \dfrac{0.667 W'/h}{W'/h + 1.444} \right) & \text{if } \dfrac{W'}{h} \geq 1, \end{cases} \tag{3.102}$$

$$F_{\Delta\mathrm{s}} = 1 + \frac{2}{\pi} \arctan\{1.4(R_{\mathrm{ss}}\Delta_{\mathrm{s}}\sigma_{\mathrm{s}})^2\}, \tag{3.103}$$

$$F_{\Delta\mathrm{g}} = 1 + \frac{2}{\pi} \arctan\{1.4(R_{\mathrm{sg}}\Delta_{\mathrm{g}}\sigma_{\mathrm{g}})^2\}, \tag{3.104}$$

$$F = 1 + \frac{2h}{W'} \left(1 - \frac{1}{\pi} + \frac{W' - W}{t} \right). \tag{3.105}$$

In equations (3.100), (3.101), (3.103) and (3.104), σ_{g} is the conductivity of the ground plane, σ_{s} is the conductivity of the strip, Δ_{g} is the skin depth of the ground plane and Δ_{s} is the skin depth of the strip.[7]

3.3.6.2 Results To validate the theory, a number of prototype antennas were realized in the way described at the start of section 3.3.6. To mimic mobile wireless devices, PCBs were made as shown in Figure 3.30, where monopole strip lengths of 15 mm, 30 mm and 50 mm, ground plane lengths of 30 mm, 65 mm and 100 mm and ground plane widths of 30 mm,

[7]The skin depth is the distance a wave must travel in a medium for its field amplitude values to be reduced by a factor $\mathrm{e}^{-1} = 0.37$. For good conductors, the skin depth may be approximated by $\Delta \approx \sqrt{2/(\omega\mu\sigma)}$ [17]. So, for copper ($\sigma = 5.8 \times 10^7$ S m^{-1}) at 1 GHz, the skin depth Δ is approximately 2.1×10^{-6} m.

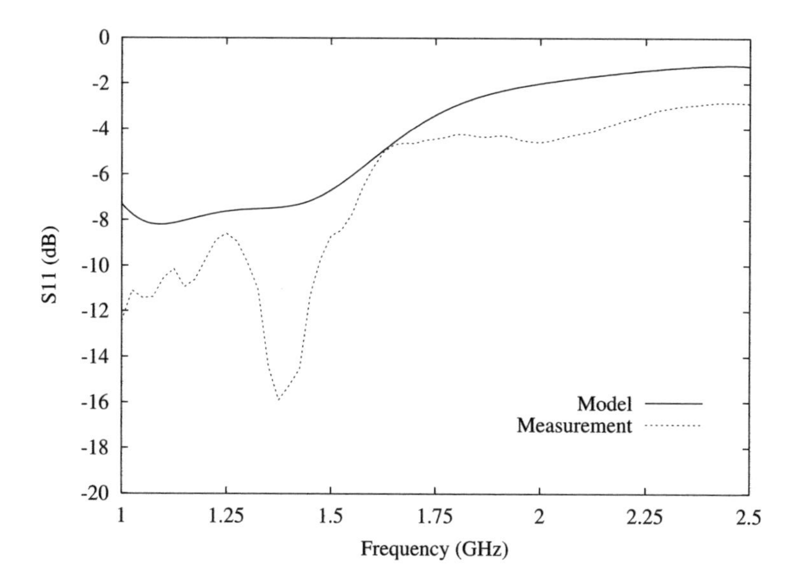

Figure 3.32 Calculated and measured return loss as a function of frequency for a strip monopole antenna of length 50 mm. The ground plane length was 100 mm, and the ground plane width was 50 mm. The strip width was 3.0 mm. The relative permittivity of the substrate was 4.28, the loss tangent was 0.016 and the thickness was 1.6 mm.

40 mm and 50 mm were used. The width of the monopole strip and microstrip was 3 mm in all prototypes. The return loss as a function of frequency was calculated and measured in the frequency range from 1 GHz to 6 GHz.[8]

We start with two examples where the match between the calculations and measurements was poor. The results are shown in Figures 3.32 and 3.33. The figure captions provide details of the antenna dimensions.

Evaluation of all of the 27 combined sets of measurements and simulations reveals that:

- *For a ground plane width $W = 50$ mm, in all situations, the correspondence between calculations and measurements is poor.* If we return to the basics of the theory, i.e. the three-term dipole model, the radius should satisfy the condition that $k_0 a \ll 1$ [25], where $k_0 = 2\pi/\lambda$. Taking for the radius a quarter of the ground plane width, $k_0 a$ for $W = 30$ mm, 40 mm and 50 mm at 3 GHz equals 0.47, 0.63 and 0.79, respectively. None of these values satisfies the radius condition completely, but the violation of the condition becomes worse for wider ground planes. Wider ground planes (larger equivalent radii) also lead to larger input impedances. So, with increasing ground plane width, the simulation results become more inaccurate and more dominant in the total antenna input impedance.

[8]We anticipate use in the frequency bands for the Global System for Mobile Communications (GSM) around 0.9 GHz and around 1.8 GHz, for the Global Positioning System (GPS) around 1.5 GHz, for Bluetooth, ZigBee and wireless local area networks around 2.4 GHz, and for future WLANs around 3.7 and 5 GHz.

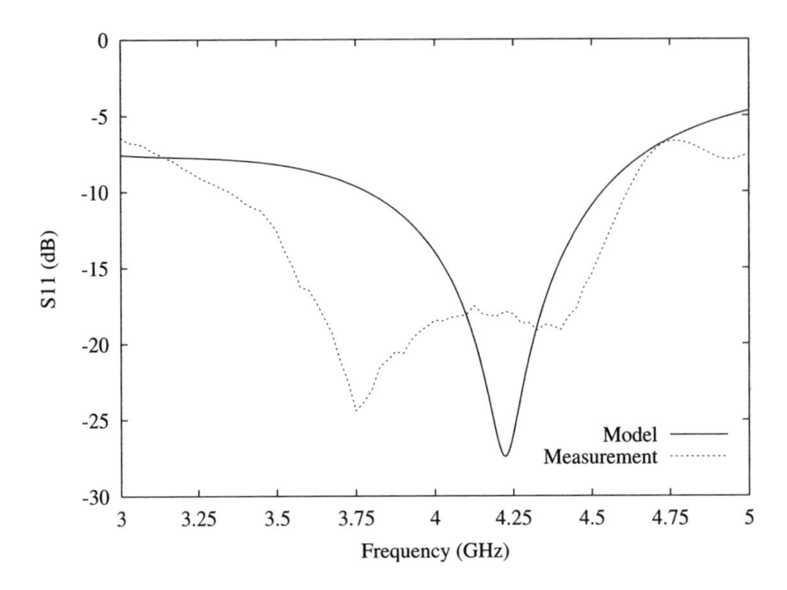

Figure 3.33 Calculated and measured return loss as a function of frequency for a strip monopole antenna of length 15 mm. The ground plane length was 30 mm, and the ground plane width was 30 mm. The strip width was 3.0 mm. The relative permittivity of the substrate was 4.28, the loss tangent was 0.016 and the thickness was 1.6 mm.

- *For a monopole length $L_{\mathrm{m}} = 15$ mm, a ground plane length $L_{\mathrm{gp}} = 65$ mm and, to an even greater extent for $L_{\mathrm{gp}} = 30$ mm, the measured minimum in the reflection occurs at a lower frequency than simulated.* The ground plane, which is wider and longer than the strip monopole antenna, has a strong contribution to the input impedance of the antenna. Owing to the unbalanced excitation of the antenna, a ground plane that is 'too short' will lead to currents flowing over the outer conductor of the coaxial cable connected to the antenna. These currents effectively lengthen the 'too short' ground plane, thus lowering the resonance frequency. The effect is hardly visible for ground planes 100 mm long, but becomes visible for ground planes 65 mm and 30 mm long.
- *For monopole antennas resonant in the frequency range from 3 to 5 GHz, a ripple in the reflection coefficient as a function of frequency may be visible.* For these frequencies, the rim of the ground plane (at the junction of the microstrip with the monopole strip) may reach a length close to half a wavelength and become resonant.

From these observations, we may distill the following set of rough design constraints for the width of the monopole strip w and the ground plane width W:

$$\begin{aligned} \frac{w}{W} &\leq 0.075, \\ k_0 \frac{W}{4} &\leq 0.65, \\ W &\leq \frac{c_0 f_{\max}}{2}, \end{aligned} \tag{3.106}$$

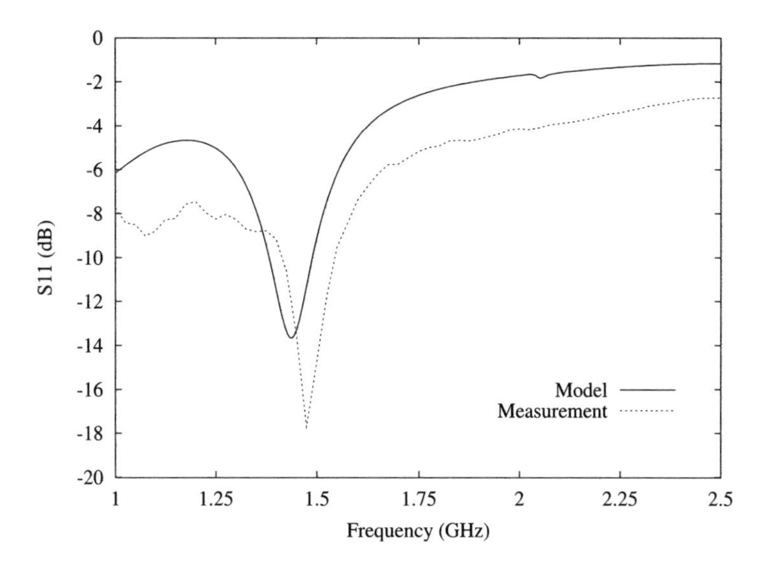

Figure 3.34 Calculated and measured return loss as a function of frequency for a strip monopole antenna of length 50 mm. The ground plane length was 100 mm, and the ground plane width was 40 mm. The strip width was 3.0 mm. The relative permittivity of the substrate was 4.28, the loss tangent was 0.016 and the thickness was 1.6 mm.

where c_0 is the speed of light in free space and $f_{\max}$ is the maximum frequency of operation for the antenna.

Some typical results (where the relative difference in resonance frequency is well below 10%), obtained following these design guidelines, are shown in Figures 3.34–3.37.[9]

In [29], the separation into monopoles was analyzed using a method of moments applied to an asymmetrically driven, bare-wire dipole antenna in free space and applied to the two monopole antennas of equal radii that the structure was separated into. From comparing the complex input impedance of the asymmetrically driven dipole antenna with the sum of the input impedances of the two monopole antennas, it was concluded that the separation led to results that were too inaccurate, and correction terms were added. The reason for the differences between the results for the dipole and the separated monopoles was considered to be due to the lack of coupling between the monopoles in the model.

The reason that the separation of the antenna structure into two monopole antennas works well for a strip monopole on a dielectric slab therefore lies in the fact that the dielectric slab confines the electromagnetic field to the dipole arms, making the coupling between the two dipole arms less critical than in the bare-wire dipole case.

[9] Observing these figures, it appears that a systematic shift in resonance frequency and return loss level has occurred in the simulations. However, other simulation and measurement results indicate that this is not true.

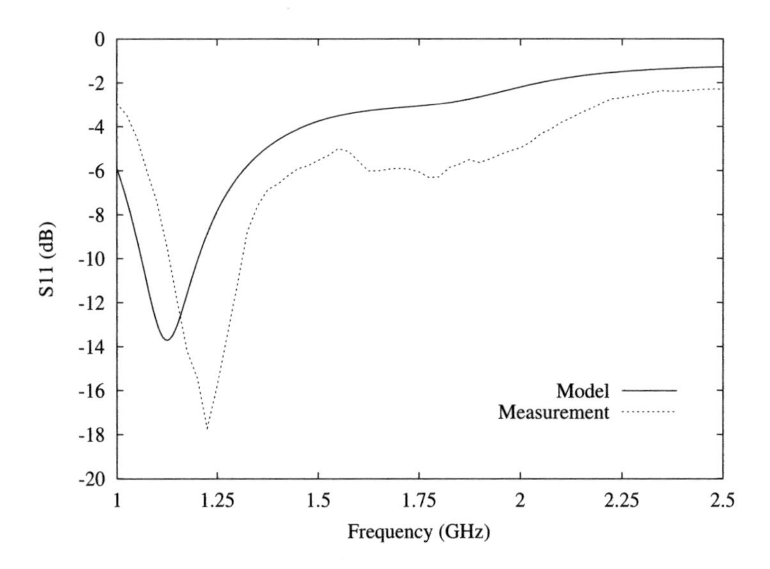

Figure 3.35 Calculated and measured return loss as a function of frequency for a strip monopole antenna of length 50 mm. The ground plane length was 65 mm, and the ground plane width was 40 mm. The strip width was 3.0 mm. The relative permittivity of the substrate was 4.28, the loss tangent was 0.016 and the thickness was 1.6 mm.

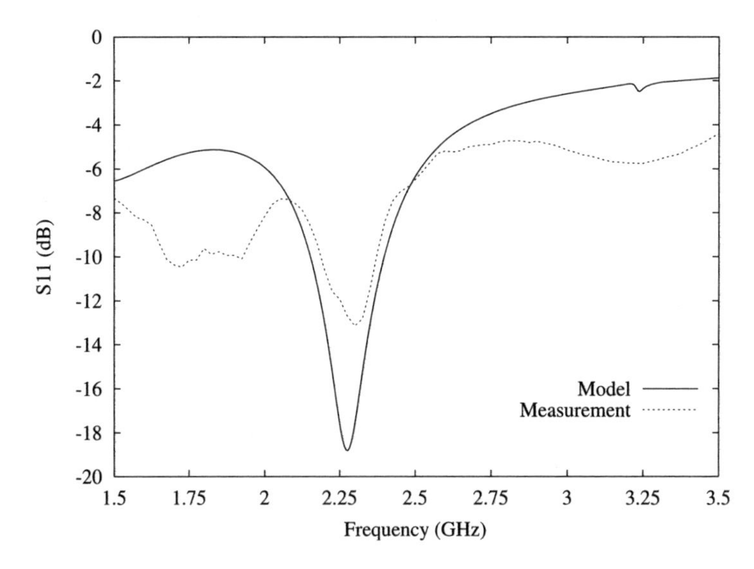

Figure 3.36 Calculated and measured return loss as a function of frequency for a strip monopole antenna of length 30 mm. The ground plane length was 65 mm, and the ground plane width was 30 mm. The strip width was 3.0 mm. The relative permittivity of the substrate was 4.28, the loss tangent was 0.016 and the thickness was 1.6 mm.

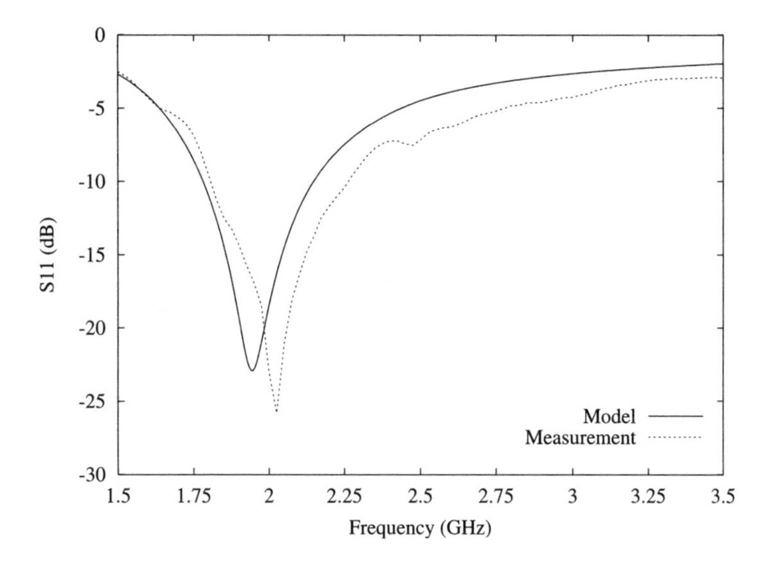

Figure 3.37 Calculated and measured return loss as a function of frequency for a strip monopole antenna of length 30 mm. The ground plane length was 30 mm, and the ground plane width was 40 mm. The strip width was 3.0 mm. The relative permittivity of the substrate was 4.28, the loss tangent was 0.016 and the thickness was 1.6 mm.

3.4 CONCLUSIONS

When physical reasoning in itself is not accurate enough for designing an antenna, this reasoning may be used as a starting point for designing an antenna employing, in an iterative way, a commercial off-the-shelf full-wave electromagnetic analysis program. The iterative use of the program replaces the trial-and-error method in which test structures are physically realized. Owing to the time-consuming character of this design strategy, it should only be applied to one-of-a-kind designs. As soon as it is understood that a class of antennas needs to be dealt with, i.e. similar antennas need to be designed for different frequency bands or made of different materials, it is worthwhile to invest in the development of an (approximate) antenna model. It has been shown that the design of a planar, printed UWB antenna may be realized by employing a full-wave analysis program, starting from the understanding that the antenna is a combination of a 'fat' dipole antenna and two tapered slot antennas. The design of a UWB antenna that incorporates filtering structures for blocking a certain frequency band may be accelerated by applying analytical models to the filtering structure. For non-UWB planar, printed dipole or monopole antennas, which may be used in various kind of wireless devices using frequencies in the range from 1 GHz to 6 GHz (GSM, GPS, Bluetooth, ZigBee, and WLAN), the development of an approximate analytical model will help in the design of these antennas. By employing a model of an equivalent-radius, magnetically coated wire dipole antenna and the separation of a microstrip-excited strip monopole antenna into a strip monopole antenna and a monopole antenna consisting of the microstrip ground plane, an approximate but still accurate analytical model has been created. Provided that the width of the ground plane does not violate the thin-wire approximation too much and that the width is also small enough to prevent resonance from the rim of the ground plane in the frequency band of interest, this model may be used to create (pre)designs of internal monopole antennas.

REFERENCES

1. H.J. Visser, 'Low-cost, compact UWB antenna with frequency band-notch function', *Proceedings of the 2nd European Conference on Antennas and Propagation, EuCAP2007*, Edinburgh, UK, pp. 1–4, November 2007.
2. C.A. Balanis, *Antenna Theory, Analysis and Design*, 2nd edition, John Wiley & Sons, New York, 1996.
3. H. Schantz, *The Art and Science of Ultrawideband Antennas*, Artech House, Boston, MA, 2005.
4. Timederivative, available at www.timederivative.com/PUBs-2cent-antenna.pdf.
5. T.W. Hertel, 'Cable-current effects of miniature UWB antennas', *Proceedings of the IEEE Antennas and Propagation Society International Symposium*, pp. 524–527, 2005.
6. E. Gueguen, F. Thudor and P. Chambelin, 'A low cost UWB printed dipole antenna with high performances', *Proceedings of the IEEE International Conference on UWB*, pp. 89–92, 2005.
7. Available at www.cst.com.
8. J. Kim, T.L. Yoon, J. Kim and J. Choi, 'Design of an ultra wide-band printed monopole antenna using FDTD and genetic algorithm', *IEEE Microwave and Wireless Components Letters*, Vol. 15, No. 6, pp. 395–397, June 2005.
9. T.S.P. See and Z.N. Chen, 'A small UWB antenna for wireless USB', *Proceedings of the IEEE International Conference on Ultra-Wideband, ICUWB2007*, pp. 198–203, 2007.
10. K.-M. Kim, S.-K. Park, I.-S. Na and C.-B. Park, 'A planar UWB antenna with band rejection characteristic', *Proceedings of the IEEE Region 10 Conference, TENCON2007*, pp. 1–4, 2007.
11. D.-H. Kwon and Y. Kim, 'Suppression of cable leakage current for edge-fed printed dipole UWB antennas using leakage-blocking slots', *IEEE Antennas and Wireless Propagation Letters*, Vol. 5, pp. 183–186, 2006.
12. Available at www.cst.com/Content/Applications/Article/Ultra-Wide-Band+Printed+Circular+ Dipole+Antenna.
13. N. Nguyen, C. Hsieh and D.W. Ball, 'Millimeter wave printed circuit spurline filters', *Proceedings of the IEEE MTT-S International Microwave Symposium*, pp. 98–100, 1983.

14. T.C. Edwards, *Foundations for Microstrip Circuit Design*, John Wiley & Sons, Chichester, 1981.

15. E. Hammerstad and Ø. Jensen, 'Accurate models for microstrip computer aided design', *Proceedings of the IEEE MTT-S International Microwave Symposium*, pp. 407–409, 1980.

16. E.M.T. Jones and J.T. Bolljahn, 'Coupled-strip-transmission-line filters and directional couplers', *IRE Transactions on Microwave Theory and Techniques*, pp. 75–81, April 1956.

17. D.M. Pozar, *Microwave Engineering*, second edition, John Wiley & Sons, New York, 1998.

18. C.-T. Tai, *Dyadic Green Functions in Electromagnetic Theory*, second edition, IEEE Press, New York, 1994.

19. K. Sawaya, T. Ishizone and Y. Mushiake, 'A simplified expression for the dyadic Green's function for a conducting halfsheet', *IEEE Transactions on Antennas and Propagation*, Vol. AP-29, pp. 749–756, September 1981.

20. D.M. Pozar and E.H. Newman, 'Analysis of a monopole mounted near an edge or a vertex', *IEEE Transactions on Antennas and Propagation*, Vol. AP-30, No. 5, pp. 401–408, May 1982.

21. R.W.P. King and T.T. Wu, 'The imperfectly conducting cylindrical transmitting antenna', *IEEE Transactions on Antennas and Propagation*, Vol. AP-14, No. 5, pp. 524–534, September 1966.

22. R.W.P. King, C.W. Harrison and E.A. Aronson, 'The imperfectly conducting cylindrical transmitting antenna, numerical results', *IEEE Transactions on Antennas and Propagation*, Vol. AP-14, No. 5, pp. 535–542, September 1966.

23. J. Moore and M.A. West, 'Simplified analysis of coated wire antennas and scatterers', *IEE Proceedings*, Vol. 142, Part H, No. 1, pp. 14–18, February 1995.

24. B.D. Popovic and A. Nesic, 'Generalisation of the concept of equivalent radius of thin cylindrical antennas', *IEE Proceedings*, Vol. 131, Part H, No. 3, pp. 153–158, June 1984.

25. R.W.P. King and T.T. Wu, 'The imperfectly conducting cylindrical transmitting antenna', *IEEE Transactions on Antennas and Propagation*, Vol. AP-14, No. 5, pp. 524–534, September 1966.

26. R.W.P. King, C.W. Harrison and E.A. Aronson, 'The imperfectly conducting cylindrical transmitting antenna, numerical results', *IEEE Transactions on Antennas and Propagation*, Vol. AP-14, No. 5, pp. 535–542, September 1966.

27. E. Ymashita and S. Yamazaki, 'Parallel-strip line embedded in or printed on a dielectric sheet', *IEEE Transactions on Microwave Theory and Techniques*, pp. 972–973, November 1968.

28. R.W.P. King, 'Asymmetrically driven antennas and the sleeve dipole', *Proceedings of the IRE*, pp. 1154–1164, October 1950.

29. K. Boyle, *Antennas for Multi-Band RF Front-End Modules*, PhD thesis, Delft University Press, 2004.

30. H.J. Visser, *Array and Phased Array Antenna Basics*, John Wiley & Sons, Chichester, UK, 2005.

31. A. van de Capelle, 'Transmission line model for rectangular microstrip antennas', in J.R. James and P.S. Hall (eds.), *Handbook of Microstrip Antennas*, Peter Peregrinus, London, pp. 527–578, 1989.

4

RFID Antennas: Folded Dipoles

Wire and strip folded-dipole antennas serve as examples of low-profile antennas that can be impedance-tuned relatively easily. This impedance tuning is necessary, since the output impedance levels of communication and RFID ICs are, in general, anything but 50 Ω. Impedance tuning of the antenna makes the application of an impedance-matching network between the transceiver and the antenna unnecessary, thus aiding in the system's compactness, power efficiency and production cost minimisation. The tuning may be accomplished by separating the transmission line mode and the dipole mode of the folded-dipole antenna and operating on the transmission line mode only. Also, the addition of a parasitic dipole element parallel to the folded-dipole antenna may be used for impedance tuning. For more directivity or a wider frequency bandwidth, several *reentrant* folded-dipole elements may be added in series to create a (log-periodic) folded-dipole array antenna.

4.1 INTRODUCTION

The half-wave dipole antenna is a much-used antenna type, owing to its structural simplicity, near-omnidirectional radiation pattern and ease of housing for applications using frequencies above 1 GHz. Some drawbacks of the half-wave dipole antenna, which may possibly prevent its use, are its narrow frequency bandwidth and almost fixed input impedance of approximately (73 + j42.5) Ω [1]. The latter means that hardly any geometrical features, except for the dipole (equivalent) radius, are at the designer's disposal for controlling the input impedance of the antenna.

By placing a short-circuited dipole antenna parallel to a dipole antenna and connecting the two dipoles at the ends (Figure 4.1(a)–(c)), a narrow loop, or *folded-dipole* antenna, is

Approximate Antenna Analysis for CAD Hubregt J. Visser

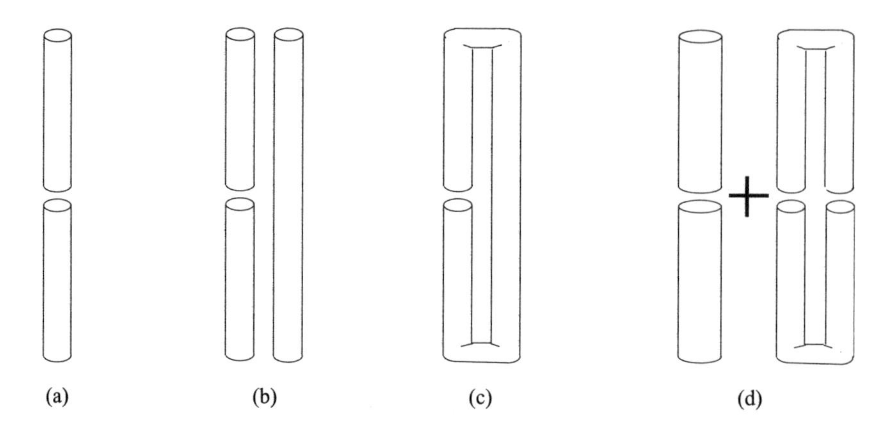

Figure 4.1 Folded-dipole antenna. (a) Single circularly cylindrical dipole antenna. (b) Dipole antenna and parallel short-circuited dipole antenna. (c) Folded-dipole antenna. (d) Equivalent-radius dipole antenna with parallel short-circuited two-wire transmission line stubs.

formed that provides more means of controlling the input impedance. At the same time, the frequency bandwidth is increased.

The need for controlling the input impedance of the antenna arises from the output impedance levels of commercially available communication and radio frequency identification (RFID) ICs, which, in general, are anything but 50 Ω.[1] If the impedance matching is accomplished directly at the antenna terminals, matching networks will not be necessary and more compact, power-efficient designs become possible.

The fact that the frequency bandwidth of a folded-dipole antenna is increased with respect to a single-dipole antenna of the same size can be easily understood by recognizing that the folded dipole in fact is a dipole antenna with parallel short-circuited two-wire transmission line stubs (Figure 4.1(d)).

To demonstrate this, we first calculate the input impedance Z_{A} of a circularly cylindrical dipole antenna of radius a and length 2ℓ as follows [3]:

$$
\begin{aligned}
Z_{\mathrm{A}} = {} & [122.65 - 204.1k\ell + 110(k\ell)^2] \\
& - \mathrm{j}\left[120\left(\ln\left(\frac{2\ell}{a}\right) - 1\right)\cot(k\ell) - 162.5 + 140k\ell - 40(k\ell)^2\right],
\end{aligned}
\tag{4.1}
$$

where $k = 2\pi/\lambda$, λ being the wavelength, $1.3 \leq k\ell \leq 1.7$ and $0.001588 \leq a/\lambda \leq 0.009525$. The input impedance Z_{T} of a short-circuited lossless transmission line stub of length l follows from the impedance transfer equation

$$
Z_{\mathrm{T}} = Z_0 \frac{Z_{\mathrm{L}} + \mathrm{j}Z_0 \tan(k\ell)}{Z_0 + \mathrm{j}Z_{\mathrm{L}} \tan(k\ell)}\bigg|_{Z_{\mathrm{L}}=0} = \mathrm{j}Z_0 \tan(k\ell),
\tag{4.2}
$$

[1] For example, the output impedance of the NORDIC nRF905 transceiver, operated at 868 MHz, is (225 – j210) Ω [2].

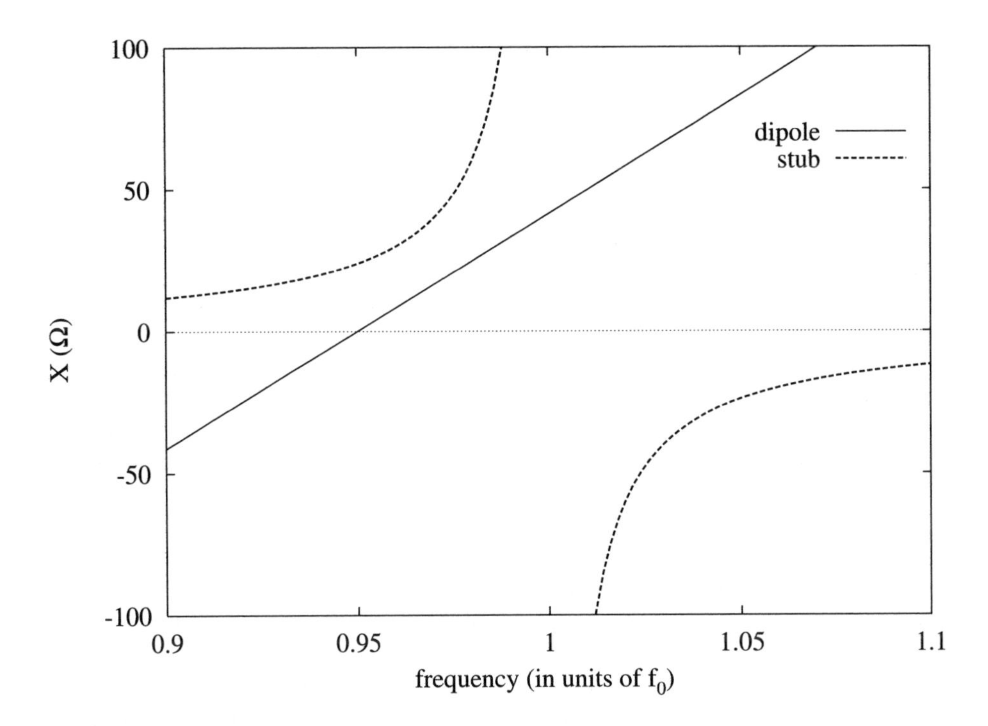

Figure 4.2 Imaginary part of the input impedance of a circularly cylindrical dipole of radius $a = 0.002\lambda_0$ and length $2\ell = \lambda_0/2$ and the input impedance divided by 100 of a short-circuited two-wire transmission line stub of length ℓ, radius a and wire separation $s = 0.01\lambda_0$, as a function of the normalized frequency.

where $Z_L = 0$ because of the short circuit, and Z_0 is the characteristic impedance of the transmission line. For a two-wire transmission line where the wire radius is a and the wire separation is s, this characteristic impedance is given by [1, 4]

$$Z_0 = \frac{\sqrt{\mu_0/\varepsilon_0}}{\pi} \ln\left[\frac{s/2 + \sqrt{(s/2)^2 - a^2}}{a}\right]. \tag{4.3}$$

The imaginary part of the input impedance of a dipole antenna of radius a and length 2ℓ and the input impedance of a shorted stub of length ℓ are shown in Figure 4.2 as a function of frequency around $f_0 = 1/(\sqrt{\varepsilon_0\mu_0}\lambda_0)$, where ε_0 and μ_0 are the permittivity and permeability, respectively, of free space. The length is such that $2\ell = \lambda_0/2$, and $a = 0.002\lambda_0$ and $s = 0.01\lambda_0$. The input impedance of the stub has been divided by a factor of 100 to improve the readability of the graph.

Although the figure does not show the input impedance of a folded-dipole antenna as a function of frequency, it does show how the behavior of a dipole as a function of frequency is – to a certain extent – compensated for by the stub's behavior. For frequencies below the dipole resonance, the dipole is capacitive while the stubs are inductive, and for frequencies above the dipole resonance, the dipole is inductive while the stubs are capacitive. Thus a frequency bandwidth that is increased with respect to the single-dipole antenna has been realized.

In the remainder of this chapter, we shall outline the analysis of the input impedance of a symmetric wire folded-dipole antenna, where we shall use the concept of separating the dipole mode and the transmission line mode. This analysis is followed by an analysis of an asymmetric wire folded-dipole antenna. The asymmetric wire folded-dipole antenna provides an additional means of impedance control through the choice of the diameters of the two wires. Then, new means of impedance control will be discussed. These means consist of placing additional short circuits into the arms of the folded dipole, placing parasitic dipole or folded-dipole elements parallel to the folded-dipole antenna or both.

Next, the general – i.e. asymmetric – coplanar-strip folded-dipole antenna will be discussed. This folded-dipole antenna is a practical printed-circuit-board (PCB) implementation of the folded-dipole antenna discussed in the previous sections. Design equations will be developed for an asymmetric coplanar-strip folded-dipole antenna on a dielectric slab. Following this analysis, *re-entrant* folded-dipole antenna elements will be combined in series-fed linear array antennas.

In the previous chapter, we desired an accuracy of a few percent in the amplitude of the reflection coefficient, since we were looking at antennas to be matched to a 50 Ω impedance level. In this chapter, we are looking at means of tuning the input impedance of an antenna to a desired complex value. So now, we require an accuracy of a few percent in both the calculated real part of the input impedance and the calculated imaginary part of the input impedance. This requirement is, as we explained in the introductory chapter, much tighter.

4.2 WIRE FOLDED-DIPOLE ANTENNAS

Before discussing the asymmetric coplanar-strip folded-dipole antenna on a dielectric slab, which is the type of most practical use for RFID and handheld-telecommunication applications, we need to understand the basic operation of the folded-dipole antenna. Therefore, we start with the wire folded-dipole antenna. First, we shall discuss the special (easier) case of the symmetric wire folded dipole, followed by a discussion of the general, i.e. asymmetric, wire folded-dipole antenna. These wire folded-dipole antennas may find applications in broadcast radio and television reception, either as stand-alone antennas or as exciting elements in a Yagi–Uda antenna.

The analyses to be presented assume that end effects may be ignored, and that the parallel wires of the folded-dipole antenna are positioned close to each other. A center-to-center spacing, assuming circularly cylindrical wires, of $0.01\lambda_0$ is frequently mentioned [5, 6, 9], and will be used later on as a rule of thumb for closely spaced wires.

4.2.1 Symmetric Folded-Dipole Antenna

To obtain the input impedance of a folded-dipole antenna of the kind shown in Figure 4.1(c), where the driven and parasitic elements are circularly cylindrical, equal-radius wires, we start by connecting a voltage source V to the input terminals of the driven element. The current flowing through the folded dipole may now be decomposed into a *transmission line*

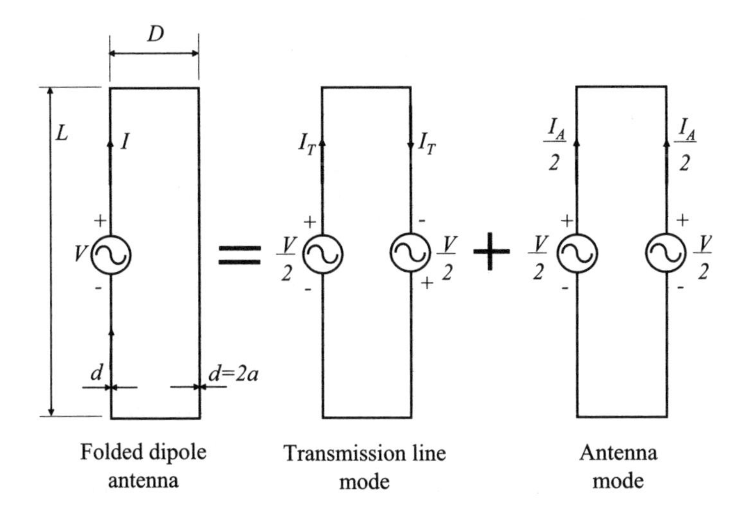

Figure 4.3 Decomposition of an equal-radius-wire folded-dipole antenna into a transmission line mode and an antenna mode.

mode current, or differential-mode current, and an *antenna mode* current, or common mode current [5, 10]. Throughout this chapter, we shall use the terms 'transmission line mode current' and 'antenna' (or 'dipole') 'mode current'. The decomposition into a transmission line mode and an antenna mode is shown in Figure 4.3 for a folded-dipole antenna of length L, wire radius a and wire separation D, $D \leq 0.01\lambda_0$.

The figure shows that the total current flowing through the left arm of the folded-dipole antenna, I, equals $I_T + (1/2)I_A$. The input impedance therefore is

$$Z_{in} = \frac{V}{I_T + (1/2)I_A}. \tag{4.4}$$

Looking at the equivalent transmission line circuit, we observe that the voltage applied to the upper short-circuited transmission line stub equals $V/2$. The transmission line mode current I_T, therefore, may be expressed as

$$I_T = \frac{V}{2Z_T}, \tag{4.5}$$

where Z_T is the impedance looking into the short-circuited two-wire transmission line stub.

Looking at the equivalent dipole antenna circuit, we observe two equal currents in parallel connected to the same voltage source (Figure 4.4). This figure reveals that the antenna mode current I_A is given by

$$I_A = \frac{V}{2Z_D}, \tag{4.6}$$

where Z_D is the impedance of an ordinary dipole of equivalent radius a_e and length L. The equivalent radius is given by [5, 11]

$$\ln(a_e) = \ln(a) + \frac{1}{2}\ln\left(\frac{D}{a}\right). \tag{4.7}$$

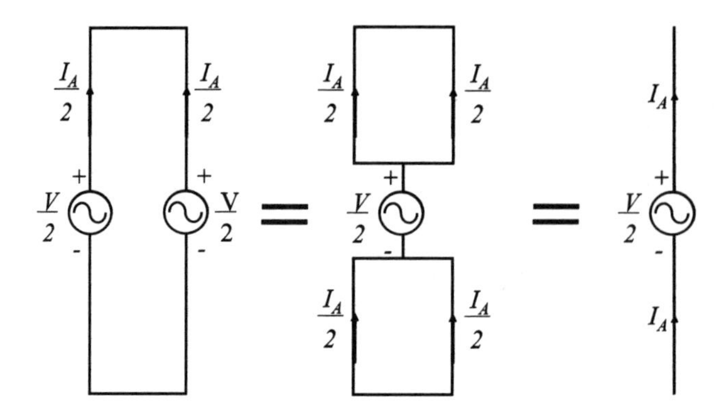

Figure 4.4 Equivalent dipole antenna circuit of a folded-dipole antenna.

The substitution of equations (4.5) and (4.6) in equation (4.4) results in the following expression for the input impedance of the folded-dipole antenna:

$$Z_{\rm in} = \frac{4Z_{\rm T}Z_{\rm D}}{Z_{\rm T} + 2Z_{\rm D}}. \tag{4.8}$$

The transmission line mode impedance $Z_{\rm T}$ is calculated using equations (4.2) and (4.3):

$$Z_{\rm T} = {\rm j}Z_{\rm c} \tan\left(\beta \frac{L}{2}\right), \tag{4.9}$$

where $\beta = 2\pi/\lambda$ and

$$Z_{\rm c} = \frac{1}{\pi}\sqrt{\frac{\mu_0}{\varepsilon_0}} \ln\left(\frac{D + \sqrt{D^2 - (2a)^2}}{2a}\right). \tag{4.10}$$

The antenna mode impedance $Z_{\rm D}$ may be calculated using equation (4.1) for very thin wires. For wires that are not very thin, one might employ the empirical double-polyfit equations for the King–Middleton second-order solution as given in [3] and, if these equations fail owing to a wire radius that is too large, one should resort to numerical methods.

An interesting feature of the symmetric folded-dipole antenna can be deduced from equations (4.8) and (4.9). If the length L of the folded-dipole antenna is chosen to be equal to half the operational wavelength, the absolute value of $Z_{\rm T}$ becomes infinite and the input impedance of the folded-dipole antenna becomes equal to four times the input impedance of a single half-wave dipole antenna of radius $a_{\rm e}$.

A two-wire transmission line, encapsulated in a ribbon-shaped isolator, having a characteristic impedance of 300 Ω is commercially available. This characteristic impedance is about four times the resistive part of the input impedance of a half-wave dipole antenna.

4.2.2 Asymmetric Folded-Dipole Antenna

If the two parallel wires or elements of a folded-dipole antenna do not have equal radii, the current in the antenna will be divided asymmetrically between the two elements. It may be

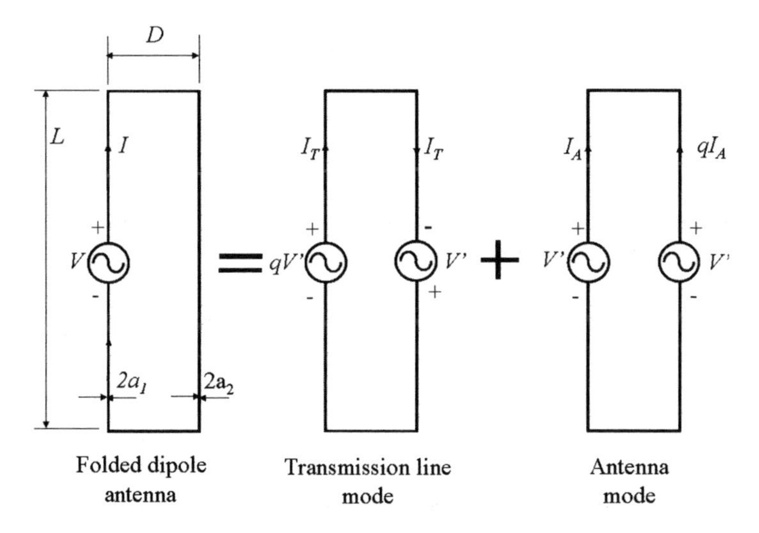

Figure 4.5 Decomposition of an asymmetric wire folded-dipole antenna into a transmission line mode and an antenna mode.

assumed that the current is divided inversely as the ratio of the characteristic impedances of the wire elements [10], i.e.

$$\frac{I_2}{I_1} = \frac{Z_{c1}}{Z_{c2}} = q. \tag{4.11}$$

The transmission line voltage will be divided according to the ratio of the characteristic impedances of the elements:

$$\frac{V_2}{V_1} = \frac{Z_{c2}}{Z_{c1}} = \frac{1}{q}. \tag{4.12}$$

The decomposition into the transmission line mode and the antenna mode is illustrated in Figure 4.5 for an asymmetric wire folded-dipole antenna of length L, current ratio q and wire separation D, $D \leq 0.01\lambda_0$. The radii of the wires are a_1 and a_2.

From this figure, we obtain the total current flowing in the left arm of the folded-dipole antenna as $I_T + I_A$. The total voltage supplied to the left arm is $(1+b)V'$. The input impedance is therefore calculated as

$$Z_{in} = \frac{(1+q)V'}{I_T + I_A}. \tag{4.13}$$

Looking at the equivalent transmission line circuit, we observe that the voltage applied to the upper short-circuited transmission line stub equals $(1/2)(1+q)V'$. The transmission line mode current I_T may therefore be expressed as

$$I_T = \frac{(1+q)V'}{2Z_T}, \tag{4.14}$$

where Z_T is the impedance looking into the short-circuited two-wire transmission line stub. Z_T is given by equation (4.9), where the characteristic impedance Z_c of the asymmetric two-wire transmission line could be calculated using equation (4.10) in which a is replaced by $\sqrt{a_1 a_2}$ [11].

The equivalent dipole antenna circuit may be regarded as two currents, I_A and qI_A, in parallel connected to the same voltage source V'. The antenna current is therefore

$$I_A = \frac{V'}{(1+q)Z_A}. \quad (4.15)$$

Here, Z_A is the input impedance of a circularly cylindrical dipole antenna of equivalent radius a_e. The equivalent radius is given by [11]

$$\ln(a_e) = \ln(a_1) + \frac{1}{(1+a_2/a_1)^2}\left[\left(\frac{a_2}{a_1}\right)^2 \ln\left(\frac{a_2}{a_1}\right) + 2\left(\frac{a_2}{a_1}\right)\ln\left(\frac{D}{a_1}\right)\right]. \quad (4.16)$$

The impedance of the antenna may be calculated as outlined in section 4.2.1.

Substitution of equations (4.14) and (4.15) in equation (4.13) results in the input impedance of an asymmetric folded-dipole antenna, expressed in terms of the stub impedance, the equivalent dipole impedance and the impedance step-up ratio:

$$Z_{in} = \frac{2(1+q)^2 Z_A Z_T}{(1+q)^2 Z_A + 2Z_T}. \quad (4.17)$$

The factor $(1+q)^2$ is known as the *impedance step-up ratio* [5, 11].

4.3 IMPEDANCE CONTROL

From a practical point of view, for *wire* implementations, symmetric folded-dipole antennas are preferred over asymmetric ones. Symmetric wire folded-dipole antennas can be constructed using a single piece of tubing, wire or electrically conducting thread. The latter allows the possibility of embroidering folded-dipole antennas into clothing [12]. Although we do have some means of controlling the input impedance (length, wire radius and wire separation), we would like to have additional means of controlling this impedance, without resorting to employing dielectric substrates and/or superstrates. These means are offered by placing short circuits in the arms of the folded-dipole antenna and/or by positioning parasitic dipole elements next to the antenna.

To test these means of input impedance control, we have made use of a symmetric wire folded-dipole antenna, half a meter long, made out of a circularly cylindrical wire of radius 0.0001 m and with a wire separation of 0.005 m. The antenna was analyzed using the transmission line theory outlined in the previous sections. Figure 4.6 shows the real and imaginary parts of the input impedance as a function of frequency, calculated using the transmission line method (TL) and (for the propose of verification) calculated with a method of moments (MoM), employing an axial-current method and delta-gap voltage excitation.

The impedance of the dipole was calculated using the empirical double-polyfit equations for the King–Middleton second-order solution as given in [3], which are valid for 0.0016 m $\leq a_e \leq$ 0.01 m and 248 MHz $\leq f \leq$ 325 MHz. We observe fair agreement between the two analysis results around resonance, even though the real part of the TL impedance differs by up to 20% from the MoM value at the highest frequency. The deviations between the TL

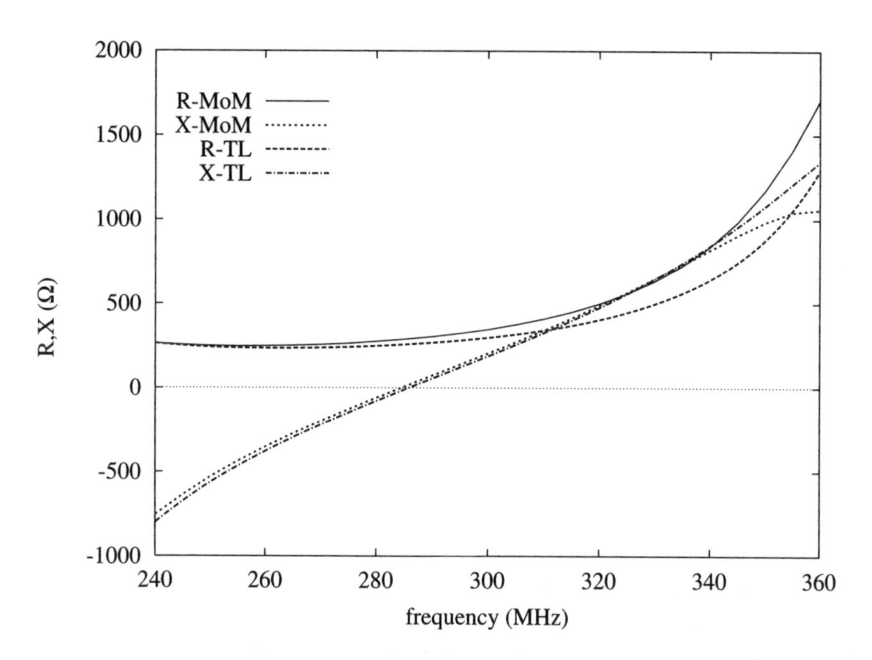

Figure 4.6 Real part (R) and imaginary part (X) of the input impedance of a symmetric wire folded-dipole antenna as a function of frequency. Antenna dimensions: $L = 0.5$ m, $a = 0.0001$ m, $D = 0.005$ m.

results and MoM results at the higher frequencies within the frequency range in which the methods are valid are attributed to inaccuracies in the method of moments used. The accuracy of the transmission line method was considered to be acceptable for design purposes, as will be explained below.

4.3.1 Power Waves

Although we compare calculated input impedance results in Figure 4.6, the parameter that really matters is another one related to the impedance; this is the amount of power delivered to the antenna in the case of transmission or the amount of power taken from the antenna in the case of reception. For a system where a generator with internal impedance Z_R is connected to a load Z_L (Figure 4.7), the standard (voltage) reflection coefficient

$$\Gamma = \frac{Z_L - Z_R}{Z_L + Z_R}, \tag{4.18}$$

is a direct measure of the reflected power only when Z_R is real [13]. When Z_R is complex and conjugate matching is applied to maximize the power transfer from the generator to the load, the standard reflection coefficient is nonzero. Therefore, the standard definition of the reflection coefficient – based on physical waves propagating along a transmission line – does not in general represent the reflection of power. Power waves – mathematical constructs

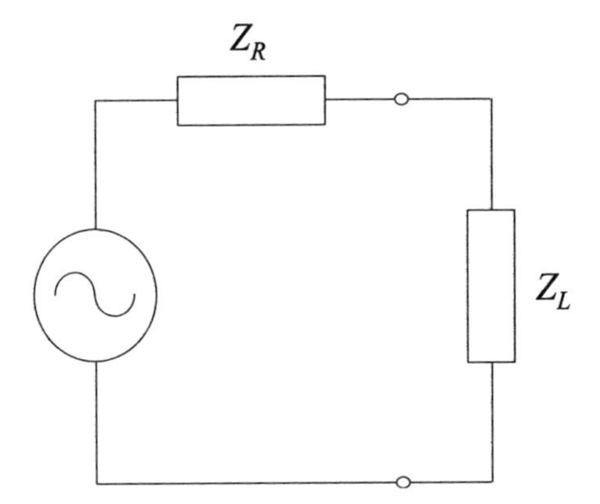

Figure 4.7 A source with internal impedance Z_R is connected to a load impedance Z_L.

from the circuit world – have been introduced to correctly describe the reflection of power in systems with complex reference impedances.

Looking at the terminals of the circuit shown in Figure 4.7, we assume a non-power-reflecting voltage wave V'^{+} propagating to the right. Associated with this voltage is a current I'^{+}. The reflected wave is defined by V'^{-} and I'^{-}. Since the wave traveling to the right is not reflected in terms of power, it must experience an impedance Z_R^* [13], where the superscript * denotes the complex conjugate. The reflected wave must experience an impedance Z_R. Thus

$$V'^{+} = Z_R^* I'^{+} \tag{4.19}$$

and

$$V'^{-} = Z_R I'^{-}. \tag{4.20}$$

The total voltage and current are given by

$$V' = V'^{+} + V'^{-} \tag{4.21}$$

and

$$I' = I'^{+} - I'^{-}, \tag{4.22}$$

respectively, and are related through the load impedance by

$$V' = Z_L I'. \tag{4.23}$$

From equations (4.19)–(4.22), it follows that

$$V'^{+} = \frac{Z_R^*}{2\Re(Z_R)}(V' + Z_R I'), \tag{4.24}$$

$$V'^{-} = \frac{Z_R}{2\Re(Z_R)}(V' - Z_R^* I'), \tag{4.25}$$

$$I'^{+} = \frac{1}{2\Re(Z_\mathrm{R})}(V' + Z_\mathrm{R} I'), \tag{4.26}$$

$$I'^{-} = \frac{1}{2\Re(Z_\mathrm{R})}(V' - Z_\mathrm{R}^{*} I'), \tag{4.27}$$

where $\Re(x)$ denotes the real part of the complex argument x.

Using the power waves defined by [13]

$$a = \frac{V' + Z_\mathrm{R} I'}{2\sqrt{\Re(Z_\mathrm{R})}}, \tag{4.28}$$

$$b = \frac{V' - Z_\mathrm{R}^{*} I'}{2\sqrt{\Re(Z_\mathrm{R})}}, \tag{4.29}$$

the total voltage and current may be written as [13]

$$V' = \frac{Z_\mathrm{R}^{*} a + Z_\mathrm{R} b}{\sqrt{\Re(Z_\mathrm{R})}}, \tag{4.30}$$

$$I' = \frac{a - b}{\sqrt{\Re(Z_\mathrm{R})}}. \tag{4.31}$$

The average power delivered to the load, P, is then

$$P = \frac{1}{2}\Re(V' I'^{*}) = \frac{|a|^2}{2} - \frac{|b|^2}{2}, \tag{4.32}$$

and the reflection coefficient for power waves is given by

$$\Gamma_\mathrm{p} = \frac{b}{a} = \frac{Z_\mathrm{L} - Z_\mathrm{R}^{*}}{Z_\mathrm{L} + Z_\mathrm{R}}. \tag{4.33}$$

Using the concept of power waves, we are now able to asses the influence of the deviation of the complex impedance calculated with the TL method from the values obtained with the MoM, as shown in Figure 4.6. Taking the MoM results as a reference, we assume, for every frequency, the reference impedance Z_R to be equal to the complex conjugate of the MoM-calculated input impedance. Thus we shall see how the deviations of the TL values are translated into power reflection factors. In Figure 4.8, we show $20\log(|\Gamma_\mathrm{p}|)$, that is, the fraction of reflected power with respect to the maximum power that can be delivered to the load, as a function of frequency.

The figure shows that the deviation from the MoM results of the input impedance calculated with the TL method results in a worst-case additional reflection of -19 dB at 325 MHz. This means that, owing to inaccuracies, an uncertainty of (much) less than 11% will be added to the power reflection. So, a deviation in the real part of the input impedance of about 20% translates to a deviation in the reflected power of about 10%. We consider this worst-case value good enough for design purposes.

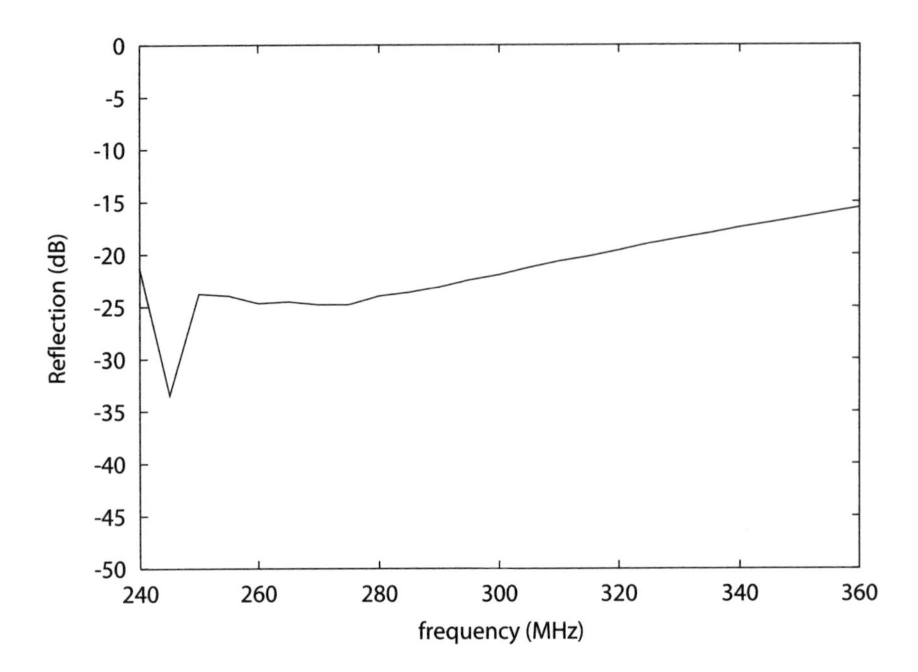

Figure 4.8 Deviation of the input impedance of a folded-dipole, calculated by the TL method, with respect to the MoM results, translated into reflected power.

4.3.2 Short Circuits

As we have seen, the folded-dipole antenna may be analyzed by decomposing the current flowing through the antenna into a transmission line mode and an antenna or dipole mode. This decomposition opens up the possibility to act either on the transmission line mode or on the antenna mode to tune the input impedance. We may act on the transmission line mode by placing short circuits in the arms of the folded-dipole antenna (Figure 4.9). In doing so, we shorten the stub lengths but keep the dipole antenna mode intact.[2] In the TL analysis, we need only to replace the stub length $L/2$ in equation (4.19) with $L'/2$ (see Figure 4.9).

To validate the above, a number of configurations were analyzed and compared with MoM analysis results. Figures 4.10 and 4.11 show two examples. Both example antennas were based on the folded-dipole antenna without additional short circuits ($L' = L$ in Figure 4.9) for which analysis results were shown in Figure 4.6.

Figures 4.6, 4.10 and 4.11 show that indeed the additional short circuits operate mainly on the transmission line mode (there is no shift in resonance frequency). Furthermore, the figures show that the input impedance can be adjusted over a large dynamic range. Comparison with the MoM analysis results shows that the TL analysis method may also be employed for these antenna structures.

[2] We shall consider only the symmetric case, i.e. the additional short circuits are placed symmetrically with respect to the horizontal folded-dipole axis.

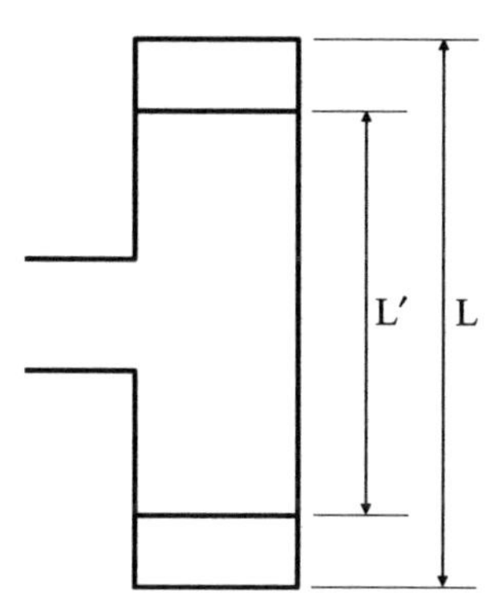

Figure 4.9 Folded-dipole antenna with additional short circuits.

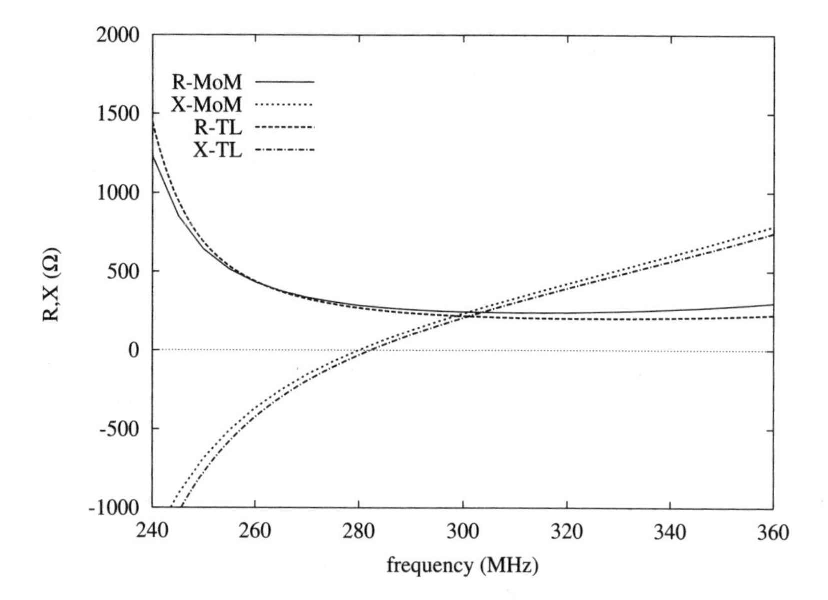

Figure 4.10 Real and imaginary parts of the input impedance of a symmetric, short-circuited wire folded-dipole antenna as a function of frequency. Antenna dimensions: $L = 0.5$ m, $a = 0.0001$ m, $D = 0.005$ m, $L' = 0.3$ m.

Note that no attempt has yet been made to tune the input impedance to any given (complex) value. The graphs are presented to show the possibility of tuning by positioning short circuits in the arms of a folded-dipole antenna.

Another tuning mechanism may be provided by acting on the antenna mode instead of on the transmission-line mode. By placing parasitic elements in close proximity to a folded dipole-antenna, one may also tune the input impedance.

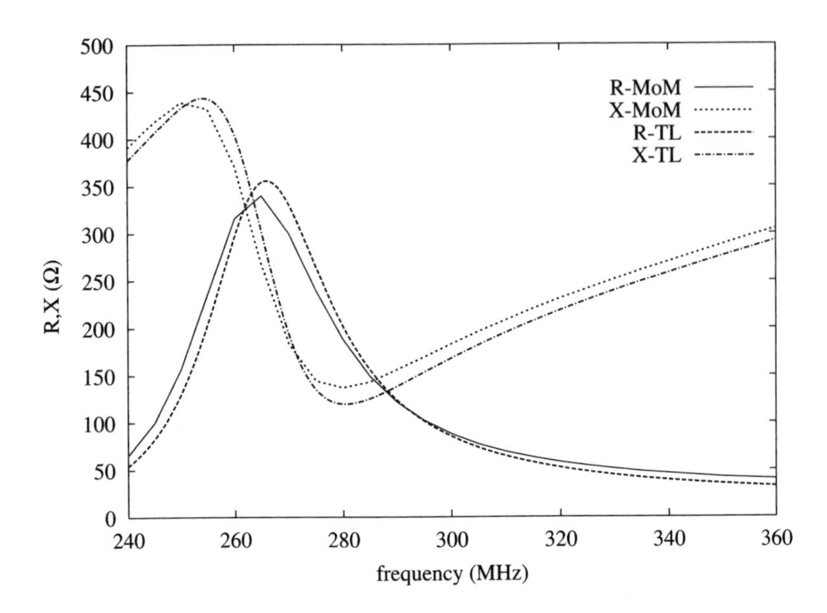

Figure 4.11 Real and imaginary parts of the input impedance of a symmetric, short-circuited wire folded-dipole antenna as a function of frequency. Antenna dimensions: $L = 0.5$ m, $a = 0.0001$ m, $D = 0.005$ m, $L' = 0.1$ m.

4.3.3 Parasitic Elements

We may tune the input impedance of a folded-dipole antenna by placing a parasitic element, for example, a short-circuited dipole or folded-dipole antenna, parallel to and in close proximity to the folded-dipole antenna as shown in Figure 4.12. The input impedance of the driven folded-dipole antenna shown in either Figure 4.12(a) or Figure 4.12(b) is[3]

$$Z_{\text{in}} = Z_{11} - \frac{Z_{12}Z_{21}}{Z_{22}}, \tag{4.34}$$

where Z_{11} is the input impedance of the isolated folded-dipole antenna on the left in Figure 4.12(a) or (b), Z_{22} is the input impedance of the isolated antenna on the right in Figure 4.12(a) or (b), and $Z_{12} = Z_{21}$ is the mutual impedance between the antenna on the left and the antenna on the right.

For the situation depicted in Figure 4.12(a), the mutual impedance is given by [14]

$$Z_{12} = Z_{21} = 2Z_{21_{\text{dipole-to-dipole}}}, \tag{4.35}$$

[3]The input impedance may be calculated from

$$V_1 = Z_{11}I_1 + Z_{12}I_2,$$
$$V_2 = Z_{21}I_1 + Z_{22}I_2,$$

where the subscripts $_1$ and $_2$ indicate the driven antenna on the left and the short-circuited antenna on the right, respectively. Thus, $V_2 = 0$. The input impedance is given by $Z_{\text{in}} = V_1/I_1$.

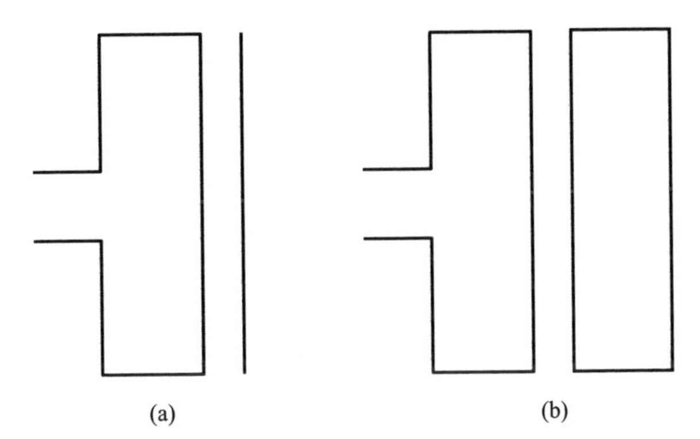

Figure 4.12 Folded-dipole antenna with parasitic elements. (a) Short-circuited dipole coupled to a folded-dipole antenna. (b) Short-circuited folded dipole coupled to a folded-dipole antenna.

where $Z21_{\text{dipole-to-dipole}}$ is the mutual impedance between two 'ordinary' dipole antennas. For thin wires, this mutual impedance may be calculated using the closed-form equations given in [15].

For the situation depicted in Figure 4.12(b), the mutual impedance is given by [14]

$$Z_{12} = Z_{21} = 4Z21_{\text{dipole-to-dipole}}, \tag{4.36}$$

where $Z21_{\text{dipole-to-dipole}}$ again is the mutual impedance between two 'ordinary' dipole antennas.

To validate the above TL analysis method, a number of configurations were analyzed and the results were compared with those of an MoM analysis. Here, the length of the driven folded-dipole antenna is L, the length of the parasitic, short-circuited dipole antenna is L_{d} and the center-to-center spacing between the folded-dipole antenna and the parasitic dipole antenna is D_{c}. All wires have an equal radius a (Figure 4.13).

For the mutual coupling between two 'ordinary' dipole antennas, the closed-form equations of [15] were used. Figures 4.14 and 4.15 show two examples of results for a folded-dipole antenna with a parasitic, short-circuited dipole antenna. Again, the folded-dipole antenna is the one for which analysis results were shown in Figure 4.6. Figures 4.14 and 4.15 therefore should be compared with Figure 4.6.

These figures show that the TL analysis method may be used for analyzing the configurations shown for frequencies around resonance. The figures also show that the input impedance of a (folded-) dipole antenna may be adjusted by positioning parasitic elements close to the driven element, using variation of the length and the distance to the driven element to obtain the required result. Note that, again, no attempt was made to tune the input impedance to a specific value.

4.4 ASYMMETRIC COPLANAR-STRIP FOLDED-DIPOLE ANTENNA ON A DIELECTRIC SLAB

The resonant folded-dipole antenna is known for its improved frequency bandwidth over that of an 'ordinary' dipole antenna [1]. We have seen, though, that its main attraction for

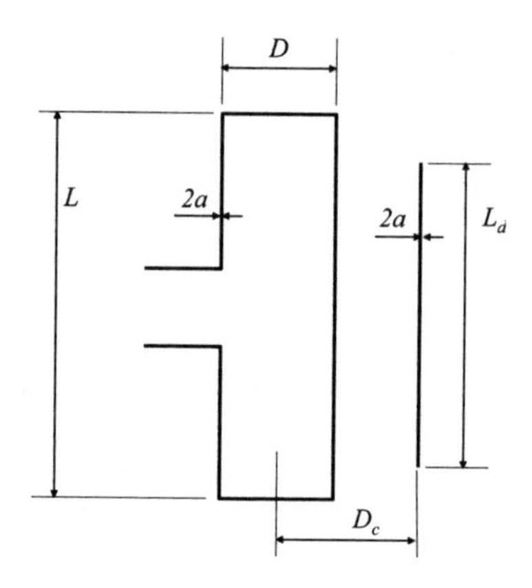

Figure 4.13 Folded-dipole antenna with parasitic, short-circuited dipole antenna.

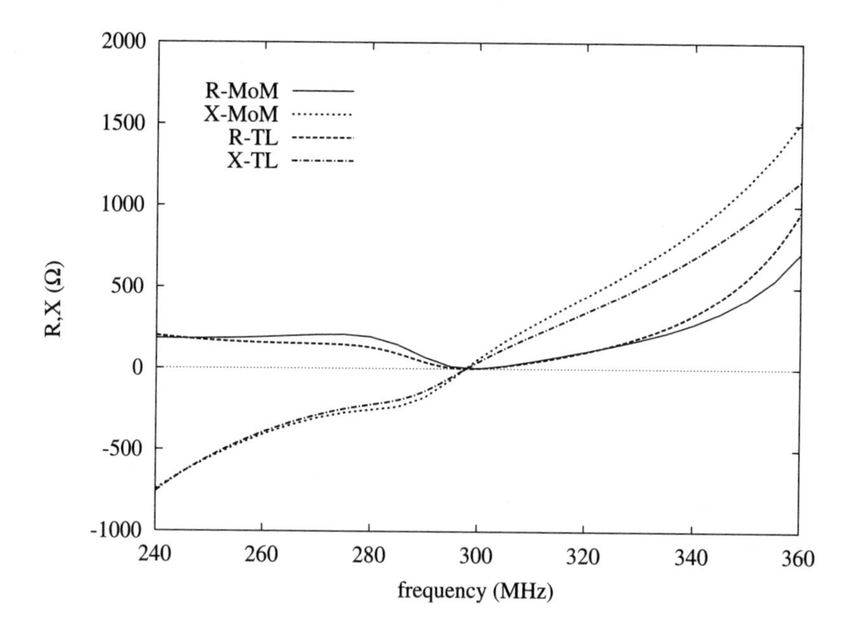

Figure 4.14 Real and imaginary parts of the input impedance of a symmetric wire folded-dipole antenna with a parasitic, short-circuited dipole antenna, as a function of frequency. Antenna dimensions: $L = 0.5$ m, $L_\mathrm{d} = 0.5$ m, $a = 0.0001$ m, $D = 0.005$ m, $D_\mathrm{c} = 0.01$ m.

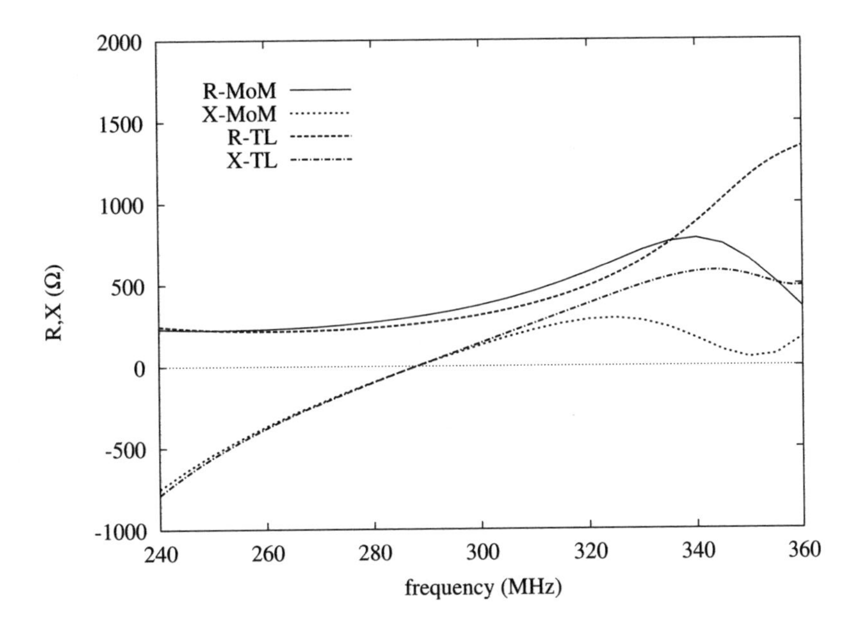

Figure 4.15 Real and imaginary parts of the input impedance of a symmetric wire folded-dipole antenna with a parasitic, short-circuited dipole antenna, as a function of frequency. Antenna dimensions: $L = 0.5$ m, $L_{\rm d} = 0.4$ m, $a = 0.0001$ m, $D = 0.005$ m, $D_{\rm c} = 0.02$ m.

wireless communication and RFID applications lies in the possibility to adjust the input impedance over a wide range of values. Adjustment of the impedance characteristics has been demonstrated for free-standing wire folded-dipole antennas. We now want to adapt the impedance control measures introduced in the preceding sections to a more practical implementation of the folded-dipole antenna. This practical implementation takes the form of an asymmetric coplanar-strip (CPS) folded-dipole antenna on a dielectric slab (Figure 4.16), which allows the integration of the antenna into a PCB design.

4.4.1 Lampe Model

Design equations for the input impedance of an asymmetric coplanar-strip folded-dipole antenna were developed by Lampe [6,7]. These equations give three means of controlling the input impedance of the antenna: the input impedance of a dipole antenna of equivalent radius; the step-up impedance ratio, which depends on the widths of the two arms of the planar folded-dipole antenna; and the characteristic impedance of the coplanar-strip transmission line formed by these two arms.

The input impedance of the antenna is given by [6,7]

$$Z_{\rm in} = \frac{2(1+q)^2 Z_{\rm D} Z_{\rm X}}{(1+q)^2 Z_{\rm D} + 2Z_{\rm X}}, \tag{4.37}$$

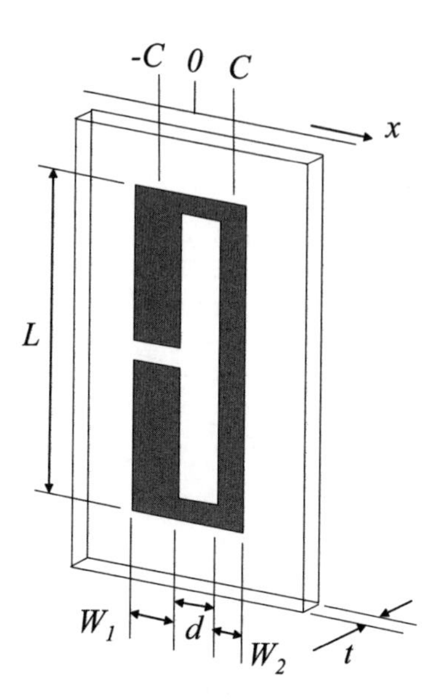

Figure 4.16 Asymmetric coplanar-strip folded-dipole antenna on a dielectric slab.

where Z_D is the input impedance of an equivalent dipole, i.e. a circular cylindrical dipole of equivalent radius ρ_e. Z_X is the input impedance of a short-circuited piece of transmission line of length $L/2$ (Figure 4.16), and $(1+q)^2$ is the step-up impedance ratio.

The impedance of the short-circuited piece of asymmetric CPS of length $L/2$ is given by

$$Z_X = \mathrm{j}\left[\frac{120\pi}{\sqrt{\varepsilon_r}}\frac{K(k)}{K'(k)}\right]\tan\left(\beta\frac{L}{2}\right). \tag{4.38}$$

The expression in the square brackets is the characteristic impedance of the CPS, embedded in a homogeneous medium of relative permittivity ε_r.[4] $K(k)$ is the complete elliptic function of the first kind, and $K'(k) = K(k')$, where $k'^2 = 1 - k^2$. β is the wave number in the medium. The ratio of complete elliptic functions in equation (4.22) may be approximated by [16]

$$\frac{K(k)}{K(k')} \approx \begin{cases} \dfrac{1}{2\pi}\ln\left[2\dfrac{\sqrt{1+k}+\sqrt[4]{k}}{\sqrt{1+k}-\sqrt[4]{k}}\right] & \text{for } 1 \le \dfrac{K}{K'} \le \infty,\ \dfrac{1}{\sqrt{2}} \le k \le 1 \\ \dfrac{2\pi}{\ln[2(\sqrt{1+k}+\sqrt[4]{k}/\sqrt{1+k}-\sqrt[4]{k})]} & \text{for } 0 \le \dfrac{K}{K'} \le 1,\ 0 \le k \le \dfrac{1}{\sqrt{2}}. \end{cases} \tag{4.39}$$

The argument k is given by, [6, 7]

$$k = \frac{(d/2)[1+e(d/2+W_1)]}{d/2+W_1+e(d/2)^2}, \tag{4.40}$$

[4]We assume, for the moment, that the material of the dielectric slab and the surrounding medium in Figure 4.16 are identical.

where

$$e = \frac{W_1 W_2 + (d/2)(W_1 + W_2) - \sqrt{W_1 W_2 (d + W_1)(d + W_2)}}{(d/2)^2 (W_1 - W_2)}. \quad (4.41)$$

The step-up impedance ratio, $(1 + b)^2$, is calculated from

$$b = \frac{\ln\{4C + 2[(2C)^2 - (W_1/2)^2]^{1/2}\} - \ln\{W_1\}}{\ln\{4C + 2[(2C)^2 - (W_2/2)^2]^{1/2}\} - \ln\{W_2\}}, \quad (4.42)$$

and the equivalent radius is given by

$$\rho_e = \left(\frac{W_1}{4}\right)^{1/(1+b)} \left(C + \sqrt{C^2 - \left(\frac{W_2}{4}\right)^2}\right)^{b/(1+b)}, \quad (4.43)$$

where C is defined in Figure 4.16.

As an example, the real and imaginary parts of the input impedance of an asymmetric coplanar-strip folded dipole are shown in Figure 4.17 as a function of frequency. The input impedance was calculated with a full-wave method (the finite integration (FI) technique, CST Microwave Studio©) and with the equations given above (the transmission line (TL) method). The dimensions of the antenna were, with reference to Figure 4.16, $W_1 = 3$ mm, $W_2 = 1$ mm, $d = 1$ mm, $L = 62.5$ mm and $\varepsilon_r = 1$. The input impedance of the equivalent circularly cylindrical dipole was calculated by applying the empirical double-polyfit equations for the King–Middleton second-order solution as given in [3].

Fair agreement is shown – bearing in mind our previous discussion concerning power waves and conjugate matching – between the two simulation results around resonance. Thus, the usefulness of the Lampe model has been demonstrated.

In Figure 4.18, we show FI-simulated results for a strip folded-dipole antenna of the same dimensions, but positioned on a dielectric slab of thickness $t = 1.6$ mm, having a relative permittivity $\varepsilon_r = 4.28$ and a loss tangent $\tan\delta = 0.016$ (FR4). The TL-simulated results for a strip folded-dipole antenna in free space are shown in the same Figure [17].

This figure shows that the input impedance of the antenna on a dielectric slab as a function of frequency is very different from that of the same antenna in free space. Therefore it is necessary to adapt the TL model to take account of the effects of the dielectric slab. The dielectric slab affects both the transmission line mode and the antenna mode. Since we expect that the dielectric slab will have a larger impact on the transmission line mode than on the antenna mode, we shall start by looking at an asymmetric coplanar-strip transmission line. The influence of the dielectric slab on the antenna or dipole mode is expected to be manifested mainly in a lower resonance frequency.

4.4.2 Asymmetric Coplanar-Strip Transmission Line

Closed-form equations for the characteristic impedance of asymmetric coplanar-strip transmission lines on a dielectric slab of finite thickness are not readily available. For symmetric CPS transmission lines, analytic formulas can be found in, for example, [18, 19].

In our first attempt to derive the required analytic equations, we shall modify the equation for the characteristic impedance Z_0 of an asymmetric CPS in a uniform medium of relative

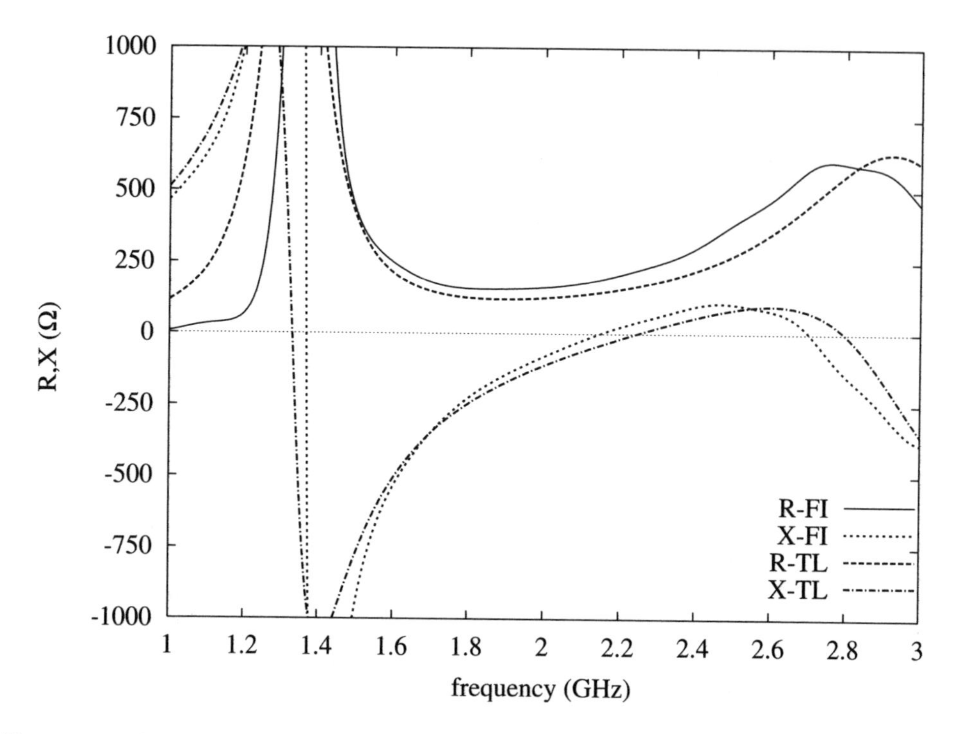

Figure 4.17 Calculated (FI and TL) real and imaginary parts of the input impedance versus frequency for an asymmetric CPS folded-dipole antenna in free space. $W_1 = 3$ mm, $W_2 = 1$ mm, $d = 1$ mm, $L = 62.5$ mm and $\varepsilon_r = 1$.

permittivity ε_r (equation (4.38)):

$$Z_0 = \frac{120\pi}{\sqrt{\varepsilon_r}} \frac{K(k)}{K'(k)}. \tag{4.44}$$

4.4.2.1 Uniform Medium First, in a very crude approximation, we shall substitute the relative permittivity of the dielectric slab for ε_r in equation (4.44). This means that we shall assume the coplanar strips to be present in a uniform medium with a relative permittivity equal to that of the (finite-thickness) dielectric slab.

The characteristic impedance was calculated in this way for several different values of the permittivity of the dielectric slab, the height, the strip separation and strip width. Results for symmetric CPS transmission lines were compared with full-wave simulation results reported in [19] (Table 4.1). In Table 4.1, t is the thickness of the dielectric slab and d is the separation of the identical strips of width $W = W_1 = W_2$ (see also Figure 4.16). The relative error is taken with respect to the full-wave value of the characteristic impedance.

The table reveals that the relative difference may be as high as 32%. The impact of the approximation chosen for the determination of the characteristic impedance of the CPS on the input impedance of an asymmetric coplanar-strip folded-dipole antenna on a dielectric slab is shown in Figure 4.19. The geometrical and electrical characteristics of this antenna

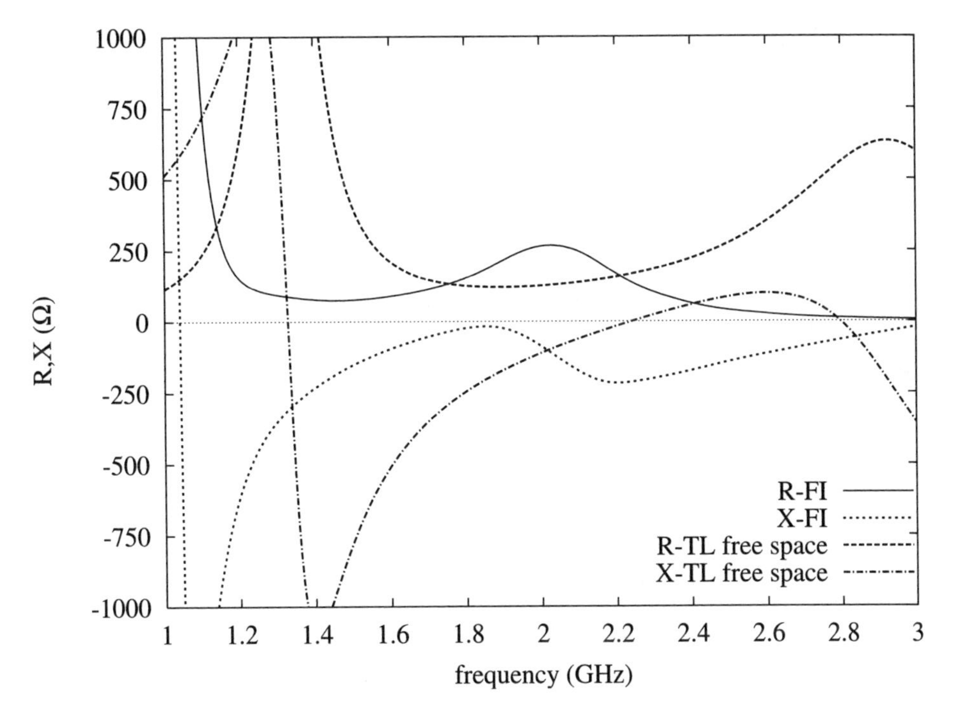

Figure 4.18 Calculated (FI and TL) real and imaginary parts of the input impedance versus frequency for an asymmetric CPS folded-dipole antenna on an FR4 dielectric slab and in free space. $W_1 = 3$ mm, $W_2 = 1$ mm, $d = 1$ mm, $L = 62.5$ mm, $\varepsilon_r = 4.28$ (FI), $\varepsilon_r = 1$ (TL) and $\tan\delta = 0.016$ (FI).

Table 4.1 Characteristic impedance of symmetric CPS transmission lines, calculated by a full-wave method and by the TL method, where, in the latter method, the dielectric slab was approximated by a uniform medium.

ε_r	t (mm)	d (mm)	W (mm)	Z_0, full wave (Ω)	Z_0, analytic (Ω)	Relative error (%)
2.20	0.79	0.10	1.52	100.07	82.60	17.46
2.20	0.79	0.30	0.76	149.79	125.56	16.18
9.90	0.64	0.04	1.27	49.91	33.99	31.90
9.90	0.64	0.37	0.51	99.98	69.84	30.15
12.90	0.25	0.026	0.38	50.00	34.23	31.54
12.90	0.25	0.15	0.13	100.05	70.51	29.53
50.00	0.25	0.031	0.20	30.03	20.71	31.04
50.00	0.25	0.030	0.025	50.03	35.63	28.78

were $W_1 = 3$ mm, $W_2 = 1$ mm, $d = 1$ mm, $L = 62.5$ mm, $t = 1.6$ mm, $\varepsilon_r = 4.28$ and $\tan\delta = 0.016$; see Figure 4.16 for the definition of the geometrical parameters.

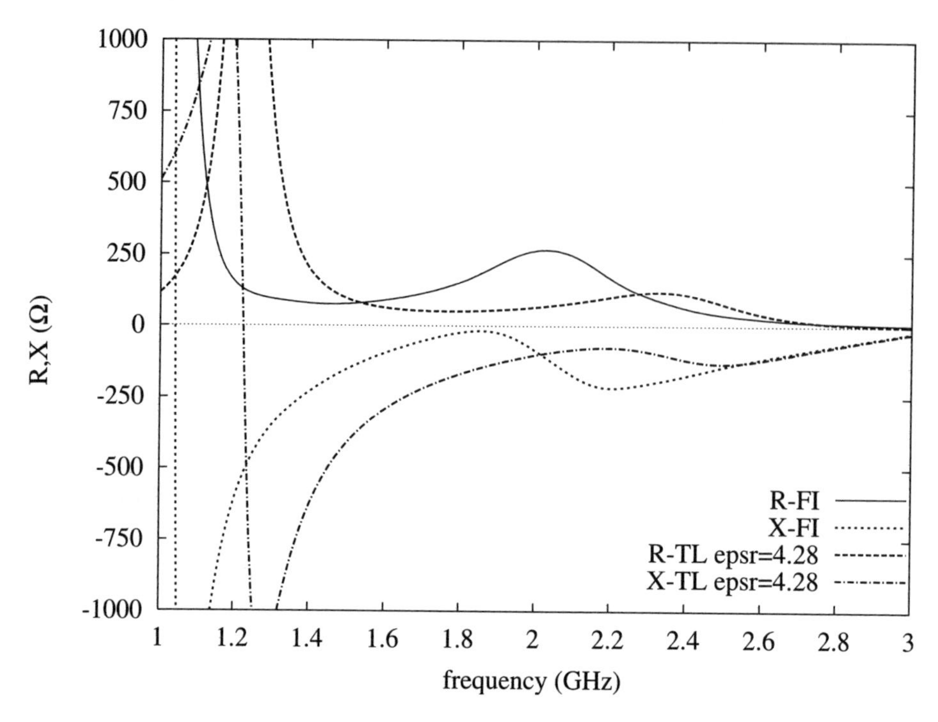

Figure 4.19 Calculated (FI and TL) real and imaginary parts of the input impedance versus frequency for an asymmetric CPS folded-dipole antenna on an FR4 dielectric slab and in a uniform medium that has the same relative permittivity as the dielectric slab. $W_1 = 3$ mm, $W_2 = 1$ mm, $d = 1$ mm, $L = 62.5$ mm, $\varepsilon_r = 4.28$ and $\tan\delta = 0.016$.

It should be noted that the dielectric slab was taken into account only for the CPS transmission line mode of the antenna. As mentioned before, the effect on the antenna mode is expected to be a shift in frequency and a scaling of the levels of the impedance curves, not a change in shape. Therefore, it suffices for the moment to take only the transmission line mode into account.

Although the impedance curves in Figure 4.19 show a distinct improvement – bearing the above in mind – with respect to those shown in Figure 4.18, there is still room for improvement. It is expected that much could be gained by improving the accuracy of the characteristic impedance of the CPS transmission line to well above 68%. Another reason for concentrating on the accuracy of the characteristic impedance of the CPS is that the antenna will most likely be connected to a transceiver by a length of CPS transmission line.

Obviously, the approximation where the entire surroundings of the antenna are considered to consist of a uniform medium with a relative permittivity equal to that of the dielectric slab is too coarse. A refinement may be made by considering half-spaces.

Table 4.2 Characteristic impedance of symmetric CPS transmission lines, calculated by a full-wave method and by the TL method, where, in the latter method, the dielectric slab was approximated by a half-space.

ε_r	t (mm)	d (mm)	W (mm)	Z_0, full wave (Ω)	Z_0, analytic (Ω)	Relative error (%)
2.20	0.79	0.10	1.52	100.07	96.86	3.21
2.20	0.79	0.30	0.76	149.79	147.24	1.70
9.90	0.64	0.04	1.27	49.91	45.80	8.23
9.90	0.64	0.37	0.51	99.98	94.13	5.85
12.90	0.25	0.026	0.38	50.00	46.63	6.74
12.90	0.25	0.15	0.13	100.05	96.06	3.99
50.00	0.25	0.031	0.20	30.03	29.00	3.43
50.00	0.25	0.030	0.025	50.03	49.89	0.28

4.4.2.2 Half-Spaces A more realistic approximation than assuming the entire space to be filled with the slab dielectric is to assume that the dielectric slab fills up a half-space. The antenna is then positioned at the interface between this dielectric half-space and free space. We then replace ε_r in equation (4.44) with the arithmetic average of the relative permittivities of the two half-spaces on the two sides of the antenna [20],

$$\varepsilon_r = \frac{\varepsilon_{r_{slab}} + 1}{2}, \tag{4.45}$$

where $\varepsilon_{r_{slab}}$ is the relative permittivity of the dielectric slab.

Characteristic impedances calculated for symmetric CPS transmission lines, where this effective relative permittivity was used in equation (5.44), were compared with full-wave analysis results reported in [19] (Table 4.2). The table shows that the relative error (with respect to the full-wave simulation results) is now less than 8.5%. The impact of the new approximation for the determination of the characteristic impedance of the CPS on the input impedance of an asymmetric coplanar-strip folded-dipole antenna on a dielectric slab is shown in Figure 4.20. The folded-dipole antenna was the same one as that analyzed in section 4.5.2.1. Again, it should be noted that the dielectric slab was taken into account only for the CPS transmission line mode of the antenna.

Upon close inspection of Figures 4.19 and 4.20, we see that the impedance curves in Figure 4.20 calculated with the transmission line model are – apart from the anticipated shift in frequency – in closer agreement than those shown in Figure 4.19 with the full-wave simulation results shown in both figures.

Our aim was to develop-closed form equations for the input impedance of asymmetric coplanar-strip folded-dipole antennas with an accuracy that is sufficient to allow antenna design. Now that we have arrived at this point, the accuracy of the transmission-line model for calculating the characteristic impedance of coplanar-strip transmission lines on dielectric slabs seems to be sufficient. However, instead of moving on to the calculation of the effects of the dielectric slab on the equivalent circularly cylindrical dipole antenna, we chose to take the calculation of the characteristic impedance of the CPS one step further and improve again on the accuracy. The reason for this exercise is that, as explained before, CPS transmission

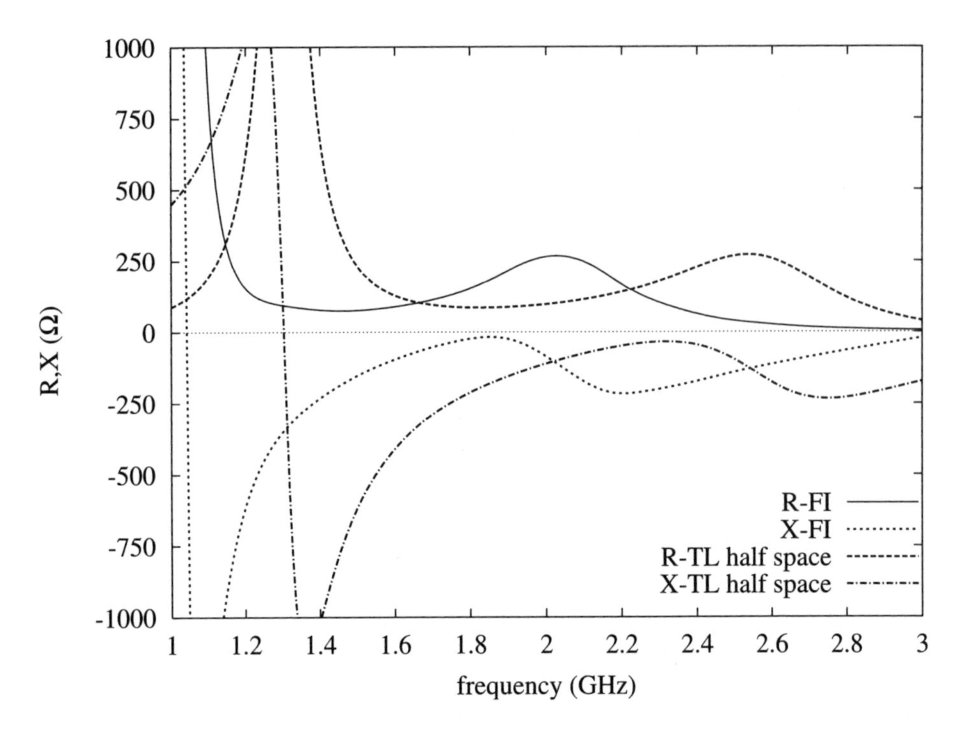

Figure 4.20 Calculated (FI and TL) real and imaginary parts of the input impedance versus frequency for an asymmetric CPS folded-dipole antenna on an FR4 dielectric slab and on a half-space that has the same relative permittivity as the dielectric slab. $W_1 = 3$ mm, $W_2 = 1$ mm, $d = 1$ mm, $L = 62.5$ mm, $\varepsilon_r = 4.28$ and $\tan\delta = 0.016$.

lines not only will be encountered within a folded-dipole antenna but also will be employed as real transmission lines between the antenna and the rest of an RFID or telecommunication system.

4.4.2.3 Analogy with Asymmetric Coplanar Waveguide In this section, we shall derive a closed-form expression for the characteristic impedance of an asymmetric CPS transmission line, based on an analogy with the asymmetric coplanar waveguide (CPW).

The characteristic impedance of a *symmetric* CPS on a dielectric slab of height t, strip width W, strip separation d and relative permittivity ε_r (Figure 4.16) is given by [18]

$$Z_0 = \frac{120\pi}{\sqrt{\varepsilon_{\text{eff}}}} \frac{K(k)}{K'(k)}, \tag{4.46}$$

where

$$\varepsilon_{\text{eff}} = 1 + \frac{\varepsilon_r - 1}{2} \frac{K(k')}{K(k)} \frac{K(k_2)}{K(k_2')} \tag{4.47}$$

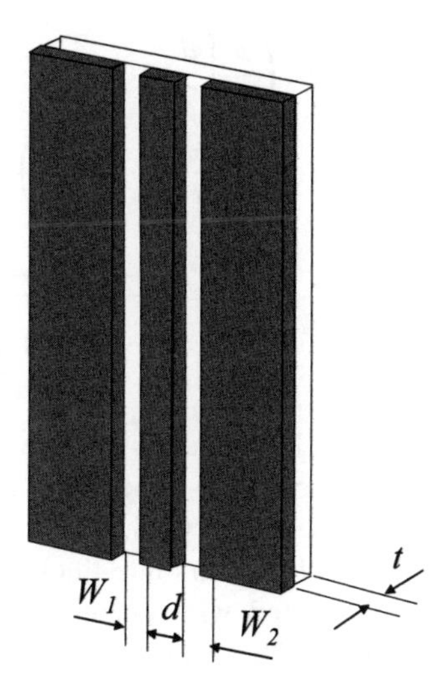

Figure 4.21 Asymmetric coplanar waveguide on a dielectric slab.

and

$$k = \frac{d}{d + 2W}, \tag{4.48}$$

$$k_2 = \frac{\sinh(\pi d/4t)}{\sinh((\pi/2t)[d/t + W])}. \tag{4.49}$$

Figure 4.21 shows an *asymmetric* CPW with slot widths W_1 and W_2 and slot separation d on a dielectric slab of thickness t. The characteristic impedance of an asymmetric CPW on a half-space dielectric slab ($t \to \infty$) of relative permittivity ε_r is given by [21]

$$Z_0 = \frac{30\pi}{\sqrt{\varepsilon_{\text{eff}}}} \frac{K'(k)}{K(k)}, \tag{4.50}$$

where k is given by equation (4.40) and ε_{eff} is given by equation (4.45). For a CPW on a dielectric slab of finite thickness t, the characteristic impedance is still given by equation (4.50), but ε_{eff} is now given by [21]

$$\varepsilon_{\text{eff}} = 1 + \frac{\varepsilon_r - 1}{2} \frac{K(k')}{K(k)} \frac{K(k_2)}{K(k_2')}, \tag{4.51}$$

where k is given by equation (4.40) and where

$$k_2 = \frac{W_A(1 + \alpha W_B)}{W_B + \alpha W_A^2}, \tag{4.52}$$

$$W_A = \sinh\left(\frac{\pi d}{4t}\right), \tag{4.53}$$

$$W_{\rm B} = \sinh\left(\frac{\pi}{2t}\left[\frac{d}{2} + W_2\right]\right), \tag{4.54}$$

$$W_{\rm E} = -\sinh\left(\frac{\pi}{2t}\left[\frac{d}{2} + W_1\right]\right) \tag{4.55}$$

and

$$\alpha = \frac{1}{W_{\rm B} + W_{\rm E}}\left[-1 - \frac{W_{\rm B} W_{\rm E}}{W_{\rm A}^2} - \sqrt{\left(\frac{W_{\rm B}^2}{W_{\rm A}^2} - 1\right)\left(\frac{W_{\rm E}^2}{W_{\rm A}^2} - 1\right)}\right]. \tag{4.56}$$

Given the analogy between a CPW and a coplanar-strip transmission line, we may easily transform the equations for the characteristic impedance of an asymmetric CPW on a finite-thickness dielectric slab to those for a CPS. The characteristic impedance of an asymmetric coplanar-strip transmission line on a finite-thickness dielectric slab is thus given by equation (4.46), where k is calculated with equations (4.40) and (4.41) and $\varepsilon_{\rm eff}$ is calculated with equations (4.51)–(4.56). The ratio of complete elliptic functions of the first kind is calculated with equation (4.39). To summarize,

$$Z_0 = \frac{120\pi}{\sqrt{\varepsilon_{\rm eff}}} \frac{K(k)}{K'(k)}, \tag{4.57}$$

where

$$\varepsilon_{\rm eff} = 1 + \frac{\varepsilon_{\rm r} - 1}{2} \frac{K(k')}{K(k)} \frac{K(k_2)}{K(k_2')}, \tag{4.58}$$

and

$$\frac{K(k)}{K(k')} \approx \begin{cases} \dfrac{1}{2\pi} \ln\left[2\dfrac{\sqrt{1+k} + \sqrt[4]{k}}{\sqrt{1+k} - \sqrt[4]{k}}\right] & \text{for } 1 \le \dfrac{K}{K'} \le \infty, \ \dfrac{1}{\sqrt{2}} \le k \le 1 \\ \dfrac{2\pi}{\ln[2(\sqrt{1+k} + \sqrt[4]{k})/(\sqrt{1+k} - \sqrt[4]{k})]} & \text{for } 0 \le \dfrac{K}{K'} \le 1, \ 0 \le k \le \dfrac{1}{\sqrt{2}}, \end{cases} \tag{4.59}$$

$$k = \frac{d/2[1 + e(d/2 + W_1)]}{d/2 + W_1 + e(d/2)^2}, \tag{4.60}$$

$$e = \frac{W_1 W_2 + (d/2)(W_1 + W_2) - \sqrt{W_1 W_2 (d + W_1)(d + W_2)}}{(d/2)^2 (W_1 - W_2)}, \tag{4.61}$$

$$k_2 = \frac{W_{\rm A}(1 + \alpha W_{\rm B})}{W_{\rm B} + \alpha W_{\rm A}^2}, \tag{4.62}$$

$$W_{\rm A} = \sinh\left(\frac{\pi d}{4t}\right), \tag{4.63}$$

$$W_{\rm B} = \sinh\left(\frac{\pi}{2t}\left[\frac{d}{2} + W_2\right]\right), \tag{4.64}$$

$$W_{\rm E} = -\sinh\left(\frac{\pi}{2t}\left[\frac{d}{2} + W_1\right]\right) \tag{4.65}$$

Table 4.3 Characteristic impedance of symmetric CPS transmission lines, calculated by a full-wave method and by the TL method, where, in the latter method, the dielectric slab was fully accounted for.

ε_r	t (mm)	d (mm)	W (mm)	Z_0, full wave (Ω)	Z_0, analytic (Ω)	Relative error (%)
2.20	0.79	0.10	1.52	100.07	100.96	0.89
2.20	0.79	0.30	0.76	149.79	151.08	0.86
9.90	0.64	0.04	1.27	49.91	49.94	0.06
9.90	0.64	0.37	0.51	99.98	99.83	0.15
12.90	0.25	0.026	0.38	50.00	49.92	0.16
12.90	0.25	0.15	0.13	100.05	99.90	0.15
50.00	0.25	0.031	0.20	30.03	29.97	0.20
50.00	0.25	0.030	0.025	50.03	49.99	0.08

and

$$\alpha = \frac{1}{W_B + W_E}\left[-1 - \frac{W_B W_E}{W_A^2} - \sqrt{\left(\frac{W_B^2}{W_A^2} - 1\right)\left(\frac{W_E^2}{W_A^2} - 1\right)}\right]. \quad (4.66)$$

Furthermore, $K'(k) = K(k')$, where $k'^2 = 1 - k^2$.

The relative error in the characteristic impedance thus calculated for symmetric CPS transmission lines compared with full-wave analysis results [19] remains well below 1%, as is demonstrated in Table 4.3.

The impact of the characteristic impedance of a CPS thus calculated on the input impedance of an asymmetric coplanar-strip folded-dipole antenna having the same dimensions and electrical characteristics as in the previous paragraphs is shown in Figure 4.22. Again, it should be noted that the dielectric slab has been taken into account only for the CPS transmission line mode of the antenna. The figure shows an improvement, although slight, over the results obtained with the two-half-spaces approximation (Figure 4.20).[5]

For designing an asymmetric coplanar-strip folded-dipole antenna on a dielectric slab, we may either use the two-half-spaces approximation for determining the characteristic impedance of the CPS or use the method described above. The latter has the disadvantage of a slightly more cumbersome implementation in software but gives excellent characteristic-impedance results; the former method has the advantage of an easy implementation in software, resulting in a fair approximation to the characteristic impedance.

In the following section, we shall briefly outline the way to analyze the dipole mode to obtain results with the best accuracy for the input impedance of a folded-dipole antenna and shall discuss in more detail a less sophisticated approximate method for getting fair results, good enough for design purposes. For the latter, we may use both methods for obtaining the characteristic impedance of the CPS. We have chosen the last method described in this section.

[5] Shifting the TL graphs and adjusting the levels would result in closer agreement over a wider range of frequencies for the TL graphs shown in Figure 4.20.

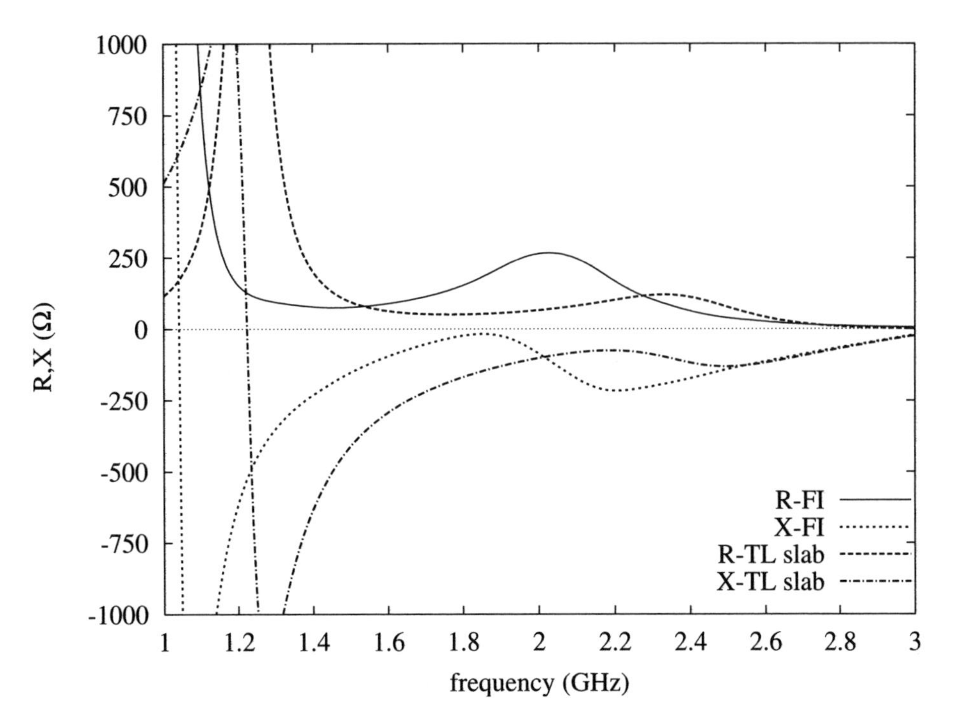

Figure 4.22 Calculated (FI and TL) real and imaginary parts of the input impedance versus frequency for an asymmetric CPS folded-dipole antenna on an FR4 dielectric slab. $W_1 = 3$ mm, $W_2 = 1$ mm, $d = 1$ mm, $L = 62.5$ mm, $\varepsilon_r = 4.28$ and $\tan\delta = 0.016$.

4.4.3 Dipole Mode Analysis

Although it is possible to account accurately for the strip dipole on the dielectric slab, we shall develop an approximate method based on correction terms applied to the free-space analysis of a folded-dipole antenna. If higher accuracy in the end results is required, an accurate analysis of the transmission line mode of the antenna needs to be used together with an accurate analysis of the dipole mode. Such an accurate method for the analysis of the dipole mode has been used in the previous chapter, on printed monopole antennas, and will be outlined again below.

4.4.3.1 Analysis of Strip Dipole In the analysis of a folded-dipole antenna, use is made of an equivalent radius ρ_e (equation (4.43)). First, this equivalent radius is transformed to an equivalent strip width, using $W_e = 4\rho_e$ [1]. To account accurately for the fact that the (equivalent) strip dipole antenna is positioned on a dielectric slab, we start with a three-term model for a cylindrical dipole antenna that models an imperfect conductor by means of a distributed impedance [22, 23]. By virtue of this distributed impedance, it is possible to model a dielectric or magnetic coating on a cylindrical dipole antenna. To that end, a distributed inductance is substituted into the distributed impedance [24]. A strip dipole on a dielectric slab is now modeled as an equivalent magnetically coated, circularly cylindrical dipole antenna [25]. In this analysis [25], the static capacitance of a coupled strip transmission line is needed, where the strip widths are equal to the equivalent dipole strip width. This capacitance value may be calculated by the method described in [26]. Further details may be found in the previous chapter.

This analysis method, however, will not be applied to the problem at hand now. Instead, we shall attempt to correct the impedance curves resulting from accounting for the dielectric slab in the transmission line mode only. This correction is accomplished by introducing correction terms applied to the free-space dipole length and equivalent radius.

4.4.3.2 Approximation for Strip Dipole We have seen that accounting for the dielectric slab in the transmission line mode of an asymmetric coplanar folded-dipole antenna led to an improvement in the impedance versus frequency curves compared with the free-space and uniform-dielectric cases. The curves resemble the ones obtained from full-wave analyses, apart from a frequency shift and an overall change in impedance level. We know that one of the main effects of a dielectric on a dipole antenna will be a lowering of the resonance frequency. Therefore, we could try, by lengthening the dipole in the antenna mode part of the analysis of the folded dipole, to make the resonance frequency coincide with that obtained by full-wave analysis.

Further, by increasing the equivalent radius, we could try to make the impedance levels coincide. Thus, in the dipole mode part of the analysis of the folded-dipole antenna, we may substitute L by L' and ρ_e by ρ'_e, where

$$L' = \kappa L \tag{4.67}$$

and

$$\rho'_e = \tau \rho_e. \tag{4.68}$$

The correction factors κ and τ have been determined for a large number of asymmetric coplanar-strip folded-dipole antennas, with different dimensions, and positioned on dielectric slabs of different heights and having different relative permittivities.

Figure 4.23 shows a *typical* example of the impedance curves thus calculated for a folded-dipole antenna together with full-wave analysis results. Figure 4.24 shows a *good* example of the impedance curves thus calculated.

The agreement between the values of the input impedance calculated as above and the values obtained through full-wave simulations tends to get better for decreasing strip widths and strip separations. For the frequency range tested (1 GHz–6 GHz), the correction factors appear to be frequency-independent and may be approximated by

$$\kappa \approx (1 + \sqrt{t} \times 10^3 \log(\varepsilon_r))^{0.45}, \tag{4.69}$$

$$\tau \approx 1.90. \tag{4.70}$$

These approximations were derived on the basis of inspection of the graphs of κ as a function of t and ε_r.

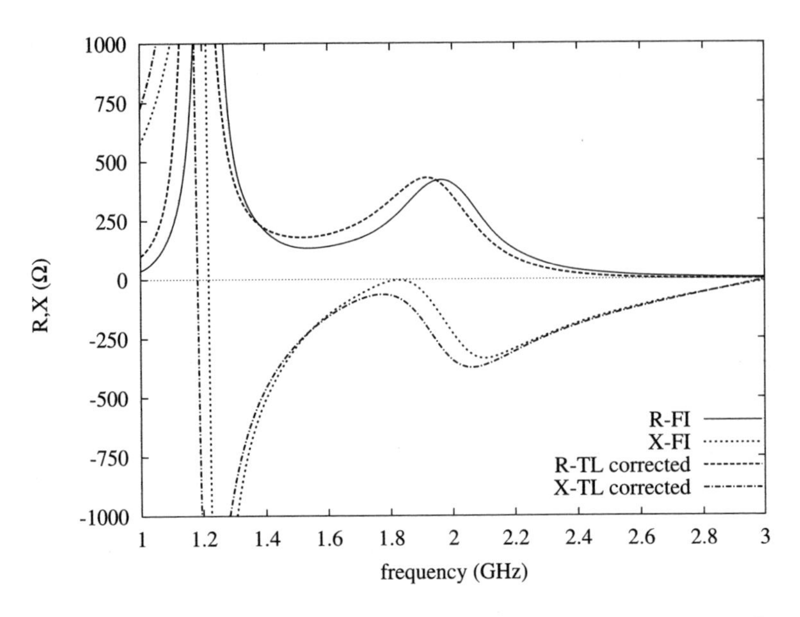

Figure 4.23 Calculated (FI and TL) real and imaginary parts of the input impedance versus frequency for an asymmetric CPS folded-dipole antenna on a dielectric slab. $W_1 = 1$ mm, $W_2 = 1$ mm, $d = 0.5$ mm, $L = 62.5$ mm, $t = 1.6$ mm, $\varepsilon_r = 4.28$, $\tan\delta = 0.016$, $\kappa = 1.30$ and $\tau = 1.90$.

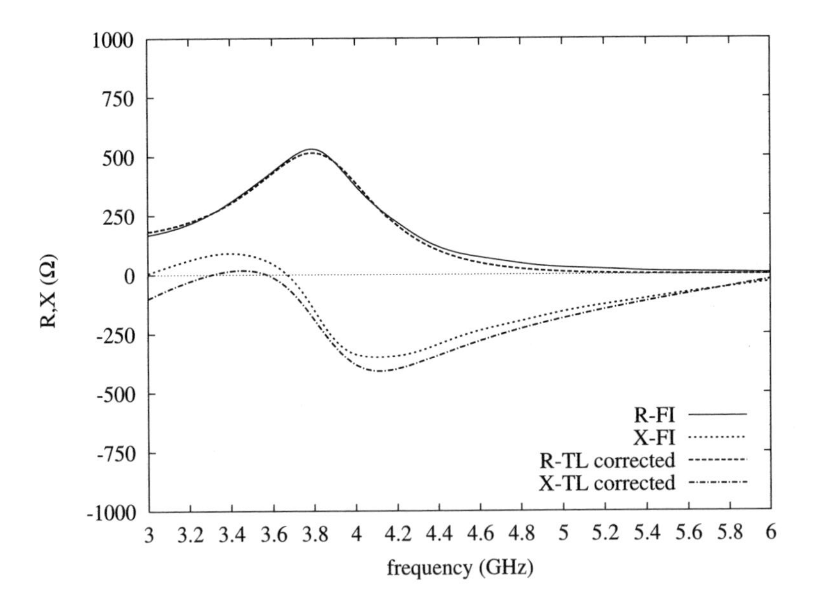

Figure 4.24 Calculated (FI and TL) real and imaginary parts of the input impedance versus frequency for an asymmetric CPS folded-dipole antenna on a dielectric slab. $W_1 = 0.75$ mm, $W_2 = 0.25$ mm, $d = 0.25$ mm, $L = 30$ mm, $t = 5.6$ mm, $\varepsilon_r = 4.28$, $\tan\delta = 0.016$, $\kappa = 1.45$ and $\tau = 1.90$.

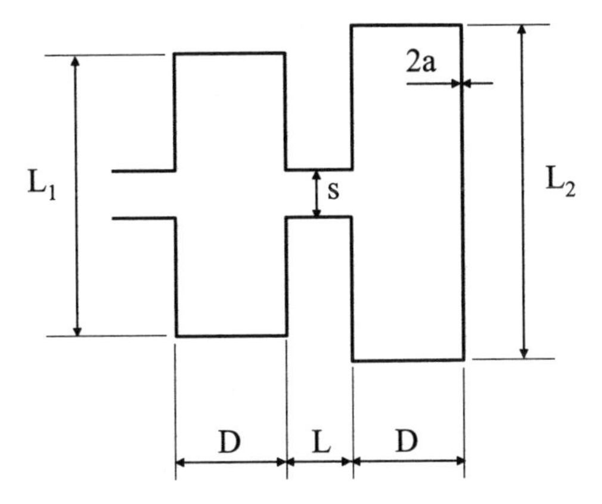

Figure 4.25 Example of a series-fed folded-dipole array antenna.

4.5 FOLDED-DIPOLE ARRAY ANTENNAS

To broaden the frequency bandwidth achievable with a single folded-dipole antenna, one can employ multiple folded-dipole elements, each of them resonant at a different frequency. When the resonance frequencies of the individual elements are close enough together, a broad bandwidth for the array may be realized. Such an antenna may be used, for example, for broadcast applications [27] or ultrawideband (UWB) applications where OFDM-like modulation schemes are used [28]. A practical way of connecting the individual elements in the array is by series feeding, as shown in the example in Figure 4.25. This figure shows a series two-element array of (wire) folded dipoles. The lengths of the folded-dipole elements are L_1 and L_2, respectively. Both of the folded-dipole elements have a wire separation D, and the spacing between the two elements is L, as indicated in the figure. The two-wire transmission line has a wire separation s, and the folded-dipole array antenna is constructed out of a single wire of radius a. The folded-dipole array antenna shown in Figure 4.25 will be used later for model verification.

With appropriate scaling and separation of the folded-dipole elements, where the required 180° phase difference is ensured through the use of the folded dipoles, the series-fed folded-dipole array antenna becomes a *log-periodic folded-dipole array antenna*, invented by John W. Greiser in 1964 [29]. Log-periodic antennas belong to the class of so-called *frequency-independent antennas*, known for their wideband characteristics, which are constructed in the case of log-periodic antennas using a particular scaling of their dimensions [11].

A linear, series-fed array of folded-dipole antennas does not consist of folded dipoles of the kind discussed before, except for the last element. The common element for these arrays is a *re-entrant folded-dipole antenna*. A reentrant folded-dipole antenna is a folded-dipole antenna equipped with an additional port (Figure 4.26). A reentrant folded-dipole antenna may be analyzed, however, using the results derived for an 'ordinary' folded-dipole antenna of the kind shown in Figure 4.1(c) [30].

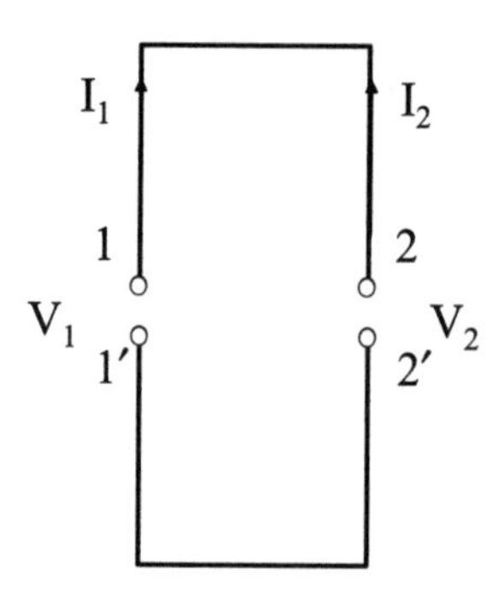

Figure 4.26 Reentrant folded-dipole antenna: geometry and definitions of voltages and currents.

4.5.1 Reentrant Folded-Dipole Antenna

Figure 4.26 shows the geometry of a reentrant folded-dipole antenna and the definitions of the voltages and currents in it. The relations between the currents and voltages are given by

$$I_1 = Y_{11}V_1 + Y_{12}V_2, \tag{4.71}$$

$$I_2 = Y_{21}V_1 + Y_{22}V_2. \tag{4.72}$$

The admittance matrix element Y_{11} is calculated from

$$Y_{11} = \left.\frac{I_1}{V_1}\right|_{V_2=0} = \frac{1}{2}Y_\mathrm{T} + \frac{1}{4}Y_\mathrm{D}, \tag{4.73}$$

where use has been made of equation (4.8), where $Y_\mathrm{T} = Z_\mathrm{T}^{-1}$ and $Y_\mathrm{D} = Z_\mathrm{D}^{-1}$. The admittance matrix element Y_{22} follows from short-circuiting port 1–1′ in Figure 4.26 ($V_1 = 0$) and exciting port 2–2′. This results in

$$Y_{22} = Y_{11}. \tag{4.74}$$

The admittance matrix element Y_{12} is calculated by equalizing the two-port voltages, i.e. $V_1 = V_2$. Only the antenna mode will be excited now. Upon substitution of equation (4.73) in equation (4.71), we find for the antenna current

$$I_1 = \left(\frac{1}{2}Y_\mathrm{T} + \frac{1}{4}Y_\mathrm{D} + Y_{12}\right)V_1. \tag{4.75}$$

From equation (4.6), we obtain the input impedance of the antenna,

$$Y_\mathrm{in} = Y_\mathrm{D} = \frac{2I_1}{V_1}. \tag{4.76}$$

The substitution of equation (4.76) in equation (4.75) results in

$$Y_{12} = Y_{21} = -\frac{1}{2}Y_\mathrm{T} + \frac{1}{4}Y_\mathrm{D}. \tag{4.77}$$

Note that we have assumed a symmetric wire folded-dipole antenna, meaning that the impedance step-up ratio is equal to 4 (in equation (4.17), $q = 1$).[6]

4.5.2 Series-Fed Linear Array of Folded Dipoles

If, for the moment, we neglect the effects of mutual coupling between the folded-dipole elements within the array antenna, the input impedance of the linear array antenna may be best calculated using *ABCD* matrices. For the nth folded-dipole element in an array consisting of N folded-dipole elements, the *ABCD* matrix is given by

$$\begin{bmatrix} A_n & B_n \\ C_n & D_n \end{bmatrix} = -\frac{1}{Y_{21_n}} \begin{bmatrix} Y_{22_n} & 1 \\ Y_{11_n} Y_{22_n} - Y_{21_n} Y_{12_n} & Y_{11_n} \end{bmatrix}, \tag{4.79}$$

where the admittance matrix elements Y_{ij_n}, $i, j = 1, 2$, $n = 1, 2, \ldots, N$, are given by equations (4.73), (4.74) and (4.77).

The *ABCD* matrix of a length of transmission line between two folded-dipole elements is given by [31]

$$\begin{bmatrix} A_p & B_p \\ C_p & D_p \end{bmatrix} = \begin{bmatrix} \cosh(\gamma_p \ell_p) & Z_{0_p} \sinh(\gamma_p \ell_p) \\ \sinh(\gamma_p \ell_p)/Z_{0_p} & \cosh(\gamma_p \ell_p) \end{bmatrix}, \tag{4.80}$$

for $p = 1, 2, \ldots, N-1$. Here ℓ_p is the length of transmission line p, Z_{0_p} is the characteristic impedance of this transmission line and γ_p is the propagation constant.

The *ABCD* matrix of a series load Y_{L} (e.g. a short circuit, but not necessarily so) at the end of the array is given by

$$\begin{bmatrix} A_{\mathrm{L}} & B_{\mathrm{L}} \\ C_{\mathrm{L}} & D_{\mathrm{L}} \end{bmatrix} = \begin{bmatrix} 1 & 0 \\ Y_{\mathrm{L}} & 1 \end{bmatrix}. \tag{4.81}$$

The complete array is characterized by

$$\begin{bmatrix} A & B \\ C & D \end{bmatrix} = \prod_{n=1}^{2N} \begin{bmatrix} A_n & B_n \\ C_n & D_n \end{bmatrix}. \tag{4.82}$$

The impedance matrix of the series-fed, linear array of folded dipoles is

$$\begin{bmatrix} Z_{11} & Z_{12} \\ Z_{21} & Z_{22} \end{bmatrix} = \frac{1}{C} \begin{bmatrix} A & AD - BC \\ 1 & D \end{bmatrix}. \tag{4.83}$$

[6] For an asymmetric wire folded-dipole antenna, the analysis is similar, but we have to use equation (4.17) instead of equation (4.8) and equation (4.15) instead of equation (4.6). The final result is

$$[Y] = \begin{bmatrix} \frac{1}{2} Y_T + \frac{1}{(1+q)^2} Y_A & -\frac{1}{2} Y_T + \frac{1}{(1+q)^2} Y_A \\ -\frac{1}{2} Y_T + \frac{1}{(1+q)^2} Y_A & \frac{1}{2} Y_T + \frac{1}{(1+q)^2} Y_A \end{bmatrix}. \tag{4.78}$$

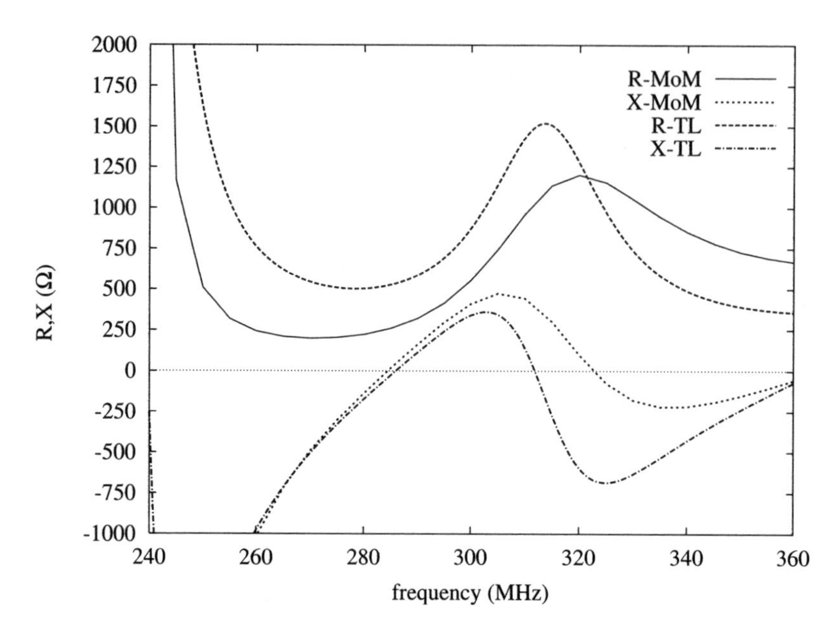

Figure 4.27 Calculated (MoM and TL) real part (R) and imaginary part (X) of the input impedance of a two-element folded-dipole array antenna versus frequency. $L_1 = 0.4$ m, $L_2 = 0.5$ m, $a = 0.0001$ m, $D = 0.005$ m, $s = 0.005$ m, $L = 0.2$ m.

4.5.3 Model Verification

To verify the model, a number of configurations of the kind shown in Figure 4.25 were analyzed using the transmission line model developed. Since we aimed only at verification of the model, no attempt was made to design an antenna to given specifications. In all calculations, the wire was assumed to be a perfect electric conductor. MoM analyses, using the axial-current method and a voltage delta-gap feed, of the same structures using copper wires gave results nearly identical to the ones shown below.

Figure 4.27 shows impedance curves calculated with the TL method and the MoM method for a two-element folded-dipole array antenna where the separation of the folded dipoles L was 0.2 m. In Figures 4.28 and 4.29, impedance curves are shown for similar folded-dipole array antennas, but now for separation of folded-dipole elements of $L = 0.4$ m and 0.8 m, respectively.

The figures show that the analysis results get better with increasing separation of the folded-dipole elements at the series-fed linear array antenna. This is consistent with the fact that the effects of mutual coupling are not accounted for in the TL analysis. Furthermore, the analysis results have been shown for rather thin wires. For thick wires, the agreement with the MoM results is not as good as that in the graphs shown here. Replacement of the dipole analysis in the TL method with an analysis better fitted to handling thicker wires, for example in MoM analysis, should be capable of curing this problem. Nevertheless, the current TL method is considered accurate enough for creating initial designs for folded-dipole array antennas.

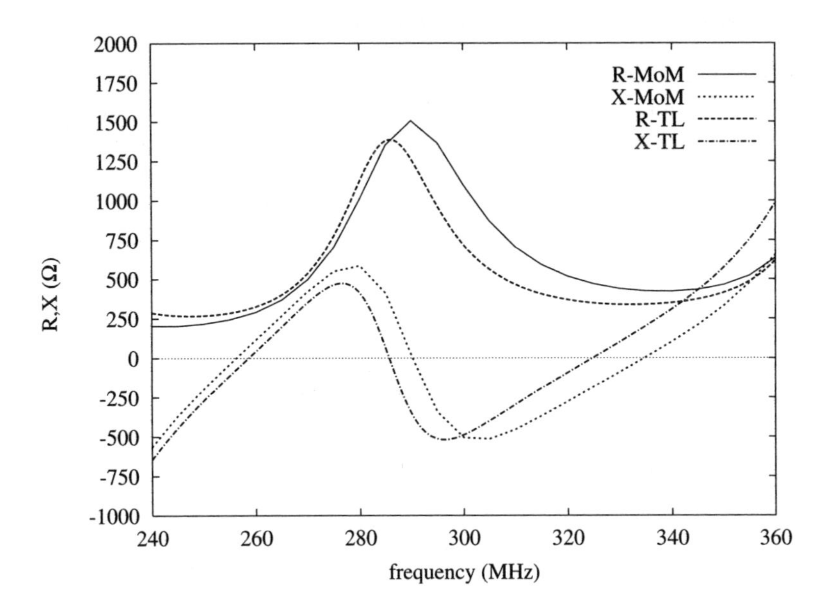

Figure 4.28 Calculated (MoM and TL) real part (R) and imaginary part (X) of the input impedance of a two-element folded-dipole array antenna versus frequency. $L_1 = 0.4$ m, $L_2 = 0.5$ m, $a = 0.0001$ m, $D = 0.005$ m, $s = 0.005$ m, $L = 0.4$ m.

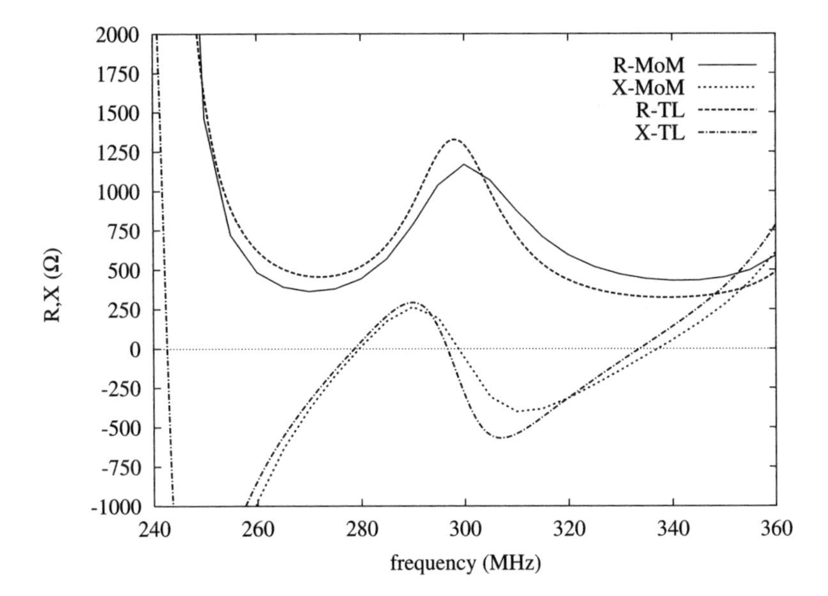

Figure 4.29 Calculated (MoM and TL) real part (R) and imaginary part (X) of the input impedance of a two-element folded-dipole array antenna versus frequency. $L_1 = 0.4$ m, $L_2 = 0.5$ m, $a = 0.0001$ m, $D = 0.005$ m, $s = 0.005$ m, $L = 0.8$ m.

4.5.4 Inclusion of Effects of Mutual Coupling

If we wish to calculate the input impedance of a linear array antenna without neglecting the effects of mutual coupling between the dipole elements, the *ABCD* matrix is no longer the preferred tool for accomplishing this task. Instead, the construction of an overall admittance matrix for the array antenna allows the inclusion of there effects in a relatively easy way.

To include the effects of mutual coupling, the radiating elements and the feeding structure are separated as described in [32, 33]. The admittance matrix of an N-element folded-dipole array may be written as a $2N \times 2N$ array $[Y]$,

$$[Y] = [Y_{\text{F}}] + [Y_{\text{A}}], \tag{4.84}$$

where $[Y_{\text{F}}]$ is the admittance matrix of the feed network, i.e. the collection of interconnecting pieces of two-wire transmission line, and where $[Y_{\text{A}}]$ is the admittance matrix of the collection of reentrant folded-dipole elements.

The admittance matrix of the feed network is given by

$$[Y_{\text{F}}] = \begin{bmatrix} 0 & 0 & 0 & \cdots & 0 & 0 \\ 0 & [Y_{\text{F1}}] & 0 & \cdots & 0 & 0 \\ 0 & 0 & [Y_{\text{F2}}] & \cdots & 0 & 0 \\ 0 & 0 & 0 & \ddots & 0 & 0 \\ \vdots & \vdots & \vdots & \cdots & \vdots & \vdots \\ 0 & 0 & 0 & \cdots & [Y_{\text{F}(N-1)}] & 0 \\ 0 & 0 & 0 & \cdots & 0 & Y_{\text{L}} \end{bmatrix}. \tag{4.85}$$

In the above equation, Y_{L} is the load admittance of the array. For the array shown in Figure 4.25, which is terminated by a short circuit, $Y_{\text{L}} = \infty$.

The 2×2 submatrices $[Y_{\text{F}i}]$, $i = 1, 2, \ldots, N-1$, are defined by [31]

$$\begin{aligned} Y_{\text{F}i_{11}} &= Y_{\text{F}i_{22}} = -\text{j}Y_{0_i}\cot(k_0 l_i), \\ Y_{\text{F}i_{12}} &= Y_{\text{F}i_{21}} = \text{j}Y_{0_i}\csc(k_0 l_i), \end{aligned} \tag{4.86}$$

where Y_{0_i} and l_i are the characteristic admittance and length, respectively, of the piece of two-wire transmission line i.

The admittance matrix of the collection of reentrant folded-dipole elements is given by

$$[Y_{\text{A}}] = \begin{bmatrix} [Y_{\text{A}_{11}}] & [C_{\text{A}_{12}}] & \cdots & [C_{\text{A}_{1N}}] \\ [C_{\text{A}_{21}}] & [Y_{\text{A}_{22}}] & \cdots & [C_{\text{A}_{2N}}] \\ \vdots & \vdots & \ddots & \vdots \\ [C_{\text{A}_{N1}}] & [C_{\text{A}_{N2}}] & \cdots & [Y_{\text{A}_{NN}}] \end{bmatrix}, \tag{4.87}$$

where the 2×2 submatrices $[Y_{\text{A}_{ii}}]$, $i = 1, 2, \ldots, N$, are defined by [30]

$$\begin{aligned} Y_{\text{A}_{ii_{11}}} &= Y_{\text{A}_{ii_{22}}} = \frac{1}{2}Y_{\text{T}_i} + \frac{1}{4}Y_{\text{D}_i}, \\ Y_{\text{A}_{ii_{12}}} &= Y_{\text{A}_{ii_{21}}} = -\frac{1}{2}Y_{\text{T}_i} + \frac{1}{4}Y_{\text{D}_i}. \end{aligned} \tag{4.88}$$

In the above equations, Y_{T_i} and Y_{D_i} are the transmission line stub admittance and the equivalent dipole admittance, respectively, of re-entrant folded dipole i. The values are calculated using equations (4.1), (4.9) and (4.10).

The 2×2 coupling submatrices $[C_{\mathrm{A}_{ij}}]$, $i, j = 1, 2, \ldots, N$, $i \neq j$, contain the mutual admittances between the reentrant folded-dipole elements. They are defined by

$$C_{\mathrm{A}_{ij11}} = C_{\mathrm{A}_{ij12}} = C_{\mathrm{A}_{ij21}} = C_{\mathrm{A}_{ij22}} = \frac{1}{4} Y_{ij}, \tag{4.89}$$

where [14]

$$Y_{ij} = Y21_{\text{dipole-to-dipole}}. \tag{4.90}$$

The mutual admittance between two thin 'ordinary' dipoles may be calculated using the closed-form equations in [15].[7]

The mutual impedance between two nonstaggered thin dipoles of half-lengths ℓ_1 and ℓ_2, separated by a distance d, is given by

$$Z_{12} = 30R_{12} + \mathrm{j}30X_{12}, \tag{4.92}$$

where

$$\begin{aligned} R_{12} = {} & \cos\{k_0(\ell_1 + \ell_2)\} \\ & \times [Ci(u_0) + Ci(v_0) - Ci(u_1) - Ci(v_1) - Ci(w_1) - Ci(y_1) + 2Ci(k_0 d)] \\ & + \cos\{k_0(\ell_1 - \ell_2)\} \\ & \times [Ci(u_0') + Ci(v_0') - Ci(u_1) - Ci(v_1) - Ci(w_1) - Ci(y_1) + 2Ci(k_0 d)] \\ & + \sin\{k_0(\ell_1 + \ell_2)\} \\ & \times [-Si(u_0) + Si(v_0) + Si(u_1) - Si(v_1) - Si(w_1) + Si(y_1)] \\ & + \sin\{k_0(\ell_1 - \ell_2)\} \\ & \times [-Si(u_0') + Si(v_0') + Si(u_1) - Si(v_1) - Si(w_1) - Si(y_1)] \end{aligned} \tag{4.93}$$

[7]For re-entrant folded dipoles with different wire radii, the coupling submatrices are defined by

$$\begin{aligned} C_{\mathrm{A}_{ij11}} &= \frac{1}{1+q_i} \frac{1}{1+q_j} Y_{ij}, \\ C_{\mathrm{A}_{ij12}} &= \frac{1}{1+q_i} \frac{1}{1+1/q_j} Y_{ij}, \\ C_{\mathrm{A}_{ij21}} &= \frac{1}{1+1/q_i} \frac{1}{1+q_j} Y_{ij}, \\ C_{\mathrm{A}_{ij22}} &= \frac{1}{1+1/q_i} \frac{1}{1+1/q_j} Y_{ij}, \end{aligned} \tag{4.91}$$

where $q_{i,j}$ is the impedance step-up ratio of the reentrant folded dipoles i and j. For equal wire radii, $q_{i,j} = 1$.

and

$$
\begin{aligned}
X_{12} = {} & \cos\{k_0(\ell_1 + \ell_2)\} \\
& \times [-Si(u_0) - Si(v_0) + Si(u_1) + Si(v_1) + Si(w_1) + Si(y_1) - 2Si(k_0 d)] \\
& + \cos\{k_0(\ell_1 - \ell_2)\} \\
& \times [-Si(u_0') - Si(v_0') + Si(u_1) + Si(v_1) + Si(w_1) + Si(y_1) - 2Si(k_0 d)] \\
& + \sin\{k_0(\ell_1 + \ell_2)\} \\
& \times [-Ci(u_0) + Ci(v_0) + Ci(u_1) - Ci(v_1) - Ci(w_1) + Ci(y_1)] \\
& + \sin\{k_0(\ell_1 - \ell_2)\} \\
& \times [-Ci(u_0') + Ci(v_0') + Ci(u_1) - Ci(v_1) - Ci(w_1) - Ci(y_1)].
\end{aligned} \tag{4.94}
$$

In equations (4.93) and (4.94), $Si(x)$ and $Ci(x)$ are, respectively, the sine and cosine integrals of argument x, and

$$
\begin{aligned}
u_0 &= k_0\left(\sqrt{d^2 + (\ell_1 + \ell_2)^2} - (\ell_1 + \ell_2)\right), \\
v_0 &= k_0\left(\sqrt{d^2 + (\ell_1 + \ell_2)^2} + (\ell_1 + \ell_2)\right), \\
u_0' &= k_0\left(\sqrt{d^2 + (\ell_1 - \ell_2)^2} - (\ell_1 - \ell_2)\right), \\
v_0' &= k_0\left(\sqrt{d^2 + (\ell_1 - \ell_2)^2} + (\ell_1 + \ell_2)\right), \\
u_1 &= k_0\left(\sqrt{d^2 + \ell_1^2} - \ell_1\right), \\
v_1 &= k_0\left(\sqrt{d^2 + \ell_1^2} + \ell_1\right), \\
w_1 &= k_0\left(\sqrt{d^2 + \ell_2^2} + \ell_2\right), \\
y_1 &= k_0\left(\sqrt{d^2 + \ell_2^2} - \ell_2\right).
\end{aligned} \tag{4.95}
$$

The mutual admittance is obtained from the mutual impedance through

$$
Y_{12} = \frac{Z_{12}}{Z_{D_1} Z_{D_2} - Z_{12}}, \tag{4.96}
$$

where Z_{D_1} and Z_{D_2} are the self-impedances of the two coupled dipole antennas.

4.5.5 Verification of Modeling of Mutual Coupling

The analysis of mutual coupling described in section 4.5.1 was implemented in software and several wire folded-dipole array antennas were analyzed. Comparisons were made with results obtained by a method-of-moments analysis. As a typical example, we shall show the results for a four-element folded-dipole array antenna, shown in Figure 4.30.

The array dimensions were: $L_1 = 11$ mm, $L_2 = 15$ mm, $L_3 = 19$ mm, $L_4 = 23$ mm, $D = 0.15$ mm, $TL = 10$ mm and $a = 3$ μm. The analysis was performed for frequencies

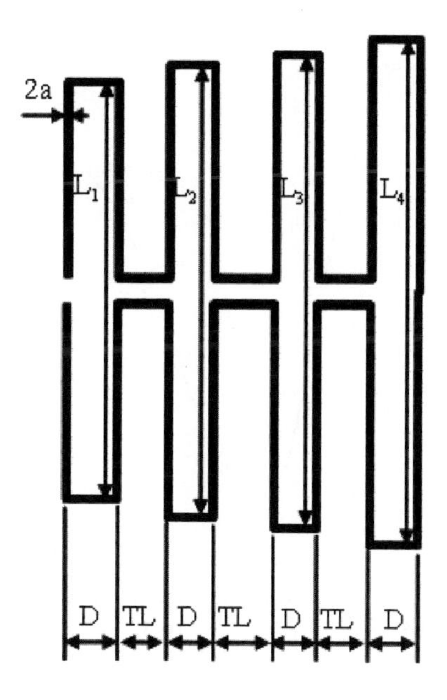

Figure 4.30 Four-element series array of reentrant folded-dipole antennas.

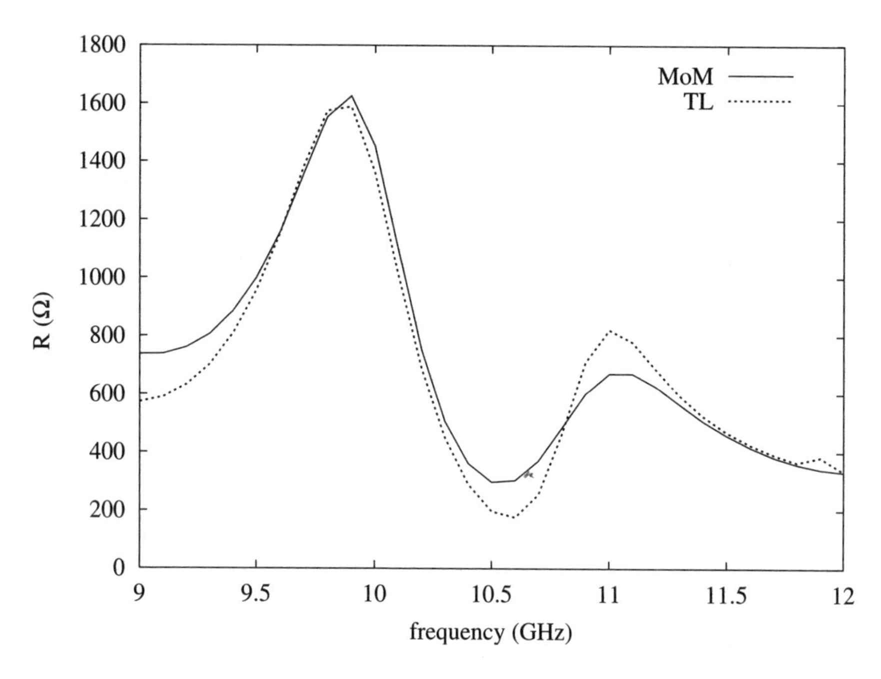

Figure 4.31 Real part of the input impedance as a function of frequency for a four-element folded-dipole array antenna.

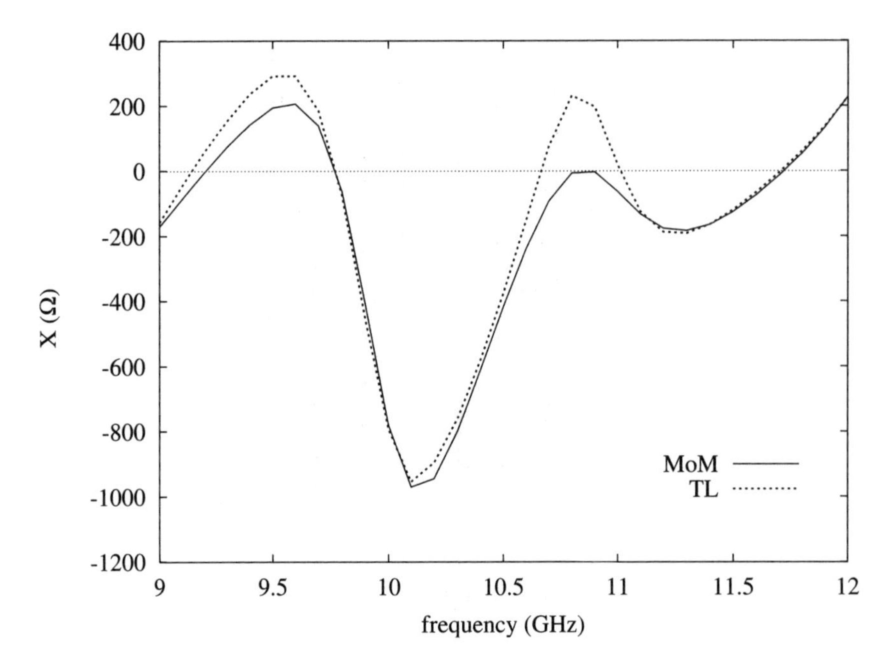

Figure 4.32 Imaginary part of the input impedance as a function of frequency for a four-element folded-dipole array antenna.

ranging from 9 GHz to 12 GHz. In Figure 4.31, the real part of the input impedance vs. frequency is shown, as calculated with the TL method. MoM analysis results are shown for comparison. In Figure 4.32, the imaginary part of the input impedance vs. frequency is shown.

These two figures show good agreement between the analysis results obtained with the TL method with inclusion of the effects of mutual coupling and the results obtained with the MoM. Although the MoM is more accurate and more versatile, the TL method, by virtue of being a dedicated method for this type of antenna, is much faster in producing analysis results. Therefore the TL method is more suited for synthesis problems, especially for generating pre-designs.

4.6 CONCLUSIONS

An accurate analytical model has been derived for a linear array of wire folded-dipole antennas. The model combines closed-form analytical equations for a folded-dipole antenna, a reentrant folded-dipole antenna, a two-wire transmission line, the mutual coupling between two folded-dipole antennas and the mutual coupling between two thin dipole antennas. A technique in which the analysis of the antennas and that of the feeding network are separated has been successfully applied in order to combine all of these closed-form equations. This analysis is expected to be extremely useful in optimization schemes to synthesize desired input impedance characteristics. To aid further in obtaining desired input impedance characteristics, measures for tuning the impedance of a folded dipole have been derived and verified. These measures consist of placing short circuits in the 'arms' of the folded dipole, placing parasitic short-circuited dipoles or folded dipoles close to the folded-dipole antenna, or both.

REFERENCES

1. C.A. Balanis, *Antenna Theory, Analysis and Design*, second edition, John Wiley & Sons, New York, 1997.
2. NORDIC Semiconductor, *Single Chip 433/868/915 MHz Transceiver nRF905*, NORDIC Semiconductor, January 2005.
3. R.S. Elliot, *Antenna Theory and Design*, revised edition, John Wiley & Sons, New York, 2003.
4. C.T.A. Johnk, *Engineering Electromagnetic Fields and Waves*, John Wiley & Sons, 1975.
5. G.A. Thiele, E.P. Ekelman Jr. and L.W. Henderson, 'On the accuracy of the transmission line model of the folded dipole', *IEEE Transactions on Antennas and Propagation*, Vol. AP-28, No. 5, pp. 700–703, September 1980.
6. R.W. Lampe, 'Design formulas for an asymmetric coplanar strip folded dipole', *IEEE Transactions on Antennas and Propagation*, Vol. AP-33, No. 9, pp. 1028–1031, September 1985.
7. R.W. Lampe, 'Correction to design formulas for an asymmetric coplanar strip folded dipole', *IEEE Transactions on Antennas and Propagation*, Vol. AP-34, No. 4, p. 611, April 1986.
8. A.R. Clark and A.P.C. Fourie, 'An improvement to the transmission line model of the folded dipole', *IEE Proceedings H*, Vol. 138, No. 6, pp. 577–579, December 1991.
9. J.A. Flint and J.C. Vardaxoglou, 'Exploitation of nonradiating modes in asymmetric coplanar strip folded dipoles', *IEE Proceedings on Microwaves, Antennas and Propagation*, Vol. 151, No. 4, pp. 307–310, August 2004.
10. E.C. Jordan, *Electromagnetic Waves and Radiating Systems*, Prentice Hall, Englewood Cliffs, 1950.
11. R.C. Johnson, *Antenna Engineering Handbook*, third edition, McGraw-Hill, New York, 1983.
12. H.J. Visser and A.C.F. Reniers, 'Textile antennas: A practical approach', *Proceedings of the 2nd European Conference on Antennas and Propagation*, Edinburgh, UK, November 2007.

13. J. Rahola, 'Power waves and conjugate matching', *IEEE Transactions on Circuits and Systems II: Express Briefs*, (in press).

14. A.R. Clark and A.P.C. Fourie, 'Mutual impedance and the folded dipole', *Proceedings of the 2nd International Conference on Computation in Electromagnetics*, pp. 347–350, April 1994.

15. H.E. King, 'Mutual impedance of unequal length antennas in echelon', *IRE Transactions on Antennas and Propagation*, pp. 306–313, July 1957.

16. W. Hilberg, 'From approximations to exact relations for characteristic impedances', *IEEE Transactions on Microwave Theory and Techniques*, Vol. MTT-17, No. 5, pp. 259–265, May 1969.

17. H.J. Visser, 'Improved design equations for asymmetric coplanar strip folded dipoles on a dielectric slab', *Proceedings of the 2nd European Conference on Antennas and Propagation*, Edinburgh, UK, November 2007.

18. M.Y. Frankel, R.H. Voelkner and J.N. Hilfiker, 'Coplanar transmission lines on thin substrates for high-speed low-loss propagation', *IEEE Transactions on Microwave Theory and Techniques*, Vol. 42, No. 3, pp. 396–402, March 1994.

19. T.Q. Deng, M.S. Leong, P.S. Kooi and T.S. Yeo, 'Synthesis formulas for coplanar lines in hybrid and monolithic MICs', *Electronics Letters*, Vol. 32, No. 24, pp. 2253–2254, November 1996.

20. S.B. Cohn, 'Slot line on a dielectric substrate', *IEEE Transactions on Microwave Theory and Techniques*, Vol. MTT-17, No. 10, pp. 768–778, October 1969.

21. V.F. Hanna and D. Thebault, 'Theoretical and experimental verification of asymmetric coplanar waveguides', *IEEE Transactions on Microwave Theory and Techniques*, Vol. MTT-32, No. 12, pp. 1649–1651, December 1984.

22. R.W.P. King and T.T. Wu, 'The imperfectly conducting cylindrical transmitting antenna', *IEEE Transactions on Antennas and Propagation*, Vol. AP-14, No. 5, pp. 524–534, September 1966.

23. R.W.P. King, C.W. Harrison and E.A. Aronson, 'The imperfectly conducting cylindrical transmitting antenna, numerical results', *IEEE Transactions on Antennas and Propagation*, Vol. AP-14, No. 5, pp. 535–542, September 1966.

24. J. Moore and M.A. West, 'Simplified analysis of coated wire antennas and scatterers', *IEE Proceedings on Microwaves, Antennas and Propagation*, Vol. 142, No. 1, pp. 14–18, February 1995.

25. B.D. Popović and A. Nesić, 'Generalisation of the concept of equivalent radius of thin cylindrical antennas', *IEE Proceedings*, Vol. 131, Part H, No. 3, pp. 153–158, June 1984.

26. E. Ymashita and S. Yamazaki, 'Parallel strip line embedded in or printed on a dielectric sheet', *IEEE Transactions on Microwave Theory and Techniques*, pp. 972–973, November 1968.

27. R. Wilensky, 'High-frequency antennas', in R.C. Johnson (ed.), *Antenna Engineering Handbook*, third edition, McGraw-Hill, pp. 26-1–26-42, 1993.

28. W. Sögel, C. Waldschnidt and W. Wiesbeck, 'Transient responses of a Vivaldi antenna and a logarithmic periodic dipole array for ultra wideband communication', *Proceedings of the IEEE Antennas and Propagation Society International Symposium*, pp. 592–595, June 2003.

29. J.W. Greiser, 'A new class of log-periodic antennas', *Proceedings of the IEEE*, pp. 617–618, May 1964.

30. H. Shnitkin, 'Analysis of log-periodic folded dipole array', *IEEE Antennas and Propagation International Symposium Digest*, pp. 2105–2108, July 1992.

31. K.C. Gupta, R. Garg and R. Chadha, *Computer Aided Design of Microwave Circuits*, Artech House, Dedham, MA, 1981.

32. J. Mautz and R. Harrington, 'Modal analysis of loaded N-port scatterers', *IEEE Transactions on Antennas and Propagation*, Vol. 21, No. 2, pp. 188–199, March 1973.

33. E. Newman and J. Tehan, 'Analysis of a microstrip array and feed network', *IEEE Transactions on Antennas and Propagation*, Vol. 33, No. 4, pp. 397–403, April 1985.

5

Rectennas: Microstrip Patch Antennas

Rectennas are used for converting wireless RF power into DC power. The challenge lies in maximizing the power conversion efficiency for low input power and – at the same time – minimizing the dimensions of the rectenna. By conjugately matching a rectifying circuit directly to a microstrip patch antenna, a matching and filtering network between the antenna and rectifying circuit can be avoided. With the aid of analytical models for the antenna and the rectifying circuit, single-layer, internally matched and filtered PCB rectennas may be designed for low input power. An efficiency of 52% for 0 dBm input power has been realized at 2.45 GHz for a rectenna on a standard PCB material (FR4). A series connection of these rectennas is able to power a standard household electric wall clock.

If the resistor in a Wilkinson power combiner is replaced by a rectifying circuit, the simultaneous transmission of power and data, employing the same antennas for power and data transmission, becomes feasible.

A survey of expected power density levels distant from GSM-900 and GSM-1800 base stations, WLAN routers and GSM mobile phones reveals that a single GSM telephone can produce sufficient energy for wirelessly powering small applications at moderate distances.

5.1 INTRODUCTION

Within the framework of the miniaturization of autonomous sensors, wireless data transmission and power management are two of the key features that determine success or failure. For the purpose of minimizing the maintenance of sensors, the combination of wireless technology for both data transmission and power transmission would be an attractive feature. The reception of microwave power is maximized by enlarging the collecting aperture. However, even for small apertures, the reception of microwave power may be advantageous.

Approximate Antenna Analysis for CAD Hubregt J. Visser

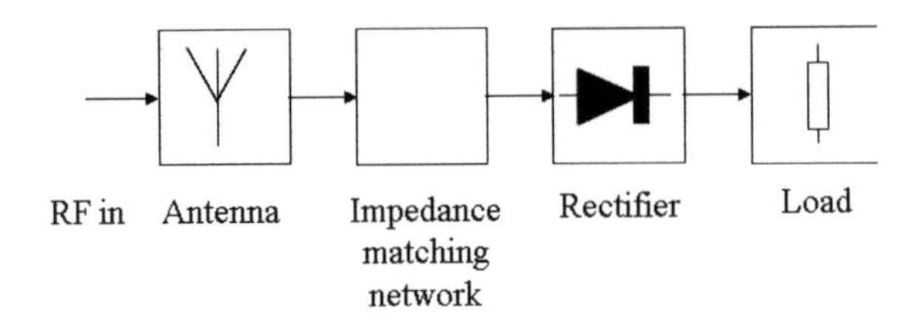

Figure 5.1 Schematic representation of a general RF power supply.

The large duty cycles that characterize a substantial class of miniature, autonomous sensors – e.g. sensors that need to communicate a temperature once every few hours – make a miniature RF power-harvesting device, used to charge a battery or capacitor, feasible.

The history of power transmission by radio waves dates back to the experiments of Heinrich Hertz in the 1880s to prove Maxwell's theory of electromagnetism. The modern history of free-space power transmission may be considered to have its origin in the late 1950s [1] with applications in microwave-powered aircraft and the Solar Power Satellite Concept.

After a period of relatively little activity in the field of free-space power transmission in the 1980s and 1990s, we can now observe a renewed interest in this field. This interest seems to have been initiated by short-range (< 2 m [2]) radio frequency identification (RFID) applications, focusing on the available industry, science and medical (ISM) frequency bands around 0.9 GHz, 2.4 GHz, 5.8 GHz and higher. Especially for the higher frequencies, the wavelengths are small enough for the realization of miniature wireless products. The power supply for these systems may consist of an antenna coupled to a rectifying component. The combination of a *rect*ifier and an ant*enna* is commonly called a *rectenna*. A rectenna is a device to be operated in the far field of a transmitting antenna. Therefore, it is substantially different from power-collecting devices employing inductively coupled coils, which need to be positioned very close together and in line with each other.

A 'traditional' RF power supply is shown schematically in Figure 5.1. RF power is collected by the antenna. The antenna is impedance-matched to the rectifier (usually a Schottky diode or diode circuit) for maximum power transfer. The diode circuit generates a voltage that is dependent of the input power and the output load [3,4]. A capacitor is placed in parallel with the load resistor. The values of the resistor and capacitor are such that significant RF feedthrough to the output port is prevented [3].

A diode becomes a more efficient rectifier at higher input power levels. In [5], a power conversion efficiency exceeding 80% for an input power level of 20 dBm was reported for a rectenna. Whenever the use of a battery is not possible or a physical connection for externally feeding an application is not available, and distance is not critical, i.e. the radiating source will be close to the rectenna, such a high-input-power-solution is preferred. For feeding wireless applications at larger distances, however, using transmission-power-restricted ISM frequency bands, we encounter three main challenges. The first challenge is to maximize the power conversion efficiency for low input power levels to the rectenna. The second challenge is to minimize the dimensions of the wireless battery.

The third challenge lies in integrating antenna and rectenna operation into the same device. We want to be able to transmit power *and data* simultaneously over a wireless link by using the same structure for both the antenna and the rectenna. Here, we shall discuss solutions to these three challenges.

In the following, we shall first discuss the design of the traditional wireless battery and same means to improve this design. Then, analytical models developed for antenna and rectifier circuits will be discussed, followed by verification of these models. A 2.45 GHz wireless-battery design will be discussed, as well as the application of eight of these batteries in series to power a household electric wall clock. Finally, we shall discuss a two-antenna array with an RF-to-DC-converting Wilkinson power-combining network, and we shall end with our conclusions.

5.2 RECTENNA DESIGN IMPROVEMENTS

In a 'traditional' rectenna or wireless battery, the receiving antenna is connected to the rectifying circuit through an impedance-matching network (e.g. [6]). The impedance-matching network, which may take the form of a transmission line and a stub, was used in [6] to match the 50 Ω antenna impedance to the rectifying circuit. In previous work on deriving analytical models for rectenna design [7], we decided to abandon the 50 Ω subsystem impedance and choose internal conjugate matching. To that end, we devised a method for analytically calculating the input impedance of the rectifying circuit. Then, by properly choosing the position of the input probe of a microstrip patch antenna, we were able to conjugately match the receiving antenna to the rectifying circuit. We aimed, in doing so, at improving the overall power conversion efficiency. The layout of the resulting structure is shown in Figure 5.2.

The antenna and the rectifying circuit are placed on both sides of a shared ground plane. Although an impedance-matching network has become unnecessary, we still observe a network between the antenna and the rectifying circuit (a diode). The purpose of this network, consisting of a transmission line and two radial stubs, is to prevent the second harmonic, generated by the diode, from being reradiated by the microstrip patch antenna. Additional radial stubs are placed between the diode and the load. One of the stubs prevents the signal at the operating frequency from being dissipated in the load, and the other prevents the signal at the second harmonic from being dissipated [7].

This rectenna may be improved with respect to complexity. The analysis may be simplified without compromising the accuracy.

The first improvement with respect to complexity consists of employing a microstrip edge feed instead of a probe feed for exciting the microstrip antenna. Thus, the whole of the antenna and the rectifying circuit may be realized on a single grounded printed circuit board (PCB). Next, the rectifying circuit (again a diode) is conjugately matched directly to the microstrip antenna. Since higher-order harmonics generated by the diode will be severely mismatched to the small-band microstrip patch antenna, the patch antenna itself will act as a higher-order-harmonic filter, making the transmission line stub network shown in Figure 5.2 superfluous. The impedance of the diode dictates the impedance of the microstrip patch antenna. The microstrip patch antenna will therefore not be operated at resonance.

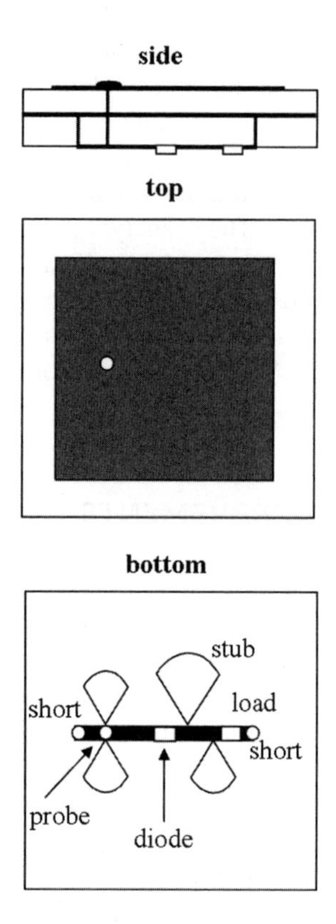

Figure 5.2 Layout of a stacked rectenna. A rectangular patch is placed on top of a grounded dielectric sheet. The ground plane of the antenna acts as the ground plane of a microstrip structure that is placed at the bottom of the antenna. The microstrip patch antenna is fed by a probe. The microstrip structure couples the rectifier to the antenna through an impedance-matching and filtering network. The rectifier output is coupled to a load through a second microstrip filtering network.

Finally, we may omit the filtering network between the diode and the load, provided that the performance of the rectenna system will not be seriously degraded owing to fundamental and higher-order harmonic frequency components. This assumption has to be verified of course.

The improvements in the analysis consist of using a robust, yet sufficiently accurate analytical model for the diode that does not depend on numerical evaluation of an integral [7], and the use of a broadband analytical model for the microstrip patch antenna. These analytical models are the subject of the next section. That section will be followed by a section where these analysis techniques are employed for the design of a compact, efficient microstrip patch rectenna.

5.3 ANALYTICAL MODELS

To conjugately match the microstrip patch antenna to the rectifying circuit, we need to be able to predict the input impedances of both devices. The input impedance of the rectifying circuit needs to be determined at the operating frequency. Once this impedance has been determined, we need to find the dimensions and excitation location of a rectangular microstrip patch antenna that will yield the conjugate value of this impedance at the operating frequency. We shall thus not use the microstrip patch antenna at resonance. As a consequence, we shall need an analytical model for the microstrip patch antenna that results in accurate impedance values over a relatively broad range of frequencies. When the analysis results for the subsystems are accurate to within a few percent, we shall be able to use our analytical models for design purposes.

5.3.1 Model of Rectangular Microstrip Patch antenna

In our recent work on rectennas [8], we applied a cavity model for a rectangular microstrip patch antenna [9, 10]. Although effective lengths and widths were employed to account for the fringe fields [10], the results proved to be too inaccurate for our needs. Therefore, in [8], we used the following method to find better effective lengths and widths.

We start by employing the cavity model of [9, 10] to find the dimensions and feed location that result in the desired (complex) input impedance. Then the dimensions and material characteristics of the microstrip patch antenna are used as input for a full-wave analysis program. The differences between the analysis results from the full-wave analysis program and from our cavity model are attributed to inaccurate length and width extensions in the cavity model. Next, for every radiating mode, the length and width extensions in the cavity model are determined such that the results of the cavity model analysis coincide with those of the full-wave analysis. With the 'corrected' length and width extensions thus found, the cavity model is applied once more to design the rectangular microstrip patch antenna that meets our requirements. As a final check, the structure thus obtained is analyzed once again with the full-wave analysis software and if differences between the results of the cavity model analysis and the full-wave analysis still exist, the procedure is repeated again. In practice, for all situations tested, one full-wave analysis iteration proved to be sufficient.

The design procedure outlined above has the advantage of speeding up the design process by employing both a cavity model and a full-wave analysis program. The fast, dedicated microstrip patch antenna model is employed for the multi-iteration design process, while the slow, general-purpose, full-wave analysis program is employed for a single-iteration – or at most dual-iteration – verification process. Notwithstanding this advantage, the need to employ a full-wave analysis program at all is regarded as undesirable. Ideally, the design should rely completely on fast analytical methods. That would create the possibility of automating the design process, where, through optimization, designs are created within an acceptable amount of time.

An analytical model that accurately predicts the input impedance of a rectangular microstrip patch antenna over a wide band of frequencies was found in [11]. This model is based on the transmission line model of [12] and involves a series combination of transmission lines for different modes. For every mode, equivalent dimensions are introduced.

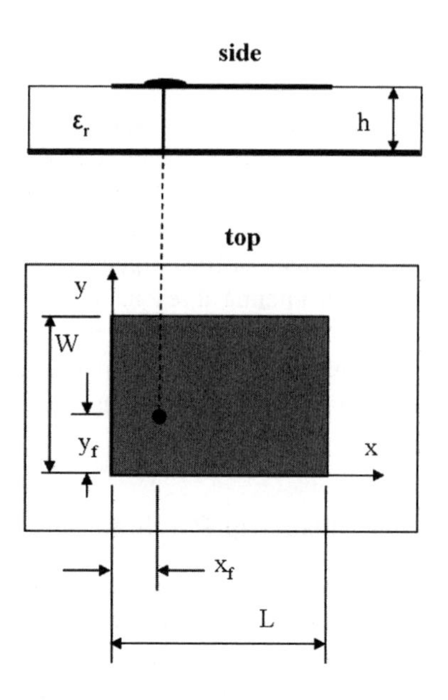

Figure 5.3 Geometry of a rectangular microstrip patch antenna.

Consider a rectangular microstrip patch antenna of length L and width W, as indicated in Figure 5.3. The feed is positioned at $x = x_\mathrm{f}$, $y = y_\mathrm{f}$, where the origin of the coordinate system is positioned at the bottom left corner of the patch. If the feed is a microstrip edge feed, one of the feed coordinates will be zero. The patch is positioned on a grounded dielectric slab of height h and relative permittivity ε_r.

The input impedance of the microstrip patch antenna is given by [11]

$$Z_\mathrm{in} = \left[\frac{1}{Y_{10}} + \frac{1}{Y_{01}} + \frac{1}{Y_{11}} + \frac{1}{Y_{21}} + \frac{1}{Y_{12}}\right]^{-1}, \tag{5.1}$$

where the Y_{mn}, $m, n = 0, 1, 2$, are the TM_{mn} mode admittances.[1] Y_{mn} is given by

$$Y_{mn} = 2Y_0 \frac{Y_0^2 + Y_\mathrm{t}^2 + 2Y_\mathrm{s}Y_0 \coth(\gamma L_\mathrm{eq}) - 2Y_\mathrm{m}Y_0 \mathrm{csch}(\gamma L_\mathrm{eq})}{(Y_0^2 + Y_\mathrm{t}^2)\coth(\gamma L_\mathrm{eq}) + (Y_0^2 - Y_\mathrm{t}^2)\cosh(2\gamma \Delta_\mathrm{eq})\mathrm{csch}(\gamma L_\mathrm{eq}) + 2Y_\mathrm{s}Y_0}, \tag{5.2}$$

where[2]

$$Y_\mathrm{t}^2 = Y_\mathrm{s}^2 - Y_\mathrm{m}^2 \tag{5.3}$$

[1]The TM_{20}, TM_{02} and TM_{22} modes are harmonics of the TM_{10}, TM_{01} and TM_{11} modes, respectively and are already accounted for in the transmission line model [11].

[2]In [11], Y_t^2 was erroneously reproduced from [12]. The plus sign should be a minus sign.

and

$$L_{eq} = \frac{LW}{\sqrt{m^2W^2 + n^2L^2}}, \tag{5.4}$$

$$\Delta_{eq} = \frac{\Delta_{xoff}\Delta_{yoff}}{\sqrt{m^2\Delta_{yoff}^2 + n^2\Delta_{xoff}^2}}, \tag{5.5}$$

$$\Delta_{xoff} = \left|\frac{L}{2} - x_f\right|, \tag{5.6}$$

$$\Delta_{yoff} = \left|\frac{W}{2} - y_f\right|. \tag{5.7}$$

In equation (5.1), Y_0 is the characteristic impedance of a microstrip transmission line of width W, and $\gamma = \alpha + \mathrm{j}\beta$ is the propagation constant. The microstrip patch antenna is regarded as a microstrip transmission line of length L, connecting two radiating slots that operate by virtue of fringe fields at the edges of the patch. Y_s in equation (6.2) is the self-admittance of such a radiating slot, and Y_m is the mutual admittance between the slots.

The characteristic admittance is given by [13]

$$Y_0 = \sqrt{\frac{\varepsilon_0}{\mu_0}}\sqrt{\varepsilon_{r_{eff}}}\frac{W_{eff}}{h}, \tag{5.8}$$

and the phase constant is

$$\beta = \omega\sqrt{\varepsilon_0\mu_0}\sqrt{\varepsilon_{r_{eff}}}. \tag{5.9}$$

To calculate the effective width as a function of frequency, W_{eff}, and the effective relative permittivity as a function of frequency, $\varepsilon_{r_{eff}}$, we first need to calculate the static effective width $W_{eff}(0)$ and static effective relative permittivity $\varepsilon_{r_{eff}}(0)$. These static parameters have been calculated as follows [13]:

$$W_{eff}(0) = \frac{2\pi h}{\ln\{hF/W' + \sqrt{1 + (2h/W')^2}\}}, \tag{5.10}$$

where

$$F = 6 + (2\pi - 6)\mathrm{e}^{-(4\pi^2/3)(h/W')^{3/4}} \tag{5.11}$$

and

$$W' = W + \frac{th}{\pi}\left\{1 + \ln\frac{4}{\sqrt{(t/h)^2 + \frac{(1/\pi)^2}{(W/t+1.1)^2}}}\right\}. \tag{5.12}$$

In the last equation, th is the thickness of the metal patch. Also,

$$\varepsilon_{r_{eff}}(0) = \frac{1}{2}\{\varepsilon_r + 1 + (\varepsilon_r - 1)G\}, \tag{5.13}$$

where

$$G = \left(1 + \frac{10h}{W}\right)^{-AB} - \frac{\ln(4)}{\pi}\frac{t}{\sqrt{Wh}} \tag{5.14}$$

and

$$A = 1 + \frac{1}{49}\ln\left\{\frac{(W/h)^4 + (W/52h)^2}{(W/h)^4 + 0.432}\right\} + \frac{1}{18.7}\ln\left\{1 + \left(\frac{W}{18.1h}\right)^3\right\}, \tag{5.15}$$

$$B = 0.564\mathrm{e}^{-0.2/(\varepsilon_\mathrm{r}+0.3)}. \tag{5.16}$$

The frequency-dependent effective relative permittivity is then calculated as [13]

$$\varepsilon_{\mathrm{r_{eff}}}(f) = \varepsilon_\mathrm{r} - \frac{\varepsilon_\mathrm{r} - \varepsilon_{\mathrm{r_{eff}}}(0)}{1 + P}, \tag{5.17}$$

where

$$P = P_1 P_2\{(0.1844 + P_3 P_4) f_\mathrm{n}\}^{1.5763}, \tag{5.18}$$

$$P_1 = 0.27488 + \left\{0.6315 + \frac{0.525}{(1 + 0.0157 f_\mathrm{n})^{20}}\right\}u - 0.065683\mathrm{e}^{-8.7513u},$$

$$P_2 = 0.33622\{1 - \mathrm{e}^{-0.03442\varepsilon_\mathrm{r}}\},$$

$$P_3 = 0.0363\mathrm{e}^{-4.6u}\{1 - \mathrm{e}^{-(f_\mathrm{n}/38.7)^{4.97}}\},$$

$$P_4 = 1 + 2.751\{1 - \mathrm{e}^{-(\varepsilon_\mathrm{r}/15.916)^8}\},$$

$$f_\mathrm{n} = fh \times 10^{-6},$$

$$u = \frac{W + (W' - W)/\varepsilon_\mathrm{r}}{h}. \tag{5.19}$$

The frequency-dependent effective width is calculated as [13]

$$W_\mathrm{eff}(f) = \frac{W}{3} + (R_W + P_W)^{1/3} - (R_W - P_W)^{1/3}, \tag{5.20}$$

where

$$P_W = \left(\frac{W}{3}\right)^3 + \frac{S_W}{2}\left[W_\mathrm{eff}(0) - \frac{W}{3}\right],$$

$$Q_W = \frac{S_W}{3} - \left(\frac{W}{3}\right)^2,$$

$$R_W = \sqrt{P_W^2 + Q_W^3},$$

$$S_W = \frac{c_0^2}{4f^2[\varepsilon_\mathrm{eff}(f) - 1]}, \tag{5.21}$$

c_0 being the speed of light in free space.

The self-admittance of the radiating slot is given by [12, 13]

$$Y_\mathrm{s} = G_\mathrm{s} + \mathrm{j}B_\mathrm{s}, \tag{5.22}$$

where

$$G_s = \frac{1}{\pi\sqrt{\mu_0/\varepsilon_0}}\left\{\left(wSi(w) + \frac{\sin(w)}{w} + \cos(w) - 2\right)\left(1 - \frac{s^2}{24}\right) + \frac{s^2}{12}\left(\frac{1}{3} + \frac{\cos(w)}{w^2} - \frac{\sin(w)}{w^3}\right)\right\} \tag{5.23}$$

and

$$B_s = Y_0 \tan(\beta \Delta l). \tag{5.24}$$

In the above equations, $w = \omega\sqrt{\varepsilon_0\mu_0}W_{\text{eff}}$, and Δl is the width of the radiating slot. Δl is given by

$$\Delta l = h\frac{\xi_1\xi_3\xi_5}{\xi_4}, \tag{5.25}$$

where

$$\begin{aligned}
\xi_1 &= 0.434907\frac{\varepsilon_{r_{\text{eff}}}^{0.81} + 0.26}{\varepsilon_{r_{\text{eff}}}^{0.81} - 0.189}\frac{(W/h)^{0.8544} + 0.236}{(W/h)^{0.8544} + 0.87},\\
\xi_2 &= 1 + \frac{(W/h)^{0.371}}{2.358\varepsilon_r + 1},\\
\xi_3 &= 1 + \frac{0.5274\arctan\{0.084(W/h)^{1.9413/\xi_2}\}}{\varepsilon_{r_{\text{eff}}}^{0.9236}},\\
\xi_4 &= 1 + 0.0377\arctan\left\{0.067\left(\frac{W}{h}\right)^{1.456}\right\}\{6 - 5e^{0.036(1-\varepsilon_r)}\},\\
\xi_5 &= 1 - 0.218e^{-7.5W/h}.
\end{aligned} \tag{5.26}$$

The mutual admittance between the radiating slots is given by [12, 13]

$$Y_m = G_m + jB_m, \tag{5.27}$$

where

$$\begin{aligned}
G_m &= G_s F_g K_g,\\
B_m &= B_s F_s K_b
\end{aligned} \tag{5.28}$$

and

$$\begin{aligned}
F_g &= J_0(l) + \frac{s^2}{24 - s^2}J_2(l),\\
F_b &= \frac{\pi}{2}\frac{Y_0(l) + (s^2/(24 - s^2))Y_2(l)}{\ln(s/2) + C_e - 3/2 + (s^2/12)/(24 - s^2)},
\end{aligned} \tag{5.29}$$

$$\begin{aligned}
K_g &= 1,\\
K_b &= 1 - e^{-0.21w}.
\end{aligned} \tag{5.30}$$

In the above, $s = \omega\sqrt{\varepsilon_0\mu_0}\Delta l$, $l = \omega\sqrt{\varepsilon_0\mu_0}(L + \Delta l)$ and C_e is Euler's constant, equal to 0.57216... $J_i(x)$ is the ith-order Bessel function of the first kind, and $Y_i(x)$ is the ith-order Bessel function of the second kind.

Finally, the attenuation constant α is given by

$$\alpha = \alpha_d + \alpha_{cs} + \alpha_{cg}, \tag{5.31}$$

where

$$\begin{aligned}
\alpha_d &= 0.5\beta\frac{\varepsilon_r}{\varepsilon_{r_{eff}}(f)}\frac{\varepsilon_{r_{eff}}(f) - 1}{\varepsilon_r - 1}\tan\delta,\\
\alpha_{cs} &= \alpha_n R_{ss} F_{\Delta_s} F_s,\\
\alpha_{cg} &= \alpha_n R_{sg} F_{\Delta_g}.
\end{aligned} \tag{5.32}$$

In the above, $\tan\delta$ is the loss tangent of the dielectric. Further,

$$\begin{aligned}
R_{ss} &= \sqrt{\frac{\pi f \mu_0}{\sigma_s}},\\
R_{sg} &= \sqrt{\frac{\pi f \mu_0}{\sigma_g}},
\end{aligned} \tag{5.33}$$

where σ_s and σ_g are the conductance of the patch and of the ground plane, respectively. In addition,

$$\begin{aligned}
\alpha_n &= \begin{cases} \dfrac{Y_0}{4\pi h}\dfrac{32 - (W'/h)^2}{32 + (W'/h)^2} & \text{for } \dfrac{W'}{h} < 1 \\[2ex] \dfrac{\sqrt{\varepsilon_{r_{eff}}(0)}}{2\sqrt{(\mu_0/\varepsilon_0)}W_{eff}(0)}\left(\dfrac{W'}{h} + \dfrac{0.667W'/h}{W'/h + 1.444}\right) & \text{for } \dfrac{W'}{h} \geq 1, \end{cases}\\
F_{\Delta_s} &= 1 + \frac{2}{\pi}\arctan\{1.4(R_{ss}\Delta_s\sigma_s)^2\},\\
F_{\Delta_g} &= 1 + \frac{2}{\pi}\arctan\{1.4(R_{sg}\Delta_g\sigma_g)^2\},\\
F_s &= 1 + \frac{2h}{W'}\left(1 - \frac{1}{\pi} + \frac{W' - W}{t}\right),
\end{aligned} \tag{5.34}$$

where Δ_s and Δ_g are the rms surface roughness of the patch and of the ground plane, respectively.

To demonstrate the accuracy of the 'modified' cavity model (i.e. using a single full-wave iteration to correct the length extension of the cavity model) and the transmission line model of [11], we calculated the input impedance of a dual-resonant microstrip patch antenna. For such an antenna, the 'standard' cavity model will fail to accurately predict the input impedance for frequencies below and above the first resonance frequency [14].

The parameters used for the microstrip patch antenna were $L = 30.8$ mm, $W = 27.7$ mm, $h = 1.6$ mm, $\varepsilon_r = 4.28$, $\tan\delta = 0.016$, $x_f = 0$, $y_f = 0.4$ mm, $th = 0.070$ mm,

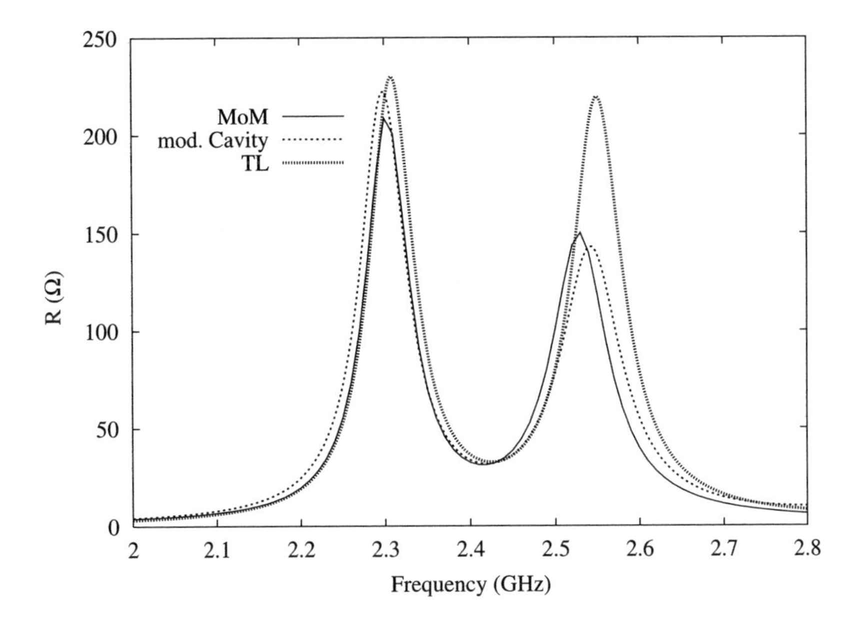

Figure 5.4 Real part of the input impedance versus frequency, calculated with three different models for a dual-resonant rectangular microstrip patch antenna.

$\sigma_s = \sigma_g = 0.556 \times 10^8$ S m^{-1} and $\Delta_s = \Delta_g = 0.0015$ mm. Figure 5.4 shows the real part of the input impedance as a function of frequency, calculated with the method of moments (MoM), with the modified cavity model and with the transmission line model. Figure 5.5 shows the calculated results for the imaginary part of the input impedance.

In comparing the calculated input impedance results, we shall take the MoM analysis results as a reference. Over recent decades, commercial off-the-shelf (COTS) MoM analysis software has become very reliable, and its results in general agree very well with measurements performed on real structures such as that under investigation here.

Figures 5.4 and 5.5 show that both for the modified cavity model and for the transmission line model, the real part of the input impedance is within a few percent of the reference value around 2.45 GHz. The imaginary part of the input impedance differs by up to 50% of the reference value around this frequency. Notwithstanding this large difference in the imaginary part of the input impedance, the power-wave reflection coefficient still is of the order of 0.20, an acceptable value for design purposes. Both the modified cavity model and the transmission line model thus generate results that are in good agreement with the full-wave analysis results. The modified cavity model is more accurate but has the drawback that it relies on a full-wave analysis program. The modified cavity model and the transmission line model have comparable accuracy in the region of interest (i.e. off resonance).

5.3.2 Model of Rectifying Circuit

The main component of a rectifying circuit is a diode. We may use a single diode in the rectifying circuit or use two diodes in a voltage doubler configuration [5]. The diode or diodes

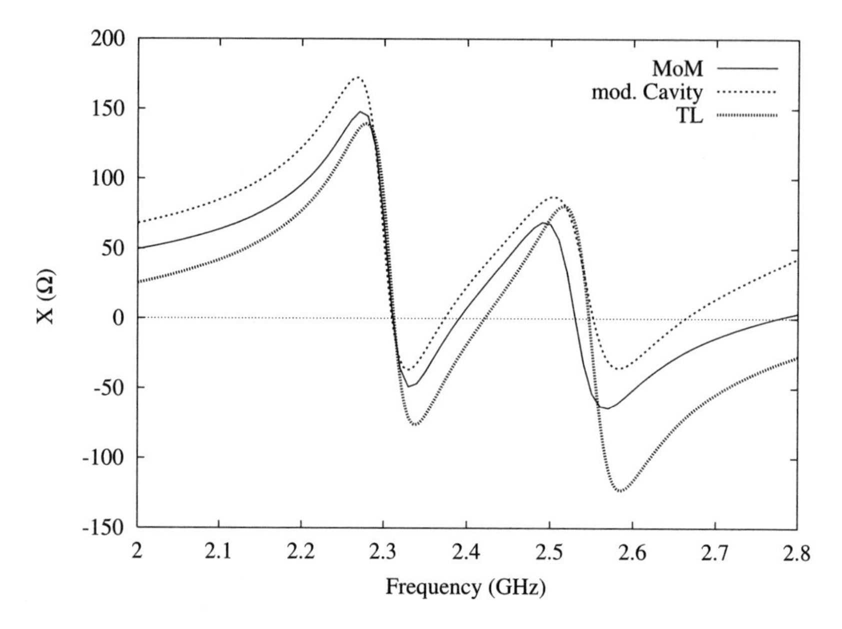

Figure 5.5 Imaginary part of the input impedance versus frequency, calculated with three different models for a dual-resonant rectangular microstrip patch antenna.

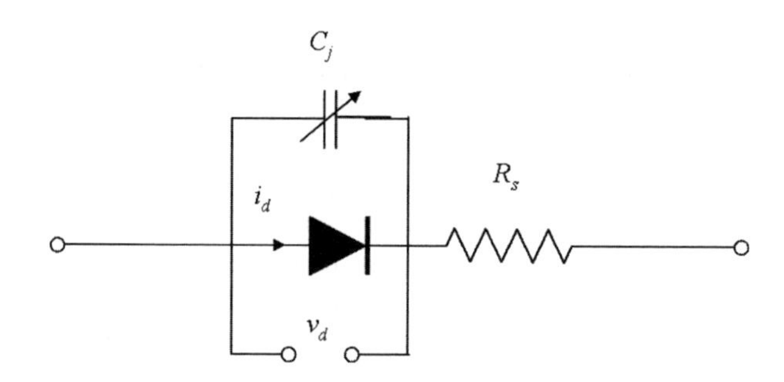

Figure 5.6 Equivalent electric circuit of a diode.

are used unbiased, since we have no power available other than the RF power delivered by the antenna. The diode considered here is of the Schottky type, because of its low forward bias voltage compared with other diode types. The diode can be represented by the equivalent electric circuit shown in Figure 5.6.

A Schottky diode consists of a semiconductor substrate attached to a metal contact. In the electric circuit in Figure 5.6, the resistance R_s represents the conduction losses in the substrate. The depletion layer in the substrate forms an insulating barrier between the two diode contacts and thus forms a junction capacitance C_j. The value of C_j depends on the

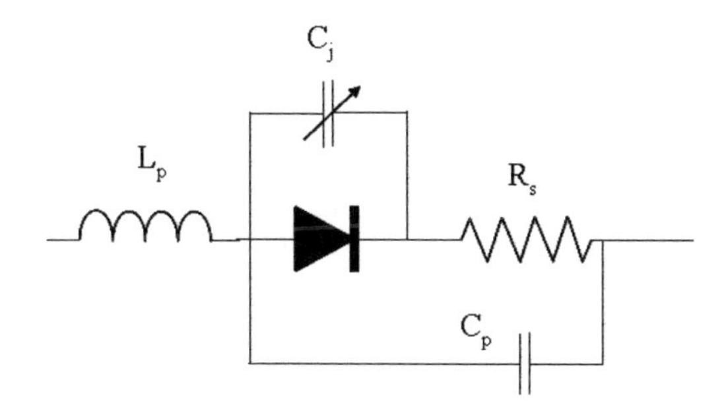

Figure 5.7 Equivalent electric circuit of a packaged diode.

applied voltage v_d according to [15]

$$C_j(v_d) = \frac{C_{j0}}{\sqrt{1 - v_d/\phi}}, \tag{5.35}$$

where v_d is the voltage across the (bare) diode (Figure 5.6), C_{j0} is the zero-bias differential barrier capacitance and ϕ is the barrier potential of the diode. Values for the latter two parameters may be found in a diode's data sheet. The nonlinear voltage–current characteristic of the diode in Figure 5.6 is given by

$$i_d = I_s(e^{(q/nkT)v_d} - 1), \tag{5.36}$$

where i_d is the current through the (nonlinear) diode, I_s is the saturation current, q is the electron charge, k is Boltzmann's constant, equal to $1.3806504 \times 10^{-23}$ J K^{-1}, T is the temperature in kelvin and n is the ideality factor of the diode. The ideality factor, a number close to one for most diodes, may be found in a diode's data sheet as well.

The above applies to the bare diode. In our rectennas, we have made use of packaged diodes. The equivalent electric circuit of a packaged diode is shown in Figure 5.7. In this circuit, we see an additional parasitic capacitance C_p and a parasitic inductance L_p, the values of which may be found in the diode's data sheet.

Several methods exist for analyzing the nonlinear behavior of a diode. Linearization of the voltage–current characteristic around a working point will not suffice for our purposes. We shall be using the diode in the large-signal region, meaning that the nonlinear behavior will show in the working region and a linearization therefore cannot be performed unambiguously. In a 'harmonic balance' analysis, the circuit is separated into a linear and a nonlinear subcircuit. The linear subcircuit is analyzed in the frequency domain, while the nonlinear subcircuit is analyzed in the time domain, after which the results of the latter analysis are Fourier transformed to the frequency domain. A solution is found if the currents through the interconnections between the linear and nonlinear subcircuits are the same for both subcircuits. The currents in the linear and nonlinear subcircuits through these interconnections have to be *balanced* at every *harmonic*. Application of an efficient harmonic balance method for analyzing a non-linear diode has proven to generate accurate results [7]. A drawback of this method is that for more than one higher harmonic, closed-form analytical solutions are no longer available.

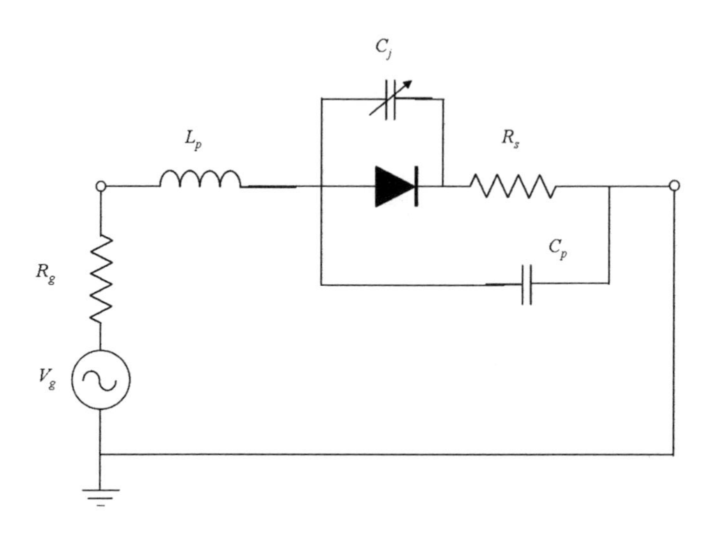

Figure 5.8 Circuit for determining the impedance of a packaged diode.

For this reason, we chose to use a 'time-marching' algorithm, which is robust and, for our purposes, where only *one* nonlinear element needs to be analyzed, still time-efficient. In this method, the system is first discretized into a set of nonlinear first-order ordinary differential equations. A starting condition is chosen and a (time-domain) input signal is impressed on the system. An integral over the differential equations, using a small time step, is then applied to determine the system state after this small time step, i.e.

$$y_{n+1} = y_n + \Delta t \cdot f(t_n, y_n), \tag{5.37}$$

where y_n is the system state at time $t = t_n$, y_{n+1} is the system state at $t = t_{n+1}$ and the function $f(t_n, y_n)$ represents the derivative of y with respect to t at $t = t_n$. Δt is the time step, i.e. $\Delta t = t_{n+1} - t_n$.

To determine the impedance of the diode, the packaged diode is embedded in the circuit shown in Figure 5.8. In this figure, R_g is the generator impedance, and the source voltage V_g is given by

$$V_g = |V_g| \cos(\omega t). \tag{5.38}$$

The voltages and currents in the circuit shown in Figure 5.8 are described by the following circuit equations:

$$\begin{aligned}
V_g &= I_g R_g + L_p \frac{dI_g}{dt} + V_{C_p}, \\
V_{C_p} &= V_d + V_{R_s}, \\
V_{R_s} &= R_s I_{R_s} = (I_{C_j} + I_d) R_s,
\end{aligned}$$

$$
\begin{aligned}
I_{C_j} &= C_j \frac{dV_\mathrm{d}}{dt}, \\
I_\mathrm{d} &= I_\mathrm{s}(\mathrm{e}^{(q/nkT)V_\mathrm{d}} - 1), \\
I_\mathrm{g} &= I_\mathrm{d} + I_{C_j} + I_{C_\mathrm{p}}, \\
I_{C_\mathrm{p}} &= C_\mathrm{p} \frac{dV_{C_\mathrm{p}}}{dt}.
\end{aligned}
\tag{5.39}
$$

From these equations, we easily obtain the following set of coupled first-order ordinary differential equations:

$$
\begin{aligned}
V_\mathrm{g} &= L_\mathrm{p}\frac{dI_\mathrm{g}}{dt} + V_\mathrm{d} + R_\mathrm{g} I_\mathrm{g} + R_\mathrm{s} C_j \frac{dV_\mathrm{d}}{dt} + R_\mathrm{s} I_\mathrm{s}(\mathrm{e}^{(q/nkT)V_\mathrm{d}} - 1), \\
I_\mathrm{g} &= I_\mathrm{s}(\mathrm{e}^{(q/nkT)V_\mathrm{d}} - 1) + C_j \frac{dV_\mathrm{d}}{dt} + C_\mathrm{p}\frac{dV_\mathrm{g}}{dt} - C_\mathrm{p} R_\mathrm{g}\frac{dI_\mathrm{g}}{dt} - C_\mathrm{p} L_\mathrm{p} \frac{dI_\mathrm{X}}{dt},
\end{aligned}
\tag{5.40}
$$

where

$$
I_\mathrm{X} = \frac{dI_\mathrm{g}}{dt}
\tag{5.41}
$$

has been introduced to avoid the occurrence of second-order derivatives. V_d and I_g are the unknowns. The first step in solving an ordinary differential equation usually consists of rewriting the equation as a set of coupled *first-order* differential equations [16]. Having accomplished this, in the next step we rewrite the set of coupled first-order ordinary differential equations in a form wherein every unknown derivative is expressed in terms of known parameters:

$$
\begin{aligned}
\frac{dV_\mathrm{d}}{dt} &= \frac{1}{R_\mathrm{s} C_j}[V_\mathrm{g} - R_\mathrm{g} I_\mathrm{g} - V_\mathrm{d} - L_\mathrm{p} I_\mathrm{X} - R_\mathrm{s} I_\mathrm{s}(\mathrm{e}^{(q/nkT)V_\mathrm{d}} - 1)], \\
\frac{dI_\mathrm{X}}{dt} &= \frac{1}{C_\mathrm{p} L_\mathrm{p}}[I_\mathrm{s}(\mathrm{e}^{(q/nkT)V_\mathrm{d}} - 1) + C_j\frac{dV_\mathrm{d}}{dt} + C_\mathrm{p}\frac{dV_\mathrm{g}}{dt} - C_\mathrm{p} R_\mathrm{g} I_\mathrm{X} - I_\mathrm{g}], \\
\frac{dI_\mathrm{g}}{dt} &= I_\mathrm{X}.
\end{aligned}
\tag{5.42}
$$

The system parameters V_d, I_g and I_X are known at $t = t_0$, as is dV_g/dt, the time derivative of the impressed generator voltage (at $t = t_0$, we choose $V_\mathrm{d} = I_\mathrm{g} = I_\mathrm{X} = dV_\mathrm{g}/dt = 0$). A fourth-order Runge–Kutta method (RK4) [16] was applied to step through time. To make the method robust, an adaptive-step-size algorithm was applied.

Time-stepping results are shown in Figure 5.9 for an Agilent HSMS-2852 Schottky diode [3]. In this figure, the diode voltage V_d is shown, together with the generator voltage V_g. The frequency of the generator was 2.45 GHz and the input power was 0 dBm.

By applying RK4, we may find the diode junction voltage and the generator current and, from equation (6.40), the diode current. After applying a fast Fourier transform (FFT) to the diode voltage and current, the diode impedance is given, for each harmonic of the working frequency, by

$$
Z_{\mathrm{d_f}} = \frac{V_{\mathrm{d_f}}}{I_{\mathrm{d_f}}}.
\tag{5.43}
$$

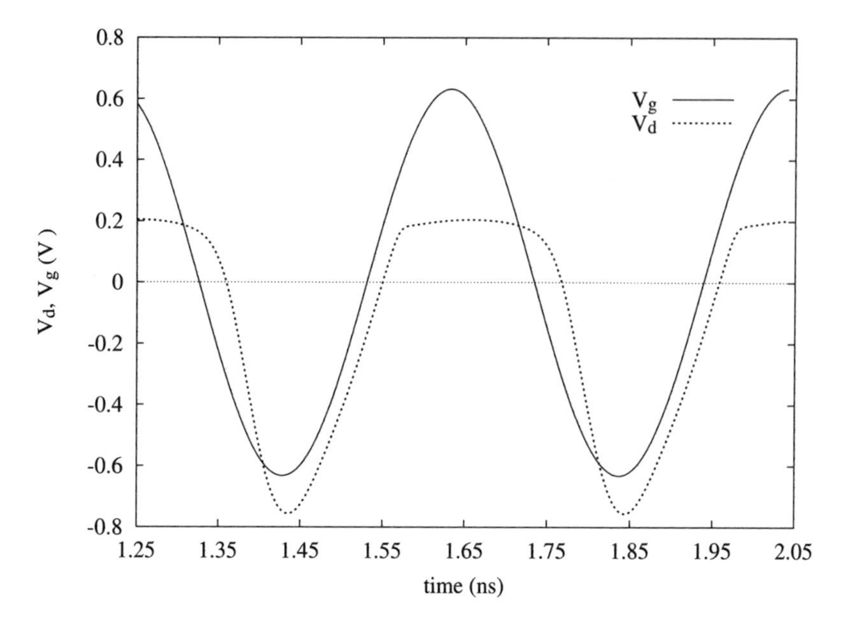

Figure 5.9 RK4 simulation results for an Agilent HSMS-2852 Schottky diode.

The incident power follows from the applied generator voltage and the generator impedance:

$$P_{\text{inc}} = \frac{|V_{\text{g}}|^2}{8R_{\text{g}}}. \tag{5.44}$$

We now have an efficient analytical time-domain method to determine the voltages and currents in a packaged diode circuit, which – after Fourier transformation – gives us the diode circuit's input impedance as a function of frequency. With an accurate transmission line model, we may determine the complex input impedance of a rectangular microstrip patch antenna as a function of frequency for any position of the excitation of the patch. So, we have the tools for, first, determining the rectifying circuit's input impedance and, second, determining the feed position of the microstrip patch for creating a conjugate impedance match to this circuit. Before we make our first design, we shall first verify the rectifier circuit model.

5.4 MODEL VERIFICATION

We have already used full-wave analysis results for a microstrip patch antenna, using (proven) COTS software, to verify the analytical models that we have developed. For the diode, the analytical models were verified by comparison with measurement results at the operating frequency and by comparison with results of COTS harmonic balance simulation for higher harmonics.

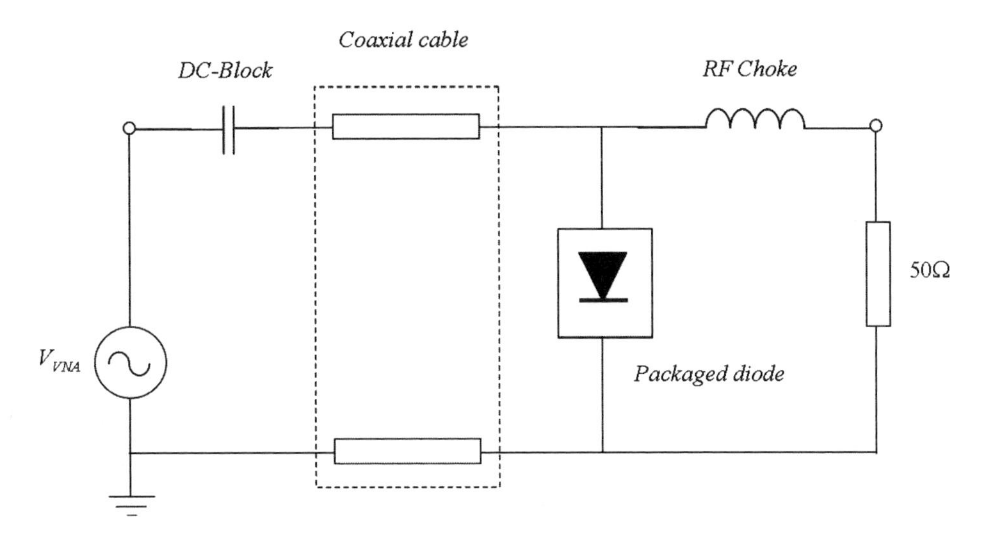

Figure 5.10 Setup for measurement of the impedance of a diode.

First, the impedance of an Agilent HSMS-2852 Schottky diode was calculated and measured, at operational frequencies in the range from 100 MHz to 4.1 GHz and for different input power levels. To that end, a packaged diode was mounted directly on a subminiature version-A (SMA) connector. This SMA connector was connected to the input port of a vector network analyzer (VNA) that was equipped with a DC block and a 50 Ω load at the DC output port, of the VNA. The measurement setup is shown in Figure 5.10.

Bearing in mind that the VNA is calibrated so that the reference plane is at the input of the packaged diode and that the RF choke isolates the DC from the AC signal, the measurement setup of Figure 5.10 is identical to that shown in Figure 5.8. Therefore, the calculated results may be compared directly with the measured results.

However, the HSMS-2852 in fact consists of two diodes, sharing a common pin of the three-pin SOT-23 package [17], where the anode of one diode and the cathode of the other are connected. When one diode is measured, the other diode, which is connected on one side and floats at the other side, will give rise to a capacitive coupling. The value of the coupled capacitor was estimated to be 0.3 pF. The value of the parasitic capacitance of a (single) diode was $C_p = 0.08$ pF. The other parameters were $C_{j0} = 0.18$ pF, $L_p = 2$ nH, $R_s = 25\ \Omega$, $I_s = 3\ \mu$A, $n = 1.06$ and $\phi = 0.35$ V [17].

The input impedance obtained from the measured reflection coefficient, corrected for the 0.3 pF capacitive coupling, is shown in Figures 5.11 and 5.12 together with the calculated results for an input power of 0 dBm. Figure 5.11 shows the real part of the input impedance, and figure 5.12 shows the imaginary part. Other input powers in the range from -10 dBm to $+10$ dBm show similar results.

The 'noise' that is visible in the measurement data can be explained by the fact that the VNA was not able to fully suppress the higher harmonics at its input. Nevertheless, the figures show that with the RK4 analysis, we are able to predict the input impedance of the packaged diode to within a few percent at frequencies around 2.45 GHz and higher.

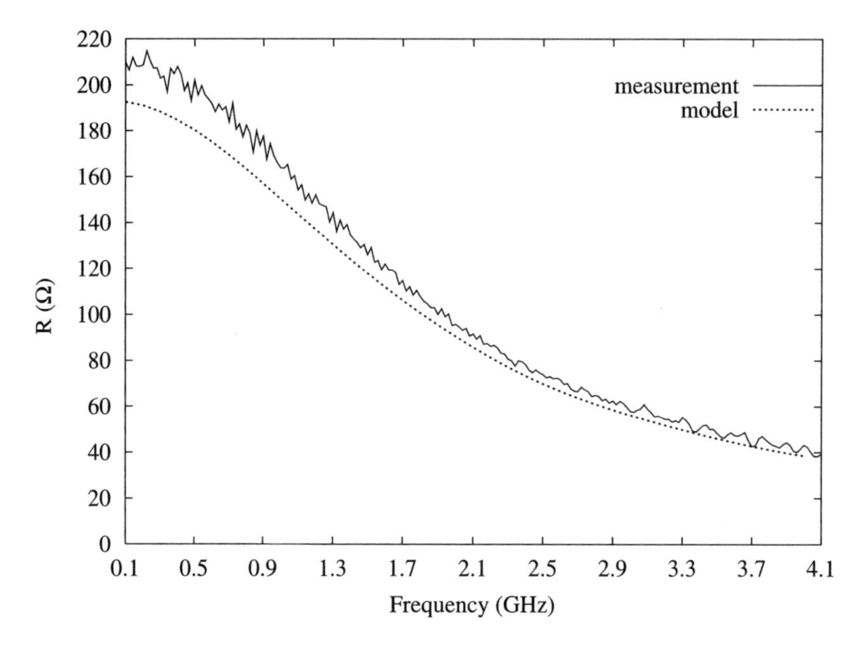

Figure 5.11 Real part of measured and simulated input impedance versus frequency for an Agilent HSMS-2852 Schottky diode, for 0 dBm input power.

Since we did not have access to equipment for measuring the input impedance of the diode at higher harmonics, we compared the results calculated with RK4 with results obtained from AWR's harmonic balance analysis program APLAC®.

In Figure 5.13, the voltage V_d is shown for the harmonic frequencies, as calculated by our RK4 method and as obtained from a harmonic balance analysis for a 0 dBm input signal at 2.45 GHz. We see, for the higher harmonics also, results that are within a few percent of the expected values.

For the problem at hand, which results in a limited set of nonlinear ordinary differential equations, the Runge–Kutta method proves to be computationally more efficient (by one order of magnitude) than the harmonic balance method, while yielding results of comparable accuracy.

Now that we have verified the models of the rectangular microstrip patch antenna and of the diode, we have completed developing the tools for designing a wireless battery.

5.5 WIRELESS BATTERY

The analysis tools described above were used to design a single rectenna, or wireless battery. After realization and characterization, this wireless battery was applied in cascade to power an electric wall clock.

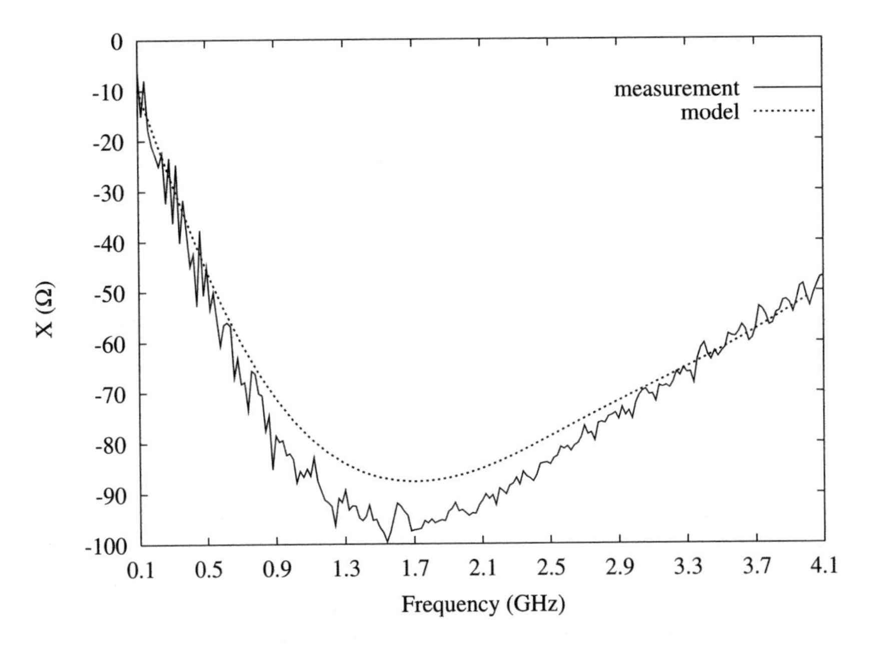

Figure 5.12 Imaginary part of measured and simulated input impedance versus frequency for an Agilent HSMS-2852 Schottky diode, for 0 dBm input power.

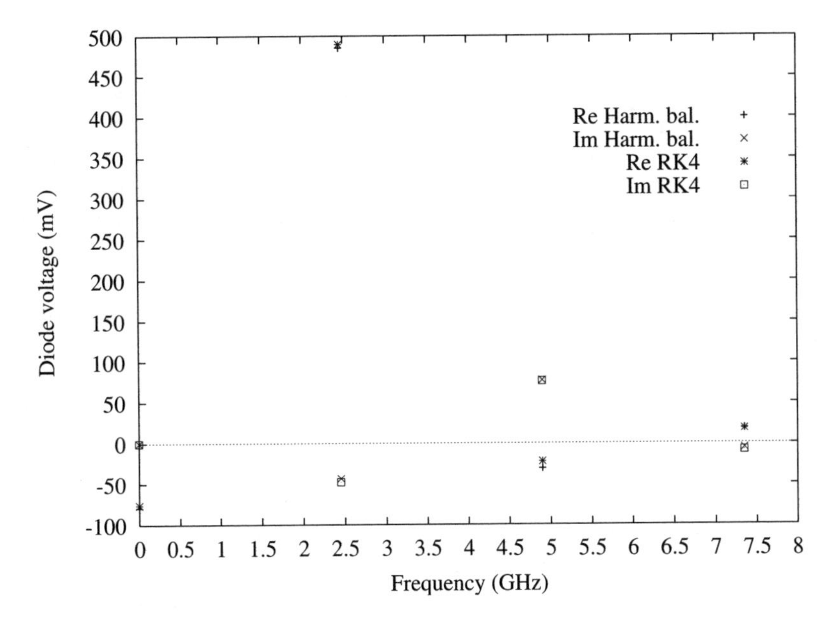

Figure 5.13 Real and imaginary parts of the diode voltage as calculated by RK4 and harmonic balance for a 0 dBm input signal at 2.45 GHz for a packaged Agilent HSMS-2852 diode.

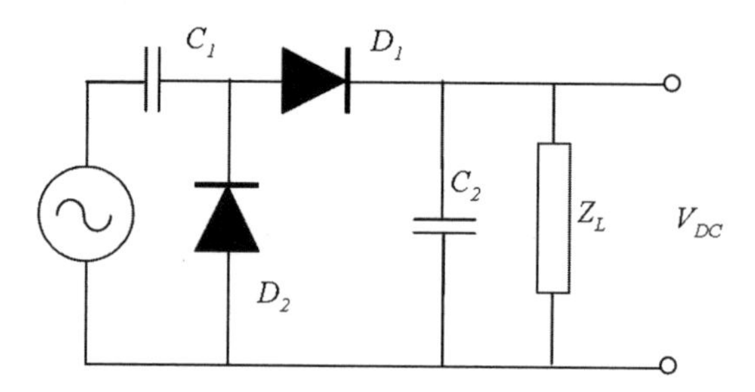

Figure 5.14 Voltage doubler diode rectifying circuit. C_1 aids in creating a DC voltage across D_2. In the rectenna, the microstrip patch antenna serves this purpose.

5.5.1 Single Rectenna

In the single rectenna designed, we made use of two diodes in a voltage doubler configuration (Figure 5.14) [15]. From Figure 5.14, we may observe that at microwave frequencies, the capacitors will behave as short circuits and, consequently, the input impedance of the rectifying circuit will be that of an antiparallel pair of diodes. Hence, the impedance will be half of that of a single diode. For DC, the capacitor will behave as an open circuit and the two diodes will now act as sources that are connected in series. The output voltage is therefore doubled in comparison with a single-diode rectifying circuit.

The design of a wireless battery starts by deciding on the operating frequency, in our case 2.45 GHz. Next, the input impedance of the diode to be used – alone or in a voltage doubler configuration – is determined for a chosen input power level, and the DC load impedance is chosen. The input power level was chosen to be 0 dBm, a value considered to be representative for the power level received by a microstrip patch antenna in an environment containing one or more ISM radio sources. With the aid of Figures 5.11 and 5.12, the input impedance of an Agilent HSMS-2852 diode under these conditions is found to be (80 – j90) Ω. The RF input impedance for a voltage doubler configuration thus equals (40 – j45) Ω.

Next, a rectangular microstrip patch antenna with a microstrip edge feed was designed for an input impedance of (40 + j45) Ω at 2.45 GHz. The real and imaginary parts of the input impedance as a function of frequency are shown in Figures 5.4 and 5.5. The values show that the dimensions of the rectangular microstrip patch antenna were such that a conjugate match with the voltage-doubling circuit had been realized. Thus the need for an impedance-matching network was circumvented. Also, a filtering structure for preventing the reradiation of higher harmonics was not necessary, since the microstrip patch antenna was mismatched to these higher harmonic signals. Finally, to suppress the presence of the fundamental frequency (and higher harmonics) in the output signal, it is common practice to use a microstrip-line stub filter between the rectifying circuit and the load (Figure 5.15(a)). Since the capacitor C in the voltage doubler circuit (Figure 5.14), takes over this role, this filtering network could be omitted here. Measurements showed that this hardly degrades the performance of the system.

So, finally, with the structure shown in Figure 5.15(b), a wireless battery that was hardly larger than the microstrip patch antenna was realized.

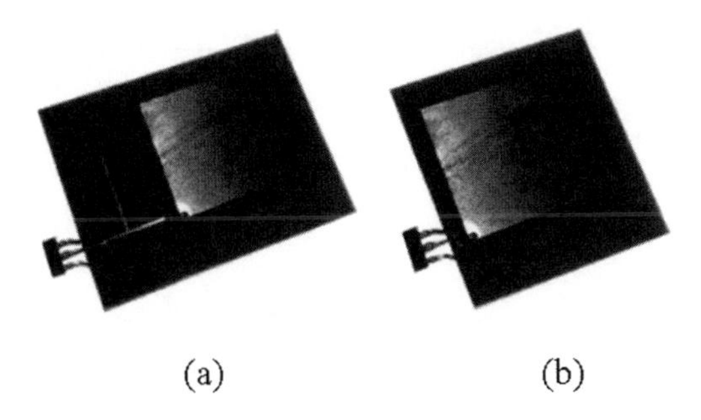

(a) (b)

Figure 5.15 Single-layer, grounded PCB rectenna consisting of a microstrip patch antenna, a diode voltage doubler and an SMD capacitor, connected directly to the edge of the patch. (a) With filtering structure, (b) without filtering structure.

5.5.2 Characterization of Rectenna

To verify the rectenna design, the unloaded output voltage as a function of the input power was calculated and measured. Since the input power coupled into the rectifying circuit was not easily accessible, a relative measurement was performed. A power-adjustable transmitter was connected to a directive antenna. The rectenna was placed at a fixed distance, and the unloaded output voltage was measured as a function of the transmitter power. For one transmitter power level, the input power to the rectifying circuit that would result in the observed unloaded output voltage was calculated. For all other transmitter power levels, the corresponding input power levels were scaled relative to this calculated value (i.e. if the transmitter power level increases by 10 dB, the input power to the rectifying circuit increases by the same amount). The results at 2.45 GHz calculated and measured in this way are shown in Figure 5.16.

First of all, the excellent agreement between the measured and calculated output voltages – the differences are within a few percent – demonstrates the validity of the design tools and of the design itself. Furthermore, it shows that for a high-impedance load, relatively high voltages may be obtained for input powers at or below 0 dBm.

To determine the efficiency of the rectenna, the distance between the transmitter and rectenna was first determined for 0 dBm input power. The Friis transmission equation was used to this end, with the gain of the transmitter antenna and the gain of the rectangular microstrip antenna as input. Then the (resistive) load was varied to maximize the dissipated power. The maximum dissipated power turned out to be 516 μW for a 900 Ω load. The efficiency η of the rectenna was then given by

$$\eta = \frac{P_{\mathrm{DC}}}{P_{\mathrm{inc}}} \times 100\% = \frac{516}{1000} \times 100\% = 52\%, \tag{5.45}$$

where P_{inc}, the input power, is 1 mW. The efficiency of this rectenna has improved by more than 10% compared with a rectenna with a filtering structure [7] evaluated at the same input power level, and the structure has become more compact.

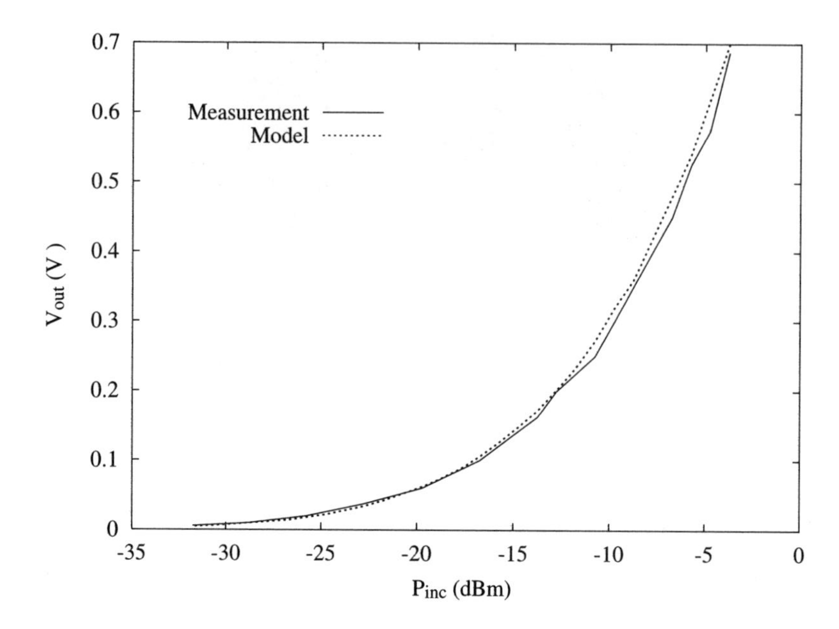

Figure 5.16 Unloaded voltage doubler output voltage as a function of input power.

5.5.3 Cascaded Rectennas

Next, it was decided to use the wireless battery in a practical application. As may be concluded from Figure 5.16, the output voltage of the wireless battery decreases with the received power and thus with distance from the RF source. By placing several batteries in series, an acceptable output voltage can be obtained at an acceptable distance from the RF source. Then, to aid further in applicability, it was decided to use a series connection of rectennas to power a low-power, low-duty-cycle application. Here, power is collected by the wireless batteries and the application employs the collected power only at discrete time intervals. Such an application was found in a common household electric wall clock, requiring 1.2 V continuously and a very small current every second. The clock was powered by eight wireless batteries connected in series, as shown in Figure 5.17.

The use of a 20 dBm transmitter, connected to a 5 dB gain horn antenna, positioned at the focal point of a 51 cm diameter parabolic reflector, enabled us to power this clock at a distance of up to 6 m.

5.6 POWER AND DATA TRANSFER

In the preceding sections, we have discussed the use of an antenna to collect RF power, and converting this RF power into DC power. In doing so, the antenna is no longer available for telecommunication purposes, i.e. the transfer of data. Instead of using two separate systems,

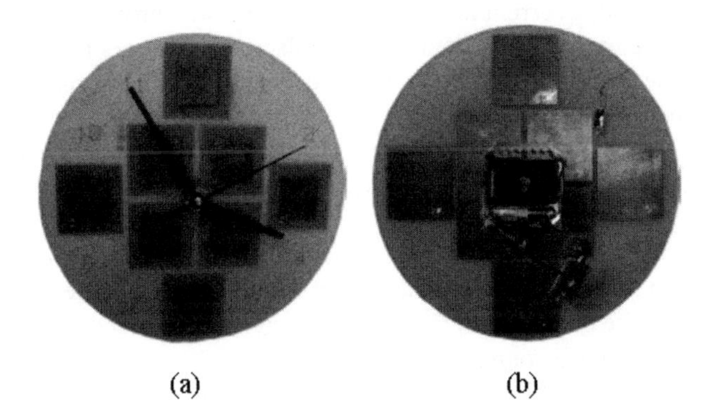

Figure 5.17 Wirelessly powered electric wall clock; (a) front view, (b) back view. The additional circuit shown in (b) is a voltage protection circuit.

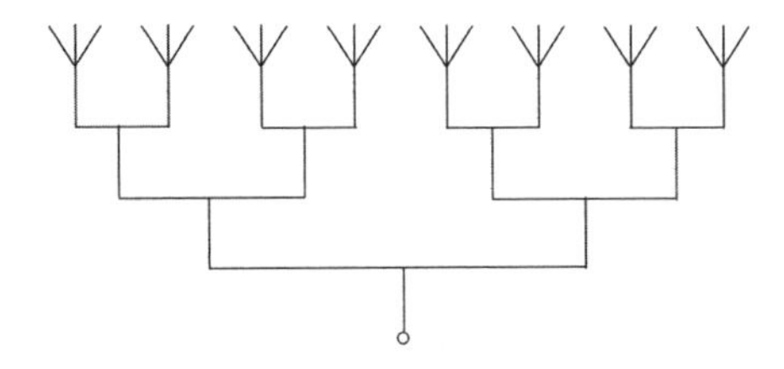

Figure 5.18 Linear array antenna with parallel (corporate) feeding network.

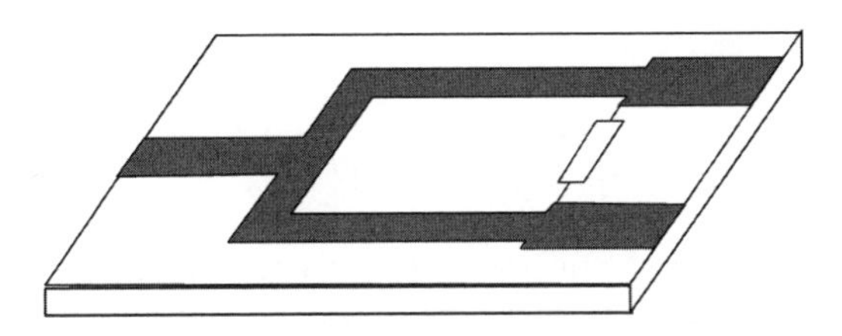

Figure 5.19 Microstrip Wilkinson power combiner.

each with their own antennas, we would prefer to have a system where the same antennas are used for both power and data transfer.

If array antennas with corporate feeds are employed (Figure 5.18), this becomes possible. To this end, we employed Wilkinson power combiners for the T-junctions in Figure 5.18 [14, 18]. A Wilkinson power combiner is a split-T combiner with a resistive element placed between the two input arms, which has all three ports matched and at the same time provides isolation between the two input ports [19, 20]. A microstrip Wilkinson power combiner is shown in Figure 5.19.

Inequalities between the antenna elements in the array, due either to tolerance effects or to phase differences in the received signals, will cause a voltage difference to occur across the resistor and thus make a current flow through it. Therefore, through dissipation, isolation between the antenna elements is created. If we replace the resistors by rectifying circuits, the inequalities in the received signals may be used for power conversion while maintaining the isolation between the antenna elements. The DC power output will be larger for larger inequalities between the antenna elements. So, the antenna elements may – or, rather, must – be imperfectly matched to the feeding network, but the elements' imperfect matches may not be identical.

Through even–odd-mode analysis for the 'standard' Wilkinson power combiner [19, 20], the lengths of the two input arms can be determined to be equal to a quarter of a wavelength at the center frequency. The characteristic impedances can be determined to be equal to $\sqrt{2}Z_0$, and the value of the resistor is found to be $2Z_0$.

In the even–odd-mode analysis [21] of the combiner shown in Figure 5.19, we start with the equivalent transmission line circuit (Figure 5.20(a)). Next, this network is drawn in a normalized, symmetric form (Figure 5.20(b)). For this equivalent circuit, two excitation modes are defined: an even or in-phase mode, for which $V_{g2} = V_{g3} = 2$ V, and an odd or out-of-phase mode, for which $V_{g2} = -V_{g3} = 2$ V. By superposition of the even and odd modes, an effective excitation of $V_{g2} = 4$ V and $V_{g3} = 0$ is created. From this excitation, the S parameters of the network can be derived [21].

If we now replace the resistor by a given parallel circuit of a resistor R and a capacitor C, we may no longer keep the lengths of the arms equal to a quarter of a wavelength at the center frequency. Figure 5.21 shows one half of the bisected circuits of the combiner for even- and odd-mode excitation.

For $\alpha = 1$, $R = Z_0$, $C = 0$, $Z_{01} = \sqrt{2}Z_0$ and $l = \lambda/4$, the standard Wilkinson power combiner is obtained. For $C \neq 0$, we need a length $\ell \neq \lambda/4 + n\lambda$, $n = 0, 1, 2, \ldots$. Then, from the even-mode analysis, we find that $\alpha = 2$ and $Z_{01} = 2Z_0$. From the odd-mode analysis, we find that $Z_0 = R/4$ and

$$\ell = \frac{1}{\beta}\arctan\left(\frac{1}{\omega RC}\right), \tag{5.46}$$

where β is the wave number of the microstrip transmission line. Quarter-lambda impedance transformers are then needed to transform the impedances of all ports to the desired value of Z_0.

If the impedance Z_{rect} of the rectifying circuit is given by

$$Z_{\text{rect}} = A - \mathrm{j}B \tag{5.47}$$

(Figures 5.11 and 5.12), the values of R and C are given by

$$R = A\left[1 + \left(\frac{B}{A}\right)^2\right], \qquad C = \frac{B/A}{\omega A[1 + (B/A)^2]}. \tag{5.48}$$

The scattering matrix of the power combiner may be obtained following the procedure outlined in [20] for the unnormalized scattering matrix, after which the results are transformed

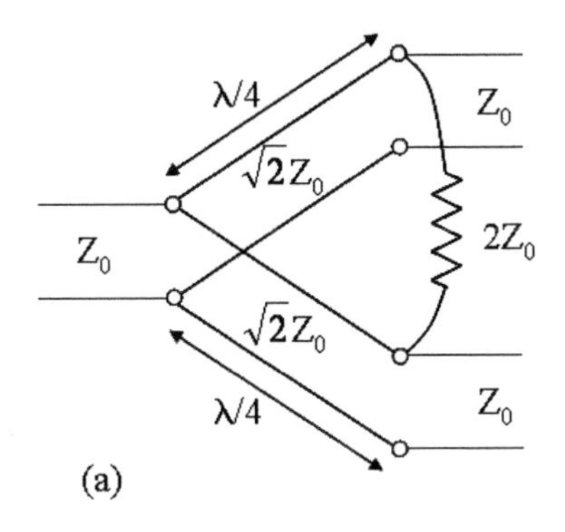

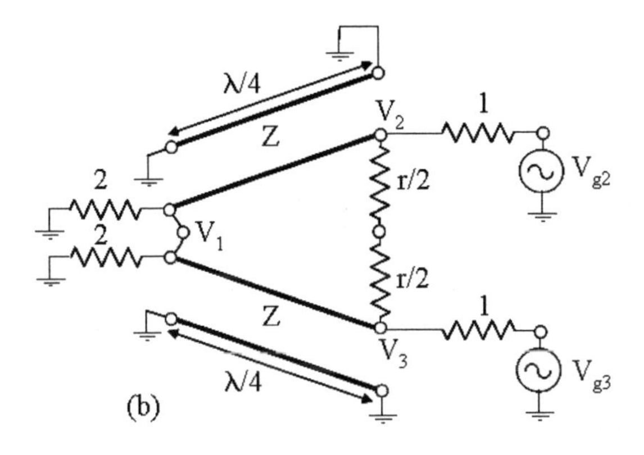

Figure 5.20 Even–odd-mode analysis of Wilkinson power combiner. (a) Equivalent transmission line circuit. (b) Equivalent transmission line circuit in normalized, symmetric form.

to the normalized scattering matrix using [14, 22]:

$$S_{11} = \Gamma_1 + (1 - \Gamma_1^2)\Gamma_2 \frac{e^{-j2\beta l}}{1 + \Gamma_1\Gamma_2 e^{-j2\beta l}}, \tag{5.49}$$

$$S_{12} = S_{21} = S_{13} = S_{31} = \frac{1}{\sqrt{\alpha}} \frac{(1+\Gamma_1)(1+\Gamma_2)e^{-j\beta l}}{1 + \Gamma_1\Gamma_2 e^{-j2\beta l}}, \tag{5.50}$$

$$S_{22} = S_{33} = \frac{Z_{2e}}{Z_{2e} + \alpha Z_0} + \frac{1}{1 + \alpha Y_{02} Z_0} - 1, \tag{5.51}$$

$$S_{23} = S_{32} = \frac{Z_{2e}}{Z_{2e} + \alpha Z_0} - \frac{1}{1 + \alpha Y_{02} Z_0}, \tag{5.52}$$

where

$$\Gamma_1 = \frac{Z_{01} - 2Z_0}{Z_{01} + 2Z_0}, \tag{5.53}$$

$$\Gamma_2 = \frac{\alpha Z_0 - Z_{01}}{\alpha Z_0 + Z_{01}}, \tag{5.54}$$

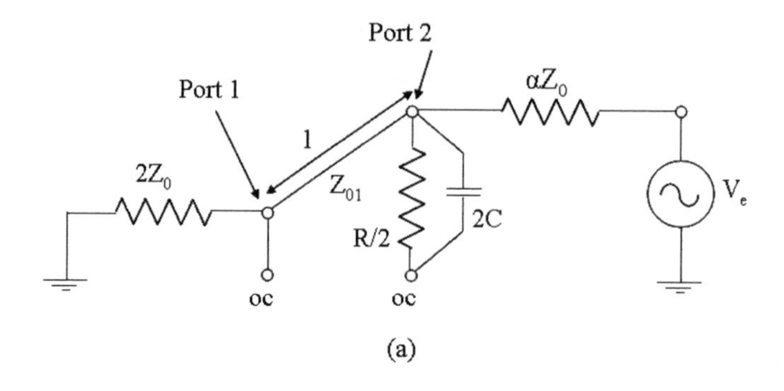

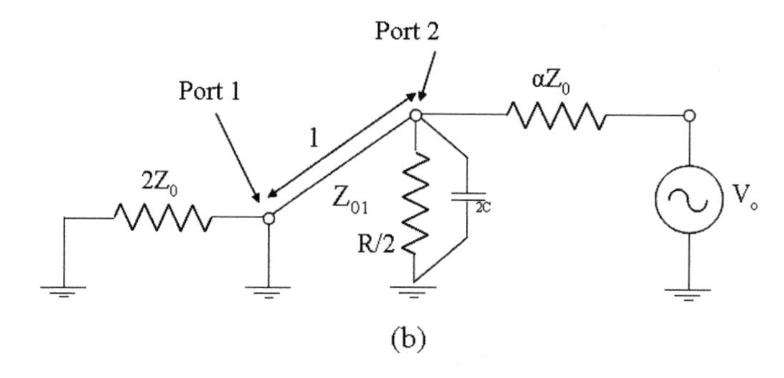

Figure 5.21 Bisections of the combiner circuit. (a) Even-mode excitation. Port 3 is excited by V_e. (b) Odd-mode excitation. Port 3 is excited by minus $-V_o$.

$$Z_{2e} = Z_{01}\frac{2Z_0 + jZ_{01}\tan(\beta l)}{Z_{01} + j2Z_0\tan(\beta l)}, \tag{5.55}$$

$$Y_{02} = \frac{1}{Z} + \frac{1}{jZ_{01}\tan(\beta l)} \tag{5.56}$$

and

$$Z = \frac{R}{2(1 + j\omega RC)}. \tag{5.57}$$

β may be calculated using the quasi-TEM equations for a microstrip given in [19].

In an initial experiment, we took a 'standard' Wilkinson power combiner and replaced the resistor by a rectifying circuit. If we replace the 100 Ω resistor in a $Z_0 = 50\ \Omega$ Wilkinson power combiner by a parallel *RC* circuit, we need to substitute $\alpha = 1$, $Z_{01} = \sqrt{2}Z_0$ and $\ell = \lambda/4$ at the center frequency in the above equations. For a power combiner *matched to the rectifying circuit*, we need to substitute $\alpha = 2$, $Z_0 = R/4$, $Z_{01} = R/2$ and the length l given by equation (5.47), calculated for the center frequency.

For the purposes of feasibility demonstration and model validation, a modulated transmission system was constructed (Figure 5.22). For this feasibility demonstration, an amplitude modulation (AM) scheme was chosen, where a low-frequency (1 kHz) block wave was superimposed on a 2.45 GHz carrier signal. The reason for this choice lay in the ease of demodulation, which could be performed with a single diode, not requiring an external power supply.

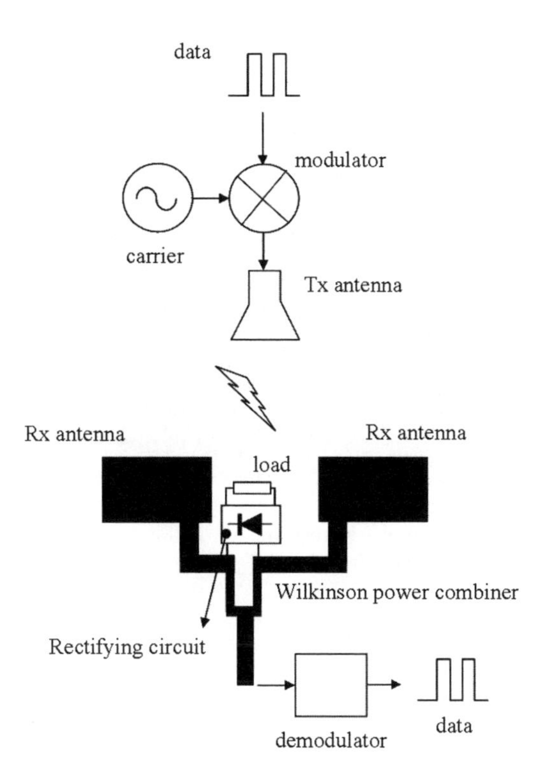

Figure 5.22 Modulated transmission system for demonstrating simultaneous transmission of power and data.

The hardware realization of the power/data receiver is shown in Figure 5.23. For this first demonstrator, a 'standard' 50 Ω microstrip Wilkinson power combiner was used, where the 100 Ω resistor was replaced by a voltage-doubling rectifying circuit using Agilent HSMS-2852 diodes.

The microstrip patch antennas, with dimensions 58.3 mm × 29.2 mm, were dual linearly polarized. The (50 Ω) feed positions were 20 mm from the corner on the long side. The width of the 50 Ω microstrip transmission lines was 3.1 mm. The patches were separated by a gap of 17 mm. The dielectric was FR4, with a thickness of 1.6 mm, a relative permittivity of 4.28 and a loss tangent of 0.016. To create an inequality between the signals at the combiner input ports, the transmission lines from the patches to the combiner input ports differed in length. The lengths of the transmission lines were, more or less arbitrarily, chosen to be 48.1 mm and 66.4 mm. For the chosen dimensions, the maximum sensitivity would be found at an angle of 26° from broadside in the plane of the array. However, the broad element patterns and the limited number of array elements had the effect that the gain variation between broadside and 26° off broadside remained within a few decibels, which was acceptable for this demonstration.

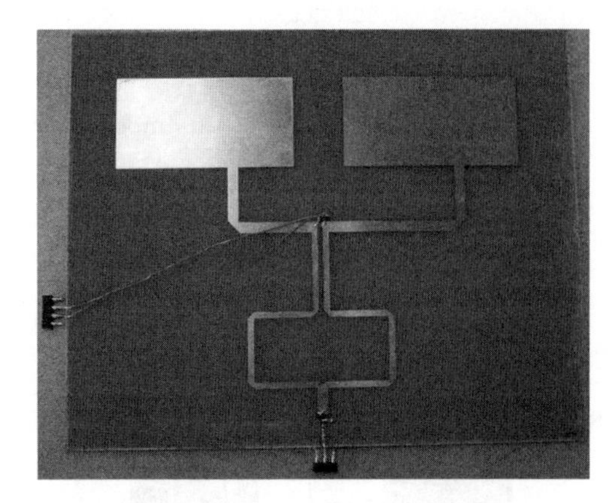

Figure 5.23 Hardware realization of AM receiver for demonstrating simultaneous transmission of power and data. The port on the left is for DC power, and the port at the bottom is for the demodulated signal.

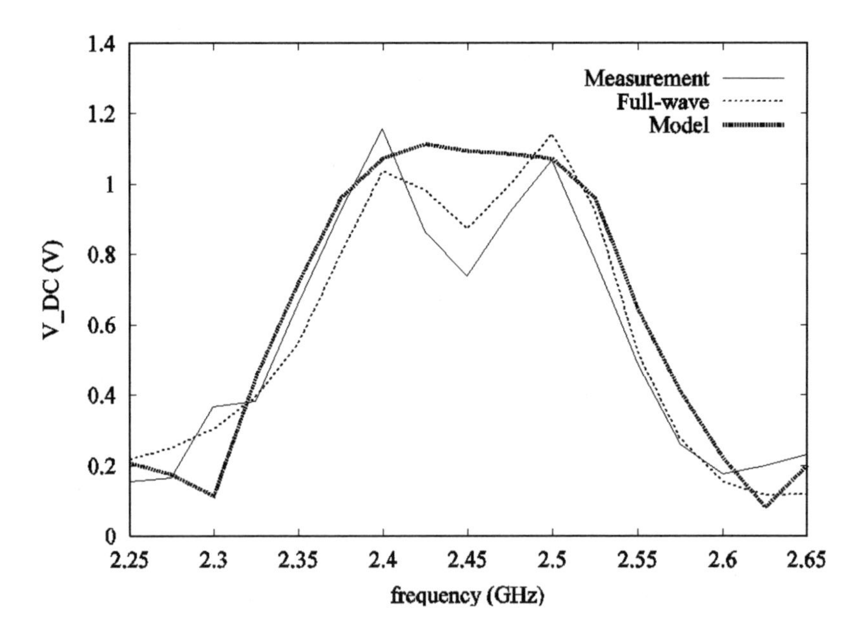

Figure 5.24 Calculated and measured unloaded output voltage of the rectifying circuit.

To calculate the unloaded output voltage of the rectifying circuit, use was made of the models developed for the power combiner and the voltage doubler circuit. Since, for the particular microstrip patch antennas used here (with a side ratio of approximately 2), the transmission line model fails to calculate accurate input impedance results, full-wave simulation results were used. Then, on the basis of the impedance of the microstrip patch antenna as a function of frequency and the frequency-dependent input impedances of the input ports of the power combiner, loaded at the output port with a single Agilent HSMS-2852 diode, the power accepted by the rectifying circuit was calculated. From that value, the unloaded output voltage of the rectifying circuit was calculated. Figure 5.24 shows the results, together with measurement results obtained when 3 mW of input power was delivered by the antennas. In the same figure, we also show simulation results based on full-wave simulations of the antennas and full-wave simulations of the power combiner with an open-circuited output port, thus neglecting the power dissipated in the demodulator. In these full-wave simulations, the impedance of the voltage doubler rectifier was linearized over a small frequency range around 2.45 GHz to create an S-parameter list for a black box to be inserted into the combiner configuration.

The full-wave simulation results are within 10% of the measured results. The results from the analytical model are within the same order of accuracy, which is adequate for generating preliminary designs.

Full-wave analysis of a matched power combiner, using the analytical model for the rectifier circuit, indicated that the unloaded output voltage level increased to a level beyond 1.2 V.

At the data output of the receiver, the 1 kHz block wave was retrieved after passing through a series capacitor and a 50 Hz band-stop filter.

5.7 RF ENERGY SCAVENGING

With the ability to analyze and design rectennas, the question arises as to whether it is possible to employ *ambient RF energy* for powering miniature wireless applications. By *ambient* RF energy, we mean RF energy not specifically introduced for wirelessly powering an application [7, 8], but RF energy available through public telecommunication services. Our main interest is in telecommunication services operating in the microwave region of the frequency spectrum, especially the Global System for Mobile Communications (GSM) and wireless local area networks (WLANs). For any of these services, printed antennas can be made with dimensions of the order of a few square centimeters, satisfying our constraint for miniature sensors.

To assess the feasibility of employing the ambient RF energy supplied by the above systems, we need to assess the power density levels in various environments (e.g. inner city areas, outer city areas, industrial areas, indoors and outdoors for GSM) and various situations (e.g. with respect to traffic, i.e. peak hours and off-peak hours, for GSM and WLANs).

5.7.1 GSM and WLAN Power Density Levels

Owing to a growing concern about a potential relation between GSM nonionizing radiation and health risks, a number of national [23] and European initiatives have been taken, dealing

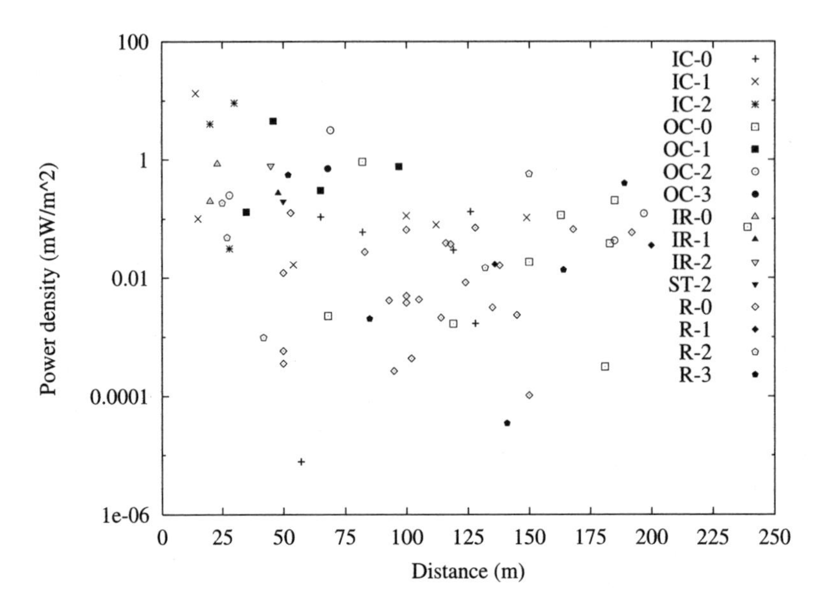

Figure 5.25 Measured GSM-900 peak power density levels as a function of distance from the nearest base station. The data is taken from [26]. The code '*XY-a*' indicates the characteristics of the area and measurement site. *XY*: IC = inner city; OC = outer country, IR = industrial area; ST = small town; R = rural or countryside area. *a*: 0 = outdoors on the ground; 1 = outdoors on roof, terrace or balcony; 2 = indoors, close to windows, 1.5 m or less; 3 = indoors, not close to windows.

with several aspects of RF exposure caused by the GSM system. The most important of the European initiatives are COST-281, 'Potential Health Implications from Mobile Communication Systems' [24], and the 'European Information System on Electromagnetic Fields Exposure and Health Impacts' [25]. Within these initiatives, ample use has been made of measurement data from several countries, gathered together in COST Action 244bis, 'Biomedical Effects of Electromagnetic Fields' [26].

We have used this data to assess the power density levels supplied by GSM base stations. We have made a distinction between the GSM-900 (downlink 935 MHz–960 MHz) and GSM-1800 (downlink 1805 MHz–1880 MHz) systems. For these communication systems, we have selected published power density measurements and made these visible in graphs showing the power density as a function of distance from the nearest GSM base station.

5.7.1.1 GSM-900 In Figure 5.25, the measured peak power densities as a function of the distance from the base station are shown. The data was measured in Austria, Germany and Hungary [26]. Measured data from France and Sweden was not used, since the distance from the nearest base station was not known for this data.

All measurements used for this figure were single-frequency spot measurements in the range 935 MHz–960 MHz. The measurements were taken in either two or three orthogonal directions, employing a *bicone* or log-periodic receiving antenna. The traffic density and the

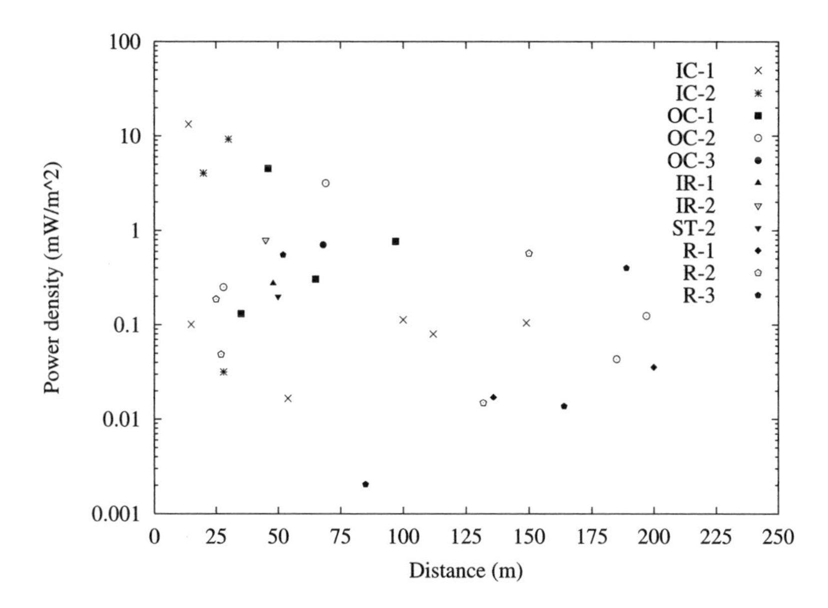

Figure 5.26 Measured GSM-900 peak power density levels as a function of distance from the nearest base station. The data is taken from [26]. The figure is the same as Figure 5.25 but with the data taken at ground level removed.

transmission power of the nearest base station were unknown [26]. Since exposure limits differ between different countries, we do not expect the base station transmission powers to be identical in all measurements. Therefore, the power density levels should be regarded as indicative. The codes '*XY-a*' associated with the various measurement sets indicate the characteristics of the area and the measurement site, as explained in the figure caption.

We may expect the largest variation in power density at ground level, on passing through the base station beam. Indoors and at an elevated level, these effects should be smaller. Indeed, in Figure 5.25 we observe a large variation in power density in the data sets IC-0, OC-0, IR-0 and R-0, all measured outdoors at ground level. If we remove these data sets from the figure, we get the results shown in Figure 5.26. From this figure, we may conclude that between 25 m and 100 m from a GSM-900 base station, we may expect – either indoors or outdoors at an elevated level – a power density between 0.1 mW m^{-2} and 1.0 mW m^{-2} (10^{-5}–10^{-4} mW cm^{-2}).

For measurements taken over the frequency band as a whole (935 MHz–960 MHz), the summed power density as a function of distance from the nearest base station is shown in Figure 5.27. From this figure, we may conclude that between 25 m and 100 m from a base station, we may expect – either indoors or outdoors at an elevated level – a summed power density between 0.3 mW m^{-2} and 3.0 mW m^{-2} ($3 \times 10^{-5} - 3 \times 10^{-4}$ mW cm^{-2}). Of course, the measurements over the entire frequency band depend strongly on the traffic density at the moment of measurement. A closer examination of the measurement data supplied by different countries [26] revealed that the power density levels for measurements over the entire frequency band may differ by a factor of one to ten from single-frequency

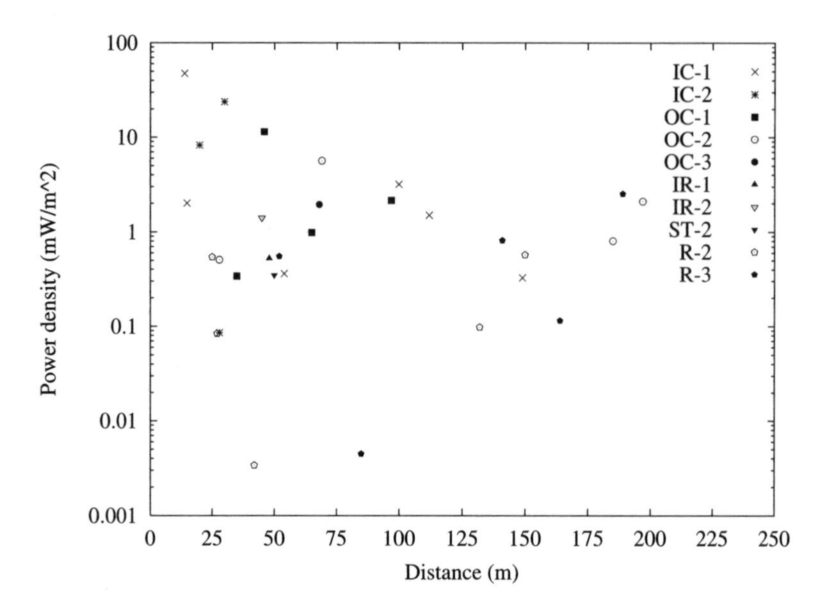

Figure 5.27 Measured GSM-900 summed power density levels as a function of distance from the nearest base station, without data taken at ground level. The data is taken from [26].

measurements. Details of GSM traffic at the moments of measurement are not available in [26].

5.7.1.2 GSM-1800 At the time the data shown in Figures 5.25–5.27 was gathered (November 1996 to November 2000 [26]), GSM-900 was dominant over GSM-1800. Therefore, more measurement data is available for GSM-900 exposure than for GSM-1800 exposure. Although the frequency of GSM-1800 is double that of GSM-900 and therefore the free-space loss has quadrupled, the ICNIRP exposure limit has only doubled [27]. Since actual exposure levels for GSM-900 are well beyond ICNIRP limits, we may expect power density levels for GSM-1800 similar to these for GSM-900.

This assumption appears to be justified by measurements conducted in the UK at 118 locations in 17 sites for a mix of GSM-900 and GSM-1800 base stations [28], and by recent measurements in Australia for 60 base stations in five cities [29] (Figure 5.28). The exposure limits in Australia follow those of ICNIRP. The measurements in [28, 29] (see Figure 5.28 again) show that the power density levels received from GSM-1800 base stations are at up to 100 m, of the same order of magnitude as those received from GSM-900 base stations at a single frequency or summed for low-traffic-density situations.

From the assessment of the power density levels produced by GSM base stations, we may conclude that it will be very difficult, at the very least, to wirelessly power or charge a small sensor.

5.7.1.3 WLANs A WLAN router (base station) will transmit less power than a GSM base station. But since WLANs are more or less confined to indoor environments and the distance

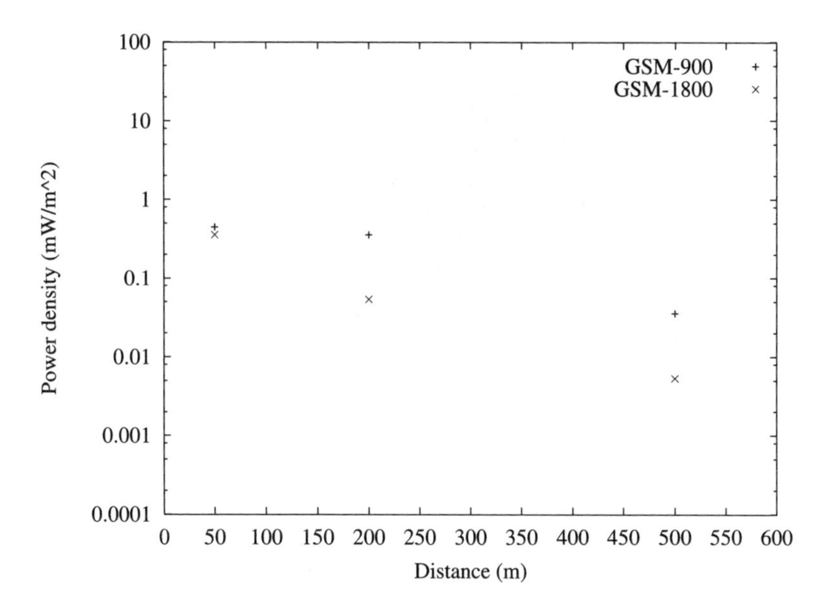

Figure 5.28 Measured summed power density levels as a function of distance from the nearest GSM-900 or GSM-1800 base station. The data is taken from [29].

from a router is usually small, reflections and a low path loss could help in establishing higher power density levels.

Initial measurements show, however, that the power density levels are at least one order of magnitude lower than those obtained close to a GSM base station. Therefore our assumptions regarding WLAN power densities prove to be wrong. It has to be noted, though, that during these measurements the traffic density was extremely low (on average two connections at a time). However, it seems unlikely that small sensors can be powered for a reasonably long time in a WLAN environment.

The idea of getting closer to an RF source, as proposed in the concept of WLAN RF energy scavenging, may be applied, though, to a GSM mobile phone.

5.7.2 GSM Mobile Phone as RF Source

In [30], the use of a GSM mobile phone was proposed as a source for RF energy scavenging. For a specific antenna, a scavenged power of 1.9 mW was predicted at a distance of 1 m from a GSM-900 mobile phone, transmitting at a power level of 2 W. This number, however, is rather optimistic and should be regarded as a theoretical maximum, since it assumes a 3 dB gain mobile phone antenna, polarization and gain alignment of the phone antenna and the rectenna, and a rectenna efficiency of 100%. Nevertheless, the concept seemed to be feasible, and therefore a microstrip patch rectenna for GSM-1800 was designed following the procedure in [8]. The rectenna was realized on FR4 and connected to an LED for demonstration purposes. A GSM-1800 mobile phone (with a maximum allowed transmission

Figure 5.29 Rectenna feeding an LED, wirelessly powered by a GSM-1800 mobile phone, transmitting at a peak power level of 1 W.

power of 1 W) was able to make the LED illuminate at up to a distance of 20 cm, proving the concept. The system is shown in Figure 5.29.

5.8 CONCLUSIONS

Analytical models for microstrip patch antennas, diodes and Wilkinson power combiners have been applied and derived to design efficient, compact, low-power rectenna systems. If a microstrip patch antenna is employed and this antenna is conjugately matched directly to a Schottky diode or a voltage-doubling circuit of Schottky diodes, the need for an impedance and filtering network no longer exists. This leads to an efficiency improvement of more than 10% with respect to a 'traditional' low-power rectenna. Rectennas may be regarded as wireless batteries, and by connecting them in series, low-power, low-duty-cycle devices may be wirelessly powered. Modifying a Wilkinson power combiner by replacing the resistive element with a rectifying circuit allows simultaneous wireless transfer of power and data. Although the feasibility of simultaneous wireless power and data transfer has been demonstrated, there is still room for improvement. The analytical tools developed and discussed here may be of help in this process.

For distances ranging from 25 m to 100 m from a GSM base station, power density levels ranging from 0.1 mW m^{-2} to 1.0 mW m^{-2} may be expected for single frequencies. For the total GSM downlink frequency bands, these levels may be elevated by a factor of between one and three, depending on the traffic density. Initial measurements in a WLAN environment indicate power density values that are at least one order of magnitude lower. Therefore, neither GSM nor a WLAN is likely to produce enough ambient RF energy for wirelessly powering miniature sensors. However, by employing rectenna *arrays* [31], this ambient RF energy may be transformed to usable DC energy. A single GSM telephone has proven to deliver enough energy to wirelessly power small applications at moderate distances.

REFERENCES

1. W.C. Brown, 'The history of power transmission by radio waves', *IEEE Transactions on Microwave Theory and Techniques*, Vol. MTT-32, No. 9, pp. 1230–1242, September 1984.

2. P. Green, 'RFID tracking in manufacturing plants, the promise and the reality', Bellhawk Systems Corporation White Papers. Available at http://www.bellhawk.com/PDF/white-papers/RFID.pdf.

3. Agilent Technologies, *Designing the Virtual Battery*, Application Note 1088, Agilent Technologies, Inc., 1999.

4. R.G. Harrison and X. Le Polozec, 'Nonsquarelaw behavior of diode detectors analyzed by the Ritz–Galerkin method', *IEEE Transactions on Microwave Theory and Techniques*, Vol. 42, No. 5, pp. 840–846, May 1994.

5. Y.-H. Suh and K. Chang, 'A high-efficiency dual-frequency rectenna for 2.45- and 5.8-GHz wireless power transmission', *IEEE Transactions on Microwave Theory and Techniques*, Vol. 50, No. 7, pp. 1784–1789, July 2002.

6. J. Heikkinen, P. Salonen and M. Kivikoski, 'Planar rectennas for 2.45GHz wireless power transfer', *Proceedings of the Radio and Wireless Conference*, Denver, CO, pp. 63–66, 2000.

7. J.A.G. Akkermans, M.C. van Beurden, G.J.N. Doodeman and H.J. Visser, 'Analytical models for low-power rectenna design', *IEEE Antennas and Wireless Propagation Letters*, Vol. 4, pp. 187–190, 2005.

8. J.A.C. Theeuwes, H.J. Visser, M.C. van Beurden and G.J.N. Doodeman, 'Efficient, compact, wireless battery design', *Proceedings of the European Conference on Wireless Technology, ECWT2007*, Munich, Germany, pp. 233–236, October 2007.

9. K.R. Carver and J.W. Mink, 'Microstrip antenna technology', *IEEE Transactions on Antennas and Propagation*, Vol. AP-29, No. 1, pp. 2–24, January 1981.

10. W.F. Richards, Y.T. Lo and D.D. Harrison, 'An improved theory for microstrip antennas and applications', *IEEE Transactions on Antennas and Propagation*, Vol. AP-29, No. 1, pp. 38–46, January 1981.

11. R.W. Dearnley and A.R.F. Barel, 'A broad-band transmission line model for a rectangular microstrip antenna', *IEEE Transactions on Antennas and Propagation*, Vol. 37, No. 1, pp. 6–15, January 1989.

12. H. Pues and A. Van de Capelle, 'Accurate transmission-line model for the rectangular microstrip antenna', *Proceedings of the IEE*, Vol. 131, Part H, No. 6, pp. 334–340, 1984.

13. A. Van de Capelle, 'Transmission-line model for rectangular microstrip antennas', in J. James and P.D. Hall (eds.), *Handbook of Microstrip Antennas*, Peter Peregrinus, pp. 527–578, 1990.

14. H.J. Visser, *Array and Phased Array Antenna Basics*, John Wiley & Sons, Chichester, 2005.

15. D.A. Fleri and L.D. Cohen, 'Nonlinear analysis of the Schottky-barrier mixer diode', *IEEE Transactions on Microwave Theory and Techniques*, Vol. 21, No. 1, pp. 39–43, January 1973.

16. W.H. Press, B.P. Flannery, S.A. Teukolsky and W.T. Vetterling, *Numerical Recipes: The Art of Scientific Computing*, Cambridge University Press, Cambridge, 1988.

17. Agilent Technologies, *Surface Mount Zero Bias Schottky Detector Diodes, Technical Data, HSMS-285x Series*, available at http://www.agilent.com/semiconductors.

18. R.C. Hansen, *Phased Array Antennas*, John Wiley & Sons, New York, 1998.

19. D.M. Pozar, *Microwave Engineering*, second edition, John Wiley & Sons, New York, 1998.

20. L.I. Parad and R.L. Moynihan, 'Split-tee power divider', *IEEE Transactions on Microwave Theory and Techniques*, Vol. 13, No. 1, pp. 91–95, January 1965.

21. J. Reed and G.J. Wheeler, 'A method of analysis of symmetrical four-port networks', *IRE Transactions on Microwave Theory and Techniques*, Vol. 4, No. 4, pp. 246–252, October 1956.

22. C.G. Montgomery (ed.), *Technique of Microwave Measurements*, Vol. 11 of MIT Radiation Laboratory Series, McGraw-Hill, New York, 1947.

23. A.P.M. Zwamborn, S.H.J.A. Vossen, B. van Leersum, M.A. Ouwens and W.N. Mäkel, *Effects of Global Communication System Radiofrequency Fields on Well Being and Cognitive Functions of Human Subjects with and without Subjective Complaints*, TNO Physics and Electronics Laboratory, The Hague, The Netherlands, Report FEL-03-C148, 2003.

24. COST-281, *Potential Health Implications from Mobile Communication Systems*, available at http://www.cost281.org/

25. C. del Pozo and D. Papameletiou, *European Information System on Electromagnetic Fields Exposure and Health Impacts – Country Reports on EMF and Health: Sources, Regulations, and Risk Communication Approaches*, European Commission, Joint Research Centre, Institute for Health and Consumer Protection, Physical and Chemical Exposure Unit, December 2005.

26. U. Bergqvist *et al.*, *Mobile Telecommunication Base Stations: Exposure to Electromagnetic Fields, Report of a Short Term Mission within COST-244bis*, COST-244bis Short Term Mission on Base Station Exposure, 2000.

27. ICNIRP, *Guidelines for Limiting Exposure to Time-Varying Electric, Magnetic and Electromagnetic Fields (up to 300 GHz)*, International Commission on Non-Ionizing Radiation Protection, Oberschleissheim, Germany, 1998.

28. S.M. Mann, T.G. Cooper, S.G. Allen, R.P. Blackwell and A.J. Lowe, *Exposure to Radio Waves near Mobile Phone Base Stations*, National Radiological Protection Board, Report NRPB-R321, June 2000.

29. S.I. Henderson and M.J. Bangay, 'Survey of RF exposure levels from mobile telephone base stations in Australia', *Bioelectromagnetics*, Vol. 27, No. 1, pp. 73–76, January 2006.

30. P. Ancey, 'Ambient functionality in MIMOSA from technology to services', *Proceedings of the Joint sOc-EUSAI Conference*, Grenoble, 2005.

31. J.A. Hagerty, F.B. Helmbrecht, W.H. McCalpin, R. Zane and Z.B. Popovic, 'Recycling ambient microwave energy with broad-band rectenna arrays', *IEEE Transactions on Microwave Theory and Techniques*, Vol. 52, No. 3, pp. 1014–1024, March 2004.

6

Large Array Antennas: Open-Ended Rectangular-Waveguide Radiators

The accurate analysis of large planar phased array antennas is a very time-consuming task owing to the incorporation of effects of the mutual coupling that exists between the radiating elements. For very large phased array antennas, i.e. for arrays consisting of roughly more than a few tens of radiating elements, the analysis time may be sped up considerably by approximating the phased array antenna by an infinite, uniformly excited phased array antenna. This is permissible because, in a very large phased array antenna, the majority of elements experience an environment identical to that of an element in a periodic phased array antenna. The analysis may now be restricted to a single unit cell only. The infinite-array concept will be applied to phased array antennas consisting of open-ended waveguide radiators, employing a mode-matching technique. To allow for impedance-matching structures inside the waveguide radiators, a mode-matching technique will be developed for waveguide-to-unit-cell junctions and waveguide-to-waveguide junctions. After a thorough validation of the results for cascaded waveguide-to-waveguide junctions, the results for the radiation into a unit cell will be validated by comparison with published analysis results and measurements on a large array antenna. For the evaluation of large array antennas analysis of the reflection coefficient of a unit cell suffices, since this reflection coefficient is directly related to the scan element pattern, which exhibits all features of a scanned phased array antenna. The material presented in this chapter is not state of the art, and dates back to the mid 1990s. Since then, a lot of progress in analyzing these kinds of antennas has been made. Nevertheless, we find it appropriate to present a 'classic' mode-matching approach in detail. This material may aid in understanding new developments and may be implemented relatively easily in software for analyzing rectangular-waveguide structures and infinite arrays of open-ended waveguides.

Approximate Antenna Analysis for CAD Hubregt J. Visser

6.1 INTRODUCTION

Array and phased array antennas are gaining in popularity. They seem no longer to be of interest only for military systems (radar), but are encountered today in many civilian systems, for example in mobile-communication base stations. Advances in electronics have made fast-switching phase shifters feasible, and increased computer power has made real-time signal processing feasible. These two developments have helped in the realization of large phased array antenna systems. In order to design (large) phased array antennas consisting of regularly positioned, identical radiating elements, an analysis technique is required that takes the mutual coupling between the radiating elements into account in addition to the characteristics of the radiating elements. The mutual coupling is largely responsible for the unique characteristics of phased arrays [1]. Only for relatively small array antennas, having (far) fewer than one hundred elements, is it possible to calculate coupling characteristics within a reasonable amount of time using a 'brute force' technique. For larger array antennas, a different approach must be taken.

This chapter deals with the application of a 'brute force' technique to large array antennas. The development of the model dates back to the mid 1990s. Since then, a lot of progress in analyzing large array antennas has been made [2–7]. Notwithstanding these developments, a presentation in detail of this 'brute force' technique (a 'classic' mode-matching technique) is still appropriate, since it is easy to understand, simple to implement in software and valuable for analyzing waveguide structures and infinite arrays of open-ended waveguide radiators.

To facilitate the analysis of large or periodic planar array antennas, we assume the array antenna to be infinitely large in the two planar directions and to be uniformly illuminated. Thus, edge effects due to the boundedness of the large but finite structure are neglected. Owing to the periodicity of the configuration, the effects of coupling between the radiators are essentially the same for each of the radiators or elements, except for phase differences. Therefore, it is permissible to place fictitious walls, perpendicular to the face of the array, between neighboring elements, thereby setting up *unit cells* into which each element individually radiates. A unit cell may be regarded as a rectangular waveguide with phase-shifting walls. Following this approach allows us to solve the infinite-array problem for a single unit cell only. Of course, in the case of a finite array antenna, the effects of coupling are not the same for each element. However, for arrays consisting of more than a few hundred elements, the majority of the (inner) elements behave locally as if embedded in an infinite array [8].

Open-ended waveguide array antennas are attractive owing to the inherent characteristics of open-ended waveguide radiators such as wide frequency bandwidth, good cross-polarization behavior and high-pass filtering characteristics due to the cutoff frequency of the waveguide modes [9, 10]. Therefore, large phased array antennas consisting of open-ended waveguide radiators are still being considered for new developments [4, 11, 13].

6.1.1 Mode Matching and Generalized Scattering Matrices

Pioneering work in the field of analyzing open-ended waveguide (phased) array antennas was performed by a number of researchers in the 1960s and 1970s [14–20]. As a result, mode matching is by now generally accepted as an analysis method that is very suitable

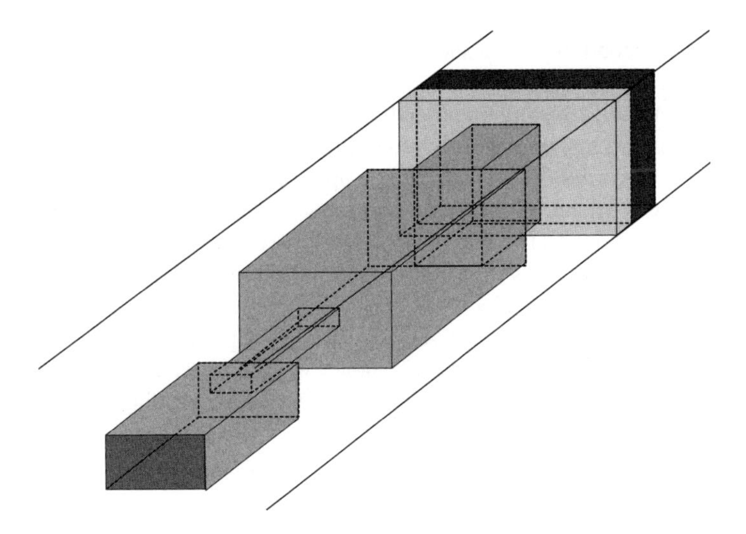

Figure 6.1 Waveguide structure consisting of cascaded rectangular-waveguide discontinuities and transmission lines.

for rectangular-waveguide structures. From the 1980s onwards, mode matching, applied for constructing generalized scattering matrices (GSMs)[1] to describe waveguide junctions, has been used with success for the characterization of filter-like structures [21–28]. Because of this success of this method with filter-like structures, we have adopted the method for the analysis of planar, infinite, open-ended waveguide (phased) array antennas.

Mode matching is employed to derive GSMs for waveguide discontinuities, including the transition from a rectangular waveguide to a rectangular unit cell, i.e. the transition from a waveguide to free space in a periodic array of waveguides.[2] A waveguide structure, as shown in Figure 6.1, is analyzed by cascading the GSMs of the discontinuities and rectangular-waveguide transmission lines, which results in the overall GSM of the structure. From this overall GSM, relevant parameters may be extracted. The dimensions of the GSM are infinite in principle, but by application of a justified truncation, desired parameters having an acceptable accuracy may be extracted.

The example shown in Figure 6.1 shows an element within a unit cell of a planar, open-ended waveguide array antenna. Going from the left to the right in this figure, we encounter a rectangular waveguide, an asymmetric two-dimensional cross-sectional reduction, a second rectangular waveguide, a symmetric two-dimensional cross-sectional enlargement, a third rectangular waveguide, a symmetric one-dimensional cross-sectional reduction, a fourth waveguide, a junction between a waveguide and a free-space unit cell, a dielectric sheet,

[1] In this chapter, a GSM is a scattering matrix, containing, besides the fundamental-mode scattering coefficients, higher-order-mode scattering coefficients.

[2] To calculate the GSM of a transition from an isolated rectangular waveguide to free space, the mode-matching technique presented in [29] may be applied.

a dielectric step, a second dielectric sheet and, finally, a dielectric step. So, for the structure shown in Figure 6.1, we need to cascade 11 GSMs to obtain the overall GSM.

The analysis method based upon cascading GSM's is very versatile. It is, for instance, capable of analyzing infinite planar arrays consisting of open-ended waveguide radiators that employ *finite-thickness* irises in the waveguide apertures (Figure 6.1). In order to obtain a wide-angle impedance match (WAIM), aperture irises and a dielectric sheet in front of the apertures may be employed [18]. Previously developed analysis methods could deal only with infinitely thin aperture irises [8,30]. More recent developments, however, have indicated that this WAIM may be improved by using one or more irises with a finite thickness [31,32].

Since our main concern is the analysis and design of planar, open-ended, rectangular-waveguide (phased) array antennas, in this chapter we shall derive only the GSM's for the building blocks necessary for analyzing these antennas. These building blocks are:

- cross-sectional reductions and enlargements of rectangular waveguides;
- junctions between a rectangular waveguide and a free-space unit cell;
- finite-length waveguides;
- finite-thickness dielectric sheets in a unit cell;
- dielectric steps in a unit cell.

Since we are dealing only with electromagnetic fields inside rectangular waveguides and inside rectangular phase-shift-wall waveguides (unit cells), we start with the derivation of field quantities within these domains. Next, we derive the GSM's of a cross-sectional reduction or enlargement in a rectangular waveguide, a junction of a rectangular waveguide to a rectangular phase-shift-wall waveguide, a dielectric step in a unit cell, and a transmission line (a waveguide with an electrically conducting wall or phase shift wall) of finite length. Every type of discontinuity will be the subject of a separate section. After the cascading of these GSM's has been discussed, a software implementation is thoroughly validated by comparing simulation results for filter-like structures with results obtained from the open literature. Simulation results for planar, open-ended waveguide array antennas are compared with results obtained from an independently developed and validated computer code [31].

6.2 WAVEGUIDE FIELDS

A semi-infinite rectangular waveguide with cross-sectional dimensions a and b is shown in Figure 6.2. A rectangular coordinate system $(\hat{\imath}_x, \hat{\imath}_y, \hat{\imath}_z)$, as indicated in the figure, is assumed. The origin of this coordinate system is positioned at the center of the waveguide cross section, following [8,33]. The waveguide walls are assumed to be perfectly electrically conducting.

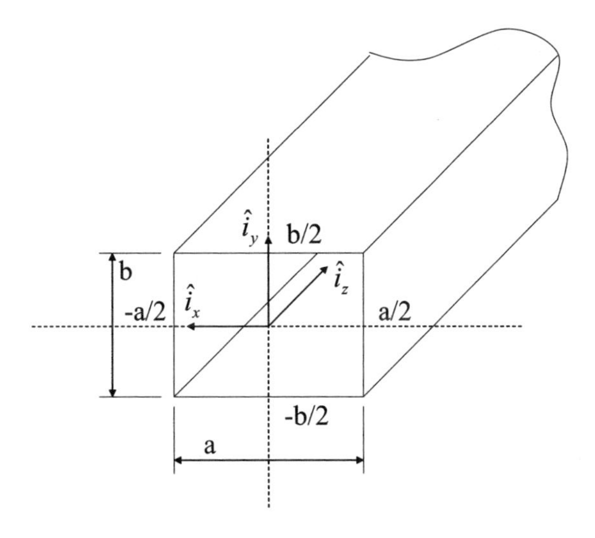

Figure 6.2 Semi-infinite waveguide of rectangular cross section, and coordinate system.

The electric field $\boldsymbol{E} = \boldsymbol{E}(\boldsymbol{r})$ and magnetic field $\boldsymbol{H} = \boldsymbol{H}(\boldsymbol{r})$ satisfy the source-free Maxwell equations

$$\nabla \times \boldsymbol{H} - \mathrm{j}\omega\varepsilon_{\mathrm{r}}\varepsilon_0\boldsymbol{E} = 0, \tag{6.1}$$

$$\nabla \times \boldsymbol{E} + \mathrm{j}\omega\mu_{\mathrm{r}}\mu_0\boldsymbol{H} = 0, \tag{6.2}$$

$$\nabla \cdot \boldsymbol{E} = 0, \tag{6.3}$$

$$\nabla \cdot \boldsymbol{H} = 0, \tag{6.4}$$

where ε_0 is the permittivity of free space and μ_0 is the permeability of free space.[3] The position vector $\mathbf{r}$ is given by

$$\boldsymbol{r} = x\hat{\boldsymbol{i}}_x + y\hat{\boldsymbol{i}}_y + z\hat{\boldsymbol{i}}_z, \tag{6.5}$$

and a harmonic time factor $\mathrm{e}^{\mathrm{j}\omega t}$ for the fields is assumed and suppressed.

The z dependence of every wave or mode is of the form $\mathrm{e}^{-\mathrm{j}\gamma z}$ for waves propagating in the direction of increasing z and of the form $\mathrm{e}^{\mathrm{j}\gamma z}$ for waves propagating in the direction of decreasing z. The wave number γ here is positive real for propagating waves and negative imaginary for evanescent waves. The evanescent waves therefore decay exponentially. To simplify the model, we restrict ourselves to loss-free media. For air-filled waveguides and loss-free or thin WAIM sheets, this is a valid restriction. Next, we shall discuss the decomposition of the electric and magnetic fields into modes. We shall consider the axial field components $e_z^{\pm}$ and $h_z^{\pm}$ as fundamental unknowns.

[3] Strictly speaking, the last two equations are not part of the Maxwell equations for the source-free situation. They follow from the first two equations.

We start by decomposing the fields into transverse (with respect to z) and axial components

$$\boldsymbol{E}(\boldsymbol{r}) = \boldsymbol{e}^{\pm}(\boldsymbol{r}_{\mathrm{T}})\mathrm{e}^{\mp \mathrm{j}\gamma z}, \tag{6.6}$$

$$\boldsymbol{H}(\boldsymbol{r}) = \boldsymbol{h}^{\pm}(\boldsymbol{r}_{\mathrm{T}})\mathrm{e}^{\mp \mathrm{j}\gamma z}, \tag{6.7}$$

where $\boldsymbol{e}^{+}(\boldsymbol{r}_{\mathrm{T}})$ and $\boldsymbol{h}^{+}(\boldsymbol{r}_{\mathrm{T}})$ are the electric field vector and magnetic field vector, respectively, for a wave traveling in the positive z direction; $\boldsymbol{e}^{-}(\boldsymbol{r}_{\mathrm{T}})$ and $\boldsymbol{h}^{-}(\boldsymbol{r}_{\mathrm{T}})$ are the electric field vector and magnetic field vector, respectively, for a wave traveling in the negative z direction. The transverse position vector $\boldsymbol{r}_{\mathrm{T}}$ is given by

$$\boldsymbol{r}_{\mathrm{T}} = x\hat{\boldsymbol{i}}_x + y\hat{\boldsymbol{i}}_y. \tag{6.8}$$

Substitution of equations (6.6) and (6.7) into the source-free Maxwell equations (6.1) and (6.2) results in

$$k_{\mathrm{T}}^2 \boldsymbol{e}_{\mathrm{T}}^{\pm} = \mathrm{j}\omega\mu_{\mathrm{r}}\mu_0(\hat{\boldsymbol{i}}_z \times \nabla_{\mathrm{T}} h_z^{\pm}) \mp \mathrm{j}\gamma \nabla_{\mathrm{T}} e_z^{\pm}, \tag{6.9}$$

$$k_{\mathrm{T}}^2 \boldsymbol{h}_{\mathrm{T}}^{\pm} = -\mathrm{j}\omega\varepsilon_{\mathrm{r}}\varepsilon_0(\hat{\boldsymbol{i}}_z \times \nabla_{\mathrm{T}} e_z^{\pm}) \mp \mathrm{j}\gamma \nabla_{\mathrm{T}} h_z^{\pm}, \tag{6.10}$$

where $\boldsymbol{e}_{\mathrm{T}}^{\pm}$ and $\boldsymbol{h}_{\mathrm{T}}^{\pm}$ are the transverse electric and magnetic field vectors;

$$\nabla_{\mathrm{T}} \equiv \frac{\partial}{\partial x}\hat{\boldsymbol{i}}_x + \frac{\partial}{\partial y}\hat{\boldsymbol{i}}_y \tag{6.11}$$

and

$$k_{\mathrm{T}}^2 = \boldsymbol{k}_{\mathrm{T}} \cdot \boldsymbol{k}_{\mathrm{T}} = \omega^2\varepsilon_0\varepsilon_{\mathrm{r}}\mu_0\mu_{\mathrm{r}} - \gamma^2 \tag{6.12}$$

where

$$\boldsymbol{k}_{\mathrm{T}} = k_x\hat{\boldsymbol{i}}_x + k_y\hat{\boldsymbol{i}}_y \tag{6.13}$$

is the transverse wave vector.

The axial wave number γ is related to the square of the transverse wave number, k_{T}^2, through

$$\gamma = \begin{cases} \sqrt{\omega^2\varepsilon_0\varepsilon_{\mathrm{r}}\mu_0\mu_{\mathrm{r}} - k_{\mathrm{T}}^2}, & \omega^2\varepsilon_0\varepsilon_{\mathrm{r}}\mu_0\mu_{\mathrm{r}} \geq k_{\mathrm{T}}^2 \\ -\mathrm{j}\sqrt{-\omega^2\varepsilon_0\varepsilon_{\mathrm{r}}\mu_0\mu_{\mathrm{r}} + k_{\mathrm{T}}^2}, & \omega^2\varepsilon_0\varepsilon_{\mathrm{r}}\mu_0\mu_{\mathrm{r}} < k_{\mathrm{T}}^2, \end{cases} \tag{6.14}$$

where the square root is nonnegative.

Equations (6.9) and (6.10) indicate that a knowledge of the axial field components of a mode suffices to determine the transverse field components of that mode. Furthermore, the problem of solving equations (6.9) and (6.10) can be separated into solving two scalar problems in the transverse plane. For transverse electric (TE) modes, we assume that $e_z^{\pm} = 0$ and $h_z^{\pm} \neq 0$. For transverse magnetic (TM) modes, we assume that $h_z^{\pm} = 0$ and $e_z^{\pm} \neq 0$. For obvious reasons, TE modes are also known as H modes and TM modes are also known as E modes. Transverse electromagnetic (TEM) waves, where $h_z^{\pm} = 0$ and $e_z^{\pm} = 0$, cannot exist in the waveguide.

6.2.1 TE Modes

For TE modes, $e_z^{\pm} = 0$ and $h_z^{\pm} \neq 0$. All field components are now determined from the scalar quantity $h_z^{\pm}$.

Taking the curl of equation (6.2), using the vector identity $\nabla\times\nabla\times\boldsymbol{A} = -\nabla\cdot\nabla\boldsymbol{A}+\nabla(\nabla\cdot\boldsymbol{A})$, results in

$$\nabla \times \nabla \times \boldsymbol{H} = -\nabla \cdot \nabla\boldsymbol{H} + \nabla(\nabla \cdot \boldsymbol{H}) = \mathrm{j}\omega\varepsilon_0\varepsilon_\mathrm{r}\nabla \times \boldsymbol{E}. \tag{6.15}$$

Next, substituting equations (6.2) and (6.4) into equation (6.15), using $\omega^2\varepsilon_0\varepsilon_\mathrm{r}\mu_0\mu_\mathrm{r} = k^2$, and rearranging terms gives

$$(\nabla \cdot \nabla + k^2)\boldsymbol{H} = 0. \tag{6.16}$$

This vectorial equation may be separated into its scalar components. So, after substitution of equation (6.7) into equation (6.16), we find for the z component of the magnetic field

$$\frac{\partial^2}{\partial x^2}h_z^{\pm}\mathrm{e}^{\mp\mathrm{j}\gamma z} + \frac{\partial^2}{\partial y^2}h_z^{\pm}\mathrm{e}^{\mp\mathrm{j}\gamma z} - \gamma^2 h_z^{\pm}\mathrm{e}^{\mp\mathrm{j}\gamma z} + k^2 h_z^{\pm}\mathrm{e}^{\mp\mathrm{j}\gamma z} = 0. \tag{6.17}$$

Using equation (6.12) and realizing that equation (6.17) should apply for all values of z leads finally to the two-dimensional Helmholtz equation,

$$(\nabla_\mathrm{T} \cdot \nabla_\mathrm{T} + k_\mathrm{T}^2)h_z^{\pm} = (\nabla_\mathrm{T}^2 + k_\mathrm{T}^2)h_z^{\pm} = 0. \tag{6.18}$$

As will be shown, nontrivial solutions to equation (6.18) exist only for certain values of k_T^2. These values depend on the transverse geometry of the waveguide. The solutions are called *modes*.

Applying the method of separation of variables [34] leads to the following solution for equation (6.18):

$$h_z^{\pm} = A^{\pm}\cos\left[k_x\left(x + \frac{a}{2}\right)\right]\cos\left[k_y\left(y + \frac{b}{2}\right)\right], \tag{6.19}$$

where $A^{\pm}$ is the mode amplitude coefficient (as yet unknown) and the *separation constants* k_x and k_y are related to the transverse and axial wave numbers through

$$k_\mathrm{T}^2 = k_x^2 + k_y^2 = \omega^2\varepsilon_0\varepsilon_\mathrm{r}\mu_0\mu_\mathrm{r} - \gamma^2, \tag{6.20}$$

as can be found by substitution of equation (6.19) into equation (6.18).

Next, we use the boundary conditions that the modulus of $h_z^{\pm}$ reaches its maximum at $x = \pm a/2$ and $y = \pm b/2$. These boundary conditions are the results of our earlier assumption of perfectly electrically conducting waveguide walls, which forces $e_z^{\pm}$ to be zero at the walls. Applying these boundary conditions leads to[4]

$$k_x = \frac{m\pi}{a}, \quad m \in \mathbb{N}, \tag{6.21}$$

[4]The natural numbers $\mathbb{N}$ are the set $\{1, 2, 3, \ldots\}$ or $\{0, 1, 2, 3, \ldots\}$. The inclusion of zero is a matter of definition [35]. Here we define $\mathbb{N}$ to include zero.

and

$$k_y = \frac{n\pi}{b}, \quad n \in \mathbb{N}. \tag{6.22}$$

Since negative values of k_x and k_y will result in the same solutions as for positive values of k_x and k_y, we shall restrict ourselves from now on to nonnegative values of m and n. The indices m and n are known as the *mode indices*. So, for the TE_{mn} mode, we have, using equation (6.20),

$$k^2_{\text{TE}_{mn}} = \left(\frac{m\pi}{a}\right)^2 + \left(\frac{n\pi}{b}\right)^2, \quad m, n \in \mathbb{N} \wedge (m, n) \neq (0, 0). \tag{6.23}$$

Substitution of equations (6.21) and (6.22) into equation (6.19) gives

$$h^{\pm}_{z_{mn}} = A^{\pm}_{mn} \cos\left[\frac{m\pi}{a}\left(x + \frac{a}{2}\right)\right] \cos\left[\frac{n\pi}{b}\left(y + \frac{b}{2}\right)\right], \tag{6.24}$$

and upon substitution of equation (6.23) into equation (6.14), the wave number of the axial mode is given by

$$\gamma_{mn} = \begin{cases} \sqrt{k_0^2 \varepsilon_r \mu_r - \left(\frac{m\pi}{a}\right)^2 - \left(\frac{n\pi}{b}\right)^2}, & k_0^2 \varepsilon_r \mu_r \geq \left(\frac{m\pi}{a}\right)^2 + \left(\frac{n\pi}{b}\right)^2 \\ -\text{j}\sqrt{-k_0^2 \varepsilon_r \mu_r + \left(\frac{m\pi}{a}\right)^2 + \left(\frac{n\pi}{b}\right)^2}, & k_0^2 \varepsilon_r \mu_r < \left(\frac{m\pi}{a}\right)^2 + \left(\frac{n\pi}{b}\right)^2, \end{cases} \tag{6.25}$$

where the square root is greater than or equal to zero and $k_0 = \omega\sqrt{\varepsilon_0 \mu_0}$ is the free-space wave number.

The modal wave admittance $Y_{\text{TE}_{mn}}$ for the TE_{mn} mode follows from substitution of equation (6.26) [36],

$$\boldsymbol{h}^{\pm}_{\text{T}_{mn}} = \pm Y_{\text{TE}_{mn}} (\hat{\boldsymbol{i}}_z \times \boldsymbol{e}^{\pm}_{\text{T}_{mn}}), \tag{6.26}$$

into equations (6.9) and (6.10). This results in

$$Y_{\text{TE}_{mn}} = \frac{\gamma_{mn}}{\omega \mu_0 \mu_r}. \tag{6.27}$$

The wave impedance for the TE_{mn} mode is then, by default, $Z_{\text{TE}_{mn}} = 1/Y_{\text{TE}_{mn}}$.

6.2.2 TM Modes

For TM modes, $h^{\pm}_z = 0$ and $e^{\pm}_z \neq 0$. All field components are now determined from a scalar quantity $e^{\pm}_z$ that satisfies the two-dimensional Helmholtz equation

$$(\nabla_\text{T} \cdot \nabla_\text{T} + k^2_\text{T}) e^{\pm}_z = 0. \tag{6.28}$$

As in the TE-mode situation, nontrivial solutions to equation (6.28) exist only for certain values – different from those for TE modes – of k^2_T. Applying the method of separation of variables leads to

$$e^{\pm}_z = A'^{\pm} \sin\left[k_x\left(x + \frac{a}{2}\right)\right] \sin\left[k_y\left(y + \frac{b}{2}\right)\right], \tag{6.29}$$

where $A'^{\pm}$ is the mode amplitude coefficient (as yet unknown).[5]

Using the boundary conditions that $e_z^{\pm}$ is zero at the waveguide boundaries $x = \pm a/2$ and $y = \pm b/2$ (we assume perfectly electrically conducting waveguide walls) leads to the following expressions for the separation constants:

$$k_x = \frac{m\pi}{a}, \quad m \in \mathbb{N} \wedge m \neq 0, \tag{6.30}$$

$$k_y = \frac{n\pi}{b}, \quad n \in \mathbb{N} \wedge n \neq 0. \tag{6.31}$$

Again, taking only positive values for k_x and k_y into account, the following relation applies for the transverse wave number of the TM_{mn} mode:

$$k_{\mathrm{TM}_{mn}}^2 = \left(\frac{m\pi}{a}\right)^2 + \left(\frac{n\pi}{b}\right)^2, \quad m, n \in \mathbb{N}, \quad m \neq 0; \quad n \neq 0. \tag{6.32}$$

Substitution of equation (6.32) into equation (6.29) yields

$$e_{z_{mn}}^{\pm} = A'^{\pm}_{mn} \sin\left[\frac{m\pi}{a}\left(x + \frac{a}{2}\right)\right] \sin\left[\frac{n\pi}{b}\left(y + \frac{b}{2}\right)\right], \tag{6.33}$$

and the axial wave number is given by equation (6.25).

The wave impedance $Z_{\mathrm{TM}_{mn}}$ for the TM_{mn} mode follows from substitution of equation (6.34) [36],

$$\boldsymbol{e}_{\mathrm{T}_{mn}}^{\pm} = \pm Z_{\mathrm{TM}_{mn}} (\boldsymbol{h}_{\mathrm{T}_{mn}}^{\pm} \times \hat{\boldsymbol{i}}_z), \tag{6.34}$$

into equations (6.9) and (6.10), which results in

$$Z_{\mathrm{TM}_{mn}} = \frac{\gamma_{mn}}{\omega \varepsilon_0 \varepsilon_{\mathrm{r}}}. \tag{6.35}$$

The wave admittance for the TM_{mn} mode is, by definition, $Y_{\mathrm{TM}_{mn}} = 1/Z_{\mathrm{TM}_{mn}}$.

6.2.3 Transverse Field Components

The complete transverse fields $\boldsymbol{E}_{\mathrm{T}}$ and $\boldsymbol{H}_{\mathrm{T}}$ in the waveguide, at a cross-sectional position z, consist of a linear superposition of modal field distributions

$$\boldsymbol{E}_{\mathrm{T}}(\mathbf{r}) = \sum_{m=0}^{\infty} \sum_{n=0}^{\infty} e_{\mathrm{T}_{mn}}^{\pm}(\mathbf{r}_{\mathrm{T}}) \mathrm{e}^{\mp \mathrm{j}\gamma_{mn} z}, \tag{6.36}$$

$$\boldsymbol{H}_{\mathrm{T}}(\mathbf{r}) = \sum_{m=0}^{\infty} \sum_{n=0}^{\infty} h_{\mathrm{T}_{mn}}^{\pm}(\mathbf{r}_{\mathrm{T}}) \mathrm{e}^{\mp \mathrm{j}\gamma_{mn} z}, \tag{6.37}$$

where it is implicitly understood that $(m, n) \neq (0, 0)$ for both TE and TM modes, and $m \neq 0$ and $n \neq 0$ for TM-mode contributions.

[5]To distinguish between TE- and TM-mode solutions in the remainder of this chapter, TE-mode amplitude coefficients will be unprimed and TM-mode amplitude coefficients will be primed.

Equations (6.36) and (6.37) can be separated into four scalar equations for the transverse field components:

$$E_x(\boldsymbol{r}) = \sum_{m=0}^{\infty}\sum_{n=0}^{\infty} e^{\pm}_{x_{mn}}(x, y)\mathrm{e}^{\mp \mathrm{j}\gamma_{mn}z}, \tag{6.38}$$

$$E_y(\boldsymbol{r}) = \sum_{m=0}^{\infty}\sum_{n=0}^{\infty} e^{\pm}_{y_{mn}}(x, y)\mathrm{e}^{\mp \mathrm{j}\gamma_{mn}z}, \tag{6.39}$$

$$H_x(\boldsymbol{r}) = \sum_{m=0}^{\infty}\sum_{n=0}^{\infty} h^{\pm}_{x_{mn}}(x, y)\mathrm{e}^{\mp j\gamma_{mn}z}, \tag{6.40}$$

$$H_y(\boldsymbol{r}) = \sum_{m=0}^{\infty}\sum_{n=0}^{\infty} h^{\pm}_{y_{mn}}(x, y)\mathrm{e}^{\mp \mathrm{j}\gamma_{mn}z}, \tag{6.41}$$

where

$$e^{\pm}_{x_{mn}} = A^{\pm}_{mn} e^{\pm}_{x\mathrm{TE}_{mn}} + A'^{\pm}_{mn} e^{\pm}_{x\mathrm{TM}_{mn}}, \tag{6.42}$$

$$e^{\pm}_{y_{mn}} = A^{\pm}_{mn} e^{\pm}_{y\mathrm{TE}_{mn}} + A'^{\pm}_{mn} e^{\pm}_{y\mathrm{TM}_{mn}}, \tag{6.43}$$

$$h^{\pm}_{x_{mn}} = A^{\pm}_{mn} h^{\pm}_{x\mathrm{TE}_{mn}} + A'^{\pm}_{mn} h^{\pm}_{x\mathrm{TM}_{mn}}, \tag{6.44}$$

$$h^{\pm}_{y_{mn}} = A^{\pm}_{mn} h^{\pm}_{y\mathrm{TE}_{mn}} + A'^{\pm}_{mn} h^{\pm}_{y\mathrm{TM}_{mn}}. \tag{6.45}$$

With the aid of equations (6.9), (6.10), (6.24) and (6.33), the mode functions can be found as

$$e^{\pm}_{x\mathrm{TE}_{mn}} = \frac{\mathrm{j}}{k^2_{\mathrm{T}_{mn}}}\omega\mu_0\mu_\mathrm{r}\left(\frac{n\pi}{b}\right)C_m\left(\frac{x}{a}\right)S_n\left(\frac{y}{b}\right), \tag{6.46}$$

$$e^{\pm}_{y\mathrm{TE}_{mn}} = -\frac{\mathrm{j}}{k^2_{\mathrm{T}_{mn}}}\omega\mu_0\mu_\mathrm{r}\left(\frac{m\pi}{a}\right)S_m\left(\frac{x}{a}\right)C_n\left(\frac{y}{b}\right), \tag{6.47}$$

$$e^{\pm}_{x\mathrm{TM}_{mn}} = \mp\frac{\mathrm{j}}{k^2_{\mathrm{T}_{mn}}}\gamma_{mn}\left(\frac{m\pi}{a}\right)C_m\left(\frac{x}{a}\right)S_n\left(\frac{y}{b}\right), \tag{6.48}$$

$$e^{\pm}_{y\mathrm{TM}_{mn}} = \mp\frac{\mathrm{j}}{k^2_{\mathrm{T}_{mn}}}\gamma_{mn}\left(\frac{n\pi}{b}\right)S_m\left(\frac{x}{a}\right)C_n\left(\frac{y}{b}\right), \tag{6.49}$$

and, using equations (6.27) and (6.35),

$$h^{\pm}_{x\mathrm{TE}_{mn}} = \mp Y_{\mathrm{TE}_{mn}} e^{\pm}_{y\mathrm{TE}_{mn}}, \tag{6.50}$$

$$h^{\pm}_{y\mathrm{TE}_{mn}} = \pm Y_{\mathrm{TE}_{mn}} e^{\pm}_{x\mathrm{TE}_{mn}}, \tag{6.51}$$

$$h^{\pm}_{x\mathrm{TM}_{mn}} = \mp Y_{\mathrm{TM}_{mn}} e^{\pm}_{y\mathrm{TM}_{mn}}, \tag{6.52}$$

$$h^{\pm}_{y\mathrm{TM}_{mn}} = \pm Y_{\mathrm{TM}_{mn}} e^{\pm}_{x\mathrm{TM}_{mn}}, \tag{6.53}$$

where

$$C_\ell(\xi) = \cos\left[\ell\pi\left(\xi + \frac{1}{2}\right)\right], \quad \ell = 0, 1, 2, \ldots, \quad -\frac{1}{2} \le \xi \le \frac{1}{2}, \tag{6.54}$$

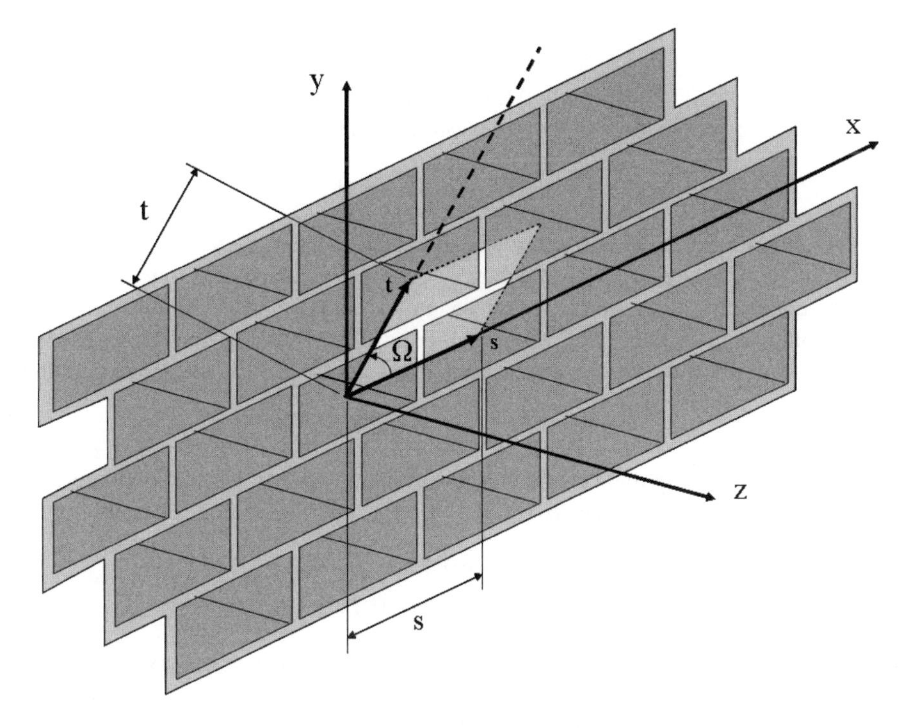

Figure 6.3 Planar, infinite open-ended waveguide array antenna with the radiators arranged in a triangular grating, plus an indication of a single unit cell.

and

$$S_\ell(\xi) = \sin\left[\ell\pi\left(\xi + \frac{1}{2}\right)\right], \quad \ell = 1, 2, 3, \ldots, \quad -\frac{1}{2} \leq \xi \leq \frac{1}{2}. \tag{6.55}$$

6.3 UNIT CELL FIELDS

By virtue of the infinite planar extent and periodicity of an open-ended waveguide array, the resulting two-dimensional grating may be decomposed into an infinite sequence of unit cells (Figure 6.3). In this figure, the radiators are arranged in a general triangular lattice that is defined by the distances s and t between elements and the angle Ω between the s and t directions.[6] One unit cell is indicated in a lighter shade.

The analysis of the fields is similar to that for a rectangular waveguide with electrically conducting side walls. So equations (6.1)–(6.14) apply equally well to the unit cell. However, in analysing the TE- and TM-mode fields, we have to take the transverse unboundedness and the periodicity of the structure into consideration.

[6]For $\Omega = \pi/2$, we have a rectangular lattice.

6.3.1 TE Modes

For TE modes, $e_z^\pm = 0$ and $h_z^\pm \neq 0$. All field components are determined from the scalar quantity $h_z^\pm$, where $h_z^\pm$ satisfies the two-dimensional Helmholtz equation

$$(\nabla_T \cdot \nabla_T + k_T^2)h_z^\pm = 0. \tag{6.56}$$

Nontrivial solutions to this equation exist only for certain values of k_T^2.

In the situation where we were dealing with a rectangular waveguide with perfectly electrically conducting side walls, we chose, in the method of separation of variables, a sample solution in the form of a product of cosines. Now, owing to the transverse unboundedness of the structure, we choose a sample solution that has an exponential form,

$$h_z^\pm = B^\pm f(x, y)\mathrm{e}^{-\mathrm{j}\boldsymbol{k}_{\mathrm{T}0}\cdot\boldsymbol{r}_\mathrm{T}} = B^\pm f(x, y)\mathrm{e}^{-\mathrm{j}(k_{x0}x + k_{y0}y)}, \tag{6.57}$$

where $B^\pm$ is the mode amplitude coefficient (as yet unknown), $f(x, y)$ is a periodic function and the separation constants are related to the spherical-coordinate angles ϑ and φ through

$$k_{x0} = k\sin(\vartheta)\cos(\varphi), \tag{6.58}$$

$$k_{y0} = k\sin(\vartheta)\sin(\varphi), \tag{6.59}$$

with $k = k_0\sqrt{\varepsilon_\mathrm{r}\mu_\mathrm{r}}$ being the wave number in the medium immediately outside the radiating apertures.

The periodicity of the grating is accounted for in the function $f(x, y)$. Since all unit cells are identical, we see from inspection of Figure 6.3 that $f(x, y)$ must satisfy [37]

$$f(x + s, y) = f(x, y) \tag{6.60}$$

and

$$f(x + t\cos(\Omega), y + t\sin(\Omega)) = f(x, y). \tag{6.61}$$

The function that satisfies these conditions is given by

$$f(x, y) = \mathrm{e}^{-\mathrm{j}p2\pi x/s}\mathrm{e}^{\mathrm{j}p2\pi y/[s\tan(\Omega)]}\mathrm{e}^{-\mathrm{j}q2\pi y/[t\sin(\Omega)]}, \quad p, q \in \mathbb{Z}, \tag{6.62}$$

which results in, for the TE_{pq} mode,

$$h_{pq}^\pm = B_{pq}^\pm \mathrm{e}^{-\mathrm{j}(u_{p0}x + v_{pq}y)}, \tag{6.63}$$

where

$$u_{p0} = k\sin(\vartheta)\cos(\varphi) + \frac{2p\pi}{s}, \quad p \in \mathbb{Z}, \tag{6.64}$$

and

$$v_{pq} = k\sin(\vartheta)\sin(\varphi) - \frac{2p\pi}{s\tan(\Omega)} + \frac{2q\pi}{t\sin(\Omega)}, \quad p, q \in \mathbb{Z}. \tag{6.65}$$

The transverse wave number of the TE_{pq} mode, $k_{\mathrm{T}_{pq}}$, which now accounts for the periodicity of the structure, is given by

$$k_{\mathrm{T}_{pq}} = \sqrt{u_{p0}^2 + v_{pq}^2}, \quad p, q \in \mathbb{Z}. \tag{6.66}$$

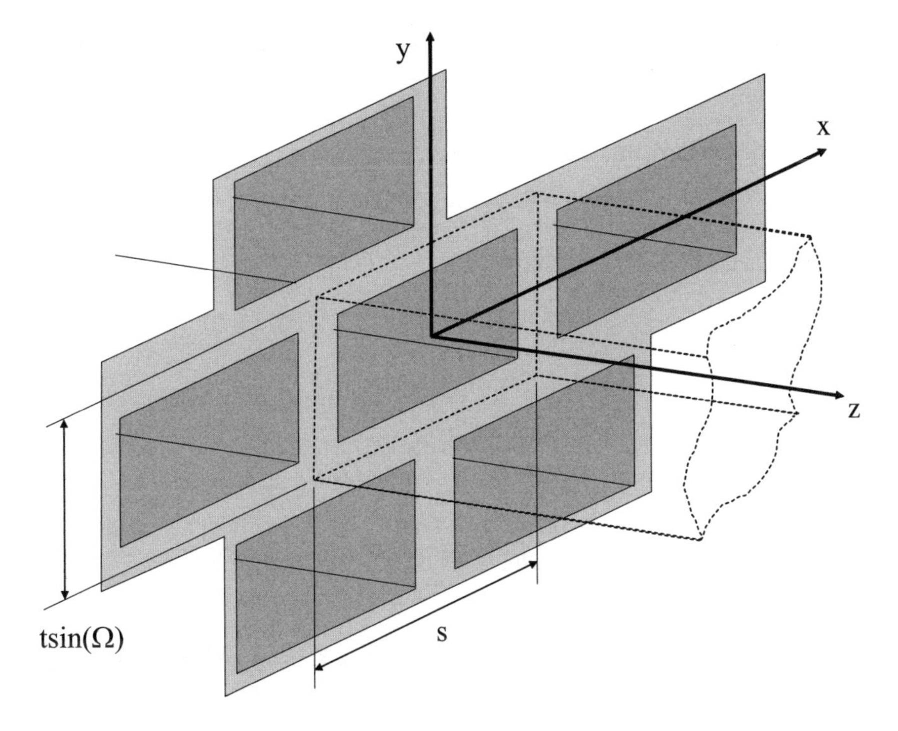

Figure 6.4 Rectangular unit cell, centered above open-ended waveguide radiator.

Each set of values (p, q) specifies a plane wave,[7] radiated into the direction (ϑ, φ), called a *space harmonic*, *grating mode* or *Floquet mode*.

As long as we do not modify the periodicity, we may deform and shift the position of the unit cell shown in Figure 6.3. For convenience, we therefore assume rectangular unit cells, centered above the open-ended waveguide apertures, as shown in Figure 6.4. The dimensions of the rectangular unit cell are s and $t\,\sin(\Omega)$ in the x and y coordinates, respectively.

It is now apparent that we may treat the unit cell as a rectangular waveguide where the walls of this waveguide do not have reflective properties, but instead phase-shifting properties, as shown by equations (6.64)–(6.66).

The axial wave number follows from equation (6.14) as

$$\gamma_{pq} = \begin{cases} \sqrt{k_0^2\varepsilon_r\mu_r - u_{p0}^2 - v_{pq}^2}, & k_0^2\varepsilon_r\mu_r \geq u_{p0}^2 + v_{pq}^2 \\ -\mathrm{j}\sqrt{-k_0^2\varepsilon_r\mu_r + u_{p0}^2 + v_{pq}^2}, & k_0^2\varepsilon_r\mu_r < u_{p0}^2 + v_{pq}^2, \end{cases} \tag{6.67}$$

where the square root is greater than or equal to zero.

[7]Contrary to the situation for a rectangular waveguide with a metallic wall, negative values need to be taken into account here, since they represent solutions that differ from their positive counterparts.

The modal wave admittance $Y_{\mathrm{TE}_{pq}}$ for the TE_{pq} mode follows from the substitution of equation (6.68) [36],

$$\boldsymbol{h}^{\pm}_{\mathrm{T}_{pq}} = \pm Y_{\mathrm{TE}_{pq}}(\hat{\boldsymbol{i}}_z \times \boldsymbol{e}^{\pm}_{\mathrm{T}_{pq}}), \tag{6.68}$$

into equations (6.9) and (6.10). This results in

$$Y_{\mathrm{TE}_{pq}} = \frac{\gamma_{pq}}{\omega\mu_0\mu_\mathrm{r}}. \tag{6.69}$$

The wave impedance for the TE_{pq} mode is $Z_{\mathrm{TE}_{pq}} = 1/Y_{\mathrm{TE}_{pq}}$.

6.3.2 TM Modes

For TM modes, $h^{\pm}_z = 0$ and $e^{\pm}_z \neq 0$. All field components are determined by $e^{\pm}_z$, which satisfies the two-dimensional wave equation

$$(\nabla_\mathrm{T} \cdot \nabla_\mathrm{T} + k^2_\mathrm{T})e^{\pm}_z = 0. \tag{6.70}$$

Solutions to equation (6.70) exist only for certain values of k^2_T. In applying the method of separation of variables, we take as a sample solution to equation (6.70)

$$e^{\pm}_{z_{pq}} = B'^{\pm}_{pq} f(x, y)\mathrm{e}^{-\mathrm{j}\boldsymbol{k}_{\mathrm{T}0}\cdot\boldsymbol{r}_\mathrm{T}}, \tag{6.71}$$

where $B'^{\pm}$ is the mode amplitude coefficient (as yet unknown) and $f(x, y)$ is a function related to the periodicity of the structure. Following the same procedure as outlined in the section 6.3.1 for TE modes, we finally find for the TM_{pq} mode

$$e^{\pm}_{z_{pq}} = B'^{\pm}_{pq}\mathrm{e}^{-\mathrm{j}(u_{p0}x + v_{pq}y)}, \tag{6.72}$$

where u_{p0} and v_{pq} are defined by equations (6.64) and (6.65), respectively. The transverse wave number is defined by equation (6.76) and the axial wave number is defined by equation (6.67).

The wave impedance $Z_{\mathrm{TM}_{pq}}$ for the TM_{pq} mode follows from substitution of equation (6.73) [36],

$$\boldsymbol{e}^{\pm}_{\mathrm{T}_{pq}} = \pm Z_{\mathrm{TM}_{pq}}(\boldsymbol{h}^{\pm}_{\mathrm{T}_{pq}} \times \hat{\boldsymbol{i}}_z), \tag{6.73}$$

into equations (6.9) and (6.10), which results in

$$Z_{\mathrm{TM}_{pq}} = \frac{\gamma_{pq}}{\omega\varepsilon_0\varepsilon_\mathrm{r}}. \tag{6.74}$$

The wave admittance for the TM_{pq} mode is $Y_{\mathrm{TM}_{pq}} = 1/Z_{\mathrm{TM}_{pq}}$.

6.3.3 Transverse Field Components

The complete transverse fields in the unit cell consist of a superposition of mode contributions

$$\boldsymbol{E}_\mathrm{T}(\boldsymbol{r}) = \sum_{p=-\infty}^{\infty}\sum_{q=-\infty}^{\infty} \boldsymbol{e}^{\pm}_{\mathrm{T}_{pq}}(\boldsymbol{r}_\mathrm{T})\mathrm{e}^{\mp\gamma_{pq}z}, \tag{6.75}$$

$$\boldsymbol{H}_\mathrm{T}(\boldsymbol{r}) = \sum_{p=-\infty}^{\infty}\sum_{q=-\infty}^{\infty} \boldsymbol{h}^{\pm}_{\mathrm{T}_{pq}}(\boldsymbol{r}_\mathrm{T})\mathrm{e}^{\mp\gamma_{pq}z}. \tag{6.76}$$

Equations (6.75) and (6.76) can be separated into four scalar equations for the transverse field components:

$$E_x(\boldsymbol{r}) = \sum_{p=-\infty}^{\infty} \sum_{q=-\infty}^{\infty} e_{x_{pq}}^{\pm}(x, y)\mathrm{e}^{\mp\gamma_{pq}z}, \tag{6.77}$$

$$E_y(\boldsymbol{r}) = \sum_{p=-\infty}^{\infty} \sum_{q=-\infty}^{\infty} e_{y_{pq}}^{\pm}(x, y)\mathrm{e}^{\mp\gamma_{pq}z}, \tag{6.78}$$

$$H_x(\boldsymbol{r}) = \sum_{p=-\infty}^{\infty} \sum_{q=-\infty}^{\infty} h_{x_{pq}}^{\pm}(x, y)\mathrm{e}^{\mp\gamma_{pq}z}, \tag{6.79}$$

$$H_y(\boldsymbol{r}) = \sum_{p=-\infty}^{\infty} \sum_{q=-\infty}^{\infty} h_{y_{pq}}^{\pm}(x, y)\mathrm{e}^{\mp\gamma_{pq}z}, \tag{6.80}$$

where

$$e_{x_{pq}}^{\pm} = B_{pq}^{\pm} e_{x\mathrm{TE}_{pq}}^{\pm} + B_{pq}^{\prime\pm} e_{x\mathrm{TM}_{pq}}^{\pm}, \tag{6.81}$$

$$e_{y_{pq}}^{\pm} = B_{pq}^{\pm} e_{y\mathrm{TE}_{pq}}^{\pm} + B_{pq}^{\prime\pm} e_{y\mathrm{TM}_{pq}}^{\pm}, \tag{6.82}$$

$$h_{x_{pq}}^{\pm} = B_{pq}^{\pm} h_{x\mathrm{TE}_{pq}}^{\pm} + B_{pq}^{\prime\pm} h_{x\mathrm{TM}_{pq}}^{\pm}, \tag{6.83}$$

$$h_{y_{pq}}^{\pm} = B_{pq}^{\pm} h_{y\mathrm{TE}_{pq}}^{\pm} + B_{pq}^{\prime\pm} h_{y\mathrm{TM}_{pq}}^{\pm}. \tag{6.84}$$

With the aid of equations (6.9), (6.10), (6.63) and (6.72), the mode functions can be found as

$$e_{x\mathrm{TE}_{pq}}^{\pm} = -\frac{1}{k_{\mathrm{T}_{pq}}^2}\omega\mu_0\mu_{\mathrm{r}}v_{pq}X_{pq}, \tag{6.85}$$

$$e_{y\mathrm{TE}_{pq}}^{\pm} = \frac{1}{k_{\mathrm{T}_{pq}}^2}\omega\mu_0\mu_{\mathrm{r}}u_{p0}X_{pq}, \tag{6.86}$$

$$e_{x\mathrm{TM}_{pq}}^{\pm} = \mp\frac{1}{k_{\mathrm{T}_{pq}}^2}\gamma_{pq}u_{p0}X_{pq}, \tag{6.87}$$

$$e_{y\mathrm{TM}_{pq}}^{\pm} = \mp\frac{1}{K_{\mathrm{T}_{pq}}^2}\gamma_{pq}v_{pq}X_{pq}, \tag{6.88}$$

and, using equations (6.69) and (6.74),

$$h_{x\mathrm{TE}_{pq}}^{\pm} = \mp Y_{\mathrm{TE}_{pq}} e_{y\mathrm{TE}_{pq}}^{\pm}, \tag{6.89}$$

$$h_{y\mathrm{TE}_{pq}}^{\pm} = \pm Y_{\mathrm{TE}_{pq}} e_{x\mathrm{TE}_{pq}}^{\pm}, \tag{6.90}$$

$$h_{x\mathrm{TM}_{pq}}^{\pm} = \mp Y_{\mathrm{TM}_{pq}} e_{y\mathrm{TM}_{pq}}^{\pm}, \tag{6.91}$$

$$h_{y\mathrm{TM}_{pq}}^{\pm} = \pm Y_{\mathrm{TM}_{pq}} e_{x\mathrm{TM}_{pq}}^{\pm}, \tag{6.92}$$

where

$$X_{pq} = \mathrm{e}^{-\mathrm{j}(u_{p0}x + v_{pq}y)} \tag{6.93}$$

and

$$k_{\mathrm{T}_{pq}}^2 = \boldsymbol{k}_{\mathrm{T}_{pq}} \cdot \boldsymbol{k}_{\mathrm{T}_{pq}} = u_{p0}^2 + v_{pq}^2. \tag{6.94}$$

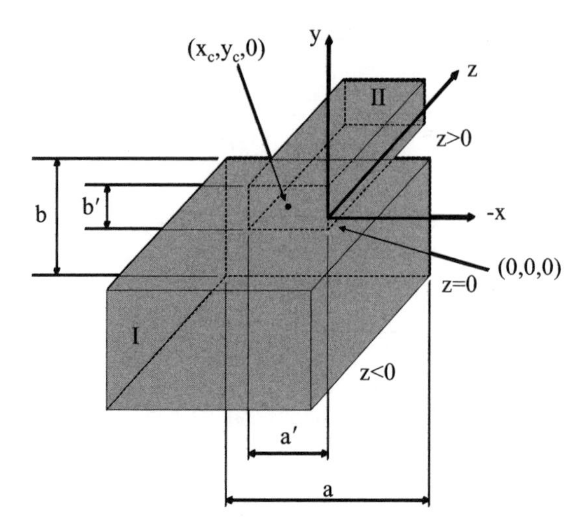

Figure 6.5 Cross-sectional step (reduction) in a rectangular waveguide, in the plane $z = 0$.

6.4 CROSS-SECTIONAL STEP IN A RECTANGULAR WAVEGUIDE

In this section, we shall derive the GSM for a one- or two-dimensional cross-sectional step in a rectangular waveguide. Seen from the input waveguide, we may regard this step as a waveguide reduction or a waveguide enlargement. When at least one of the two transverse dimensions of the output waveguide is larger than the corresponding dimension of the input waveguide, we define the step as a waveguide *enlargement*. Otherwise, we define the step as a waveguide *reduction*. Figure 6.5 shows a junction of two rectangular waveguides. The input waveguide ($z < 0$) is designated as waveguide *I*, and the output waveguide ($z > 0$) is designated as waveguide *II*. According to the definition given above, the configuration shown in the figure is a waveguide reduction.[8]

The input waveguide has cross-sectional dimensions a and b. The output waveguide has cross-sectional dimensions a' and b', where $a' \leq a$ and $b' \leq b$. The step is positioned in the plane $z = 0$. A rectangular coordinate system is centered with respect to the input waveguide. The output waveguide has the center of its cross section at $(x, y, z) = (x_c, y_c, 0)$, where

$$|x_c| + \frac{a'}{2} \leq \frac{a}{2} \tag{6.95}$$

and

$$|y_c| + \frac{b'}{2} \leq \frac{b}{2}, \tag{6.96}$$

such that the cross section of the output waveguide falls entirely within that of the input waveguide.

[8] A change in the medium filling a rectangular waveguide with a uniform cross section will also be considered a cross-sectional step.

Also, S^I is the area of the cross section of waveguide I ($-a/2 \leq x \leq a/2$, $-b/2 \leq y \leq b/2$), S^{II} is the area of the cross-section of waveguide II ($-a'/2 \leq x - x_c \leq a'/2$, $-b'/2 \leq y - y_c \leq b'/2$) and S is the area of the cross section of waveguide I, excluding the cross section of waveguide II. S is the complement of the intersection of S^I and S^{II}. S is assumed to be perfectly electrically conducting.

In accordance with the analysis in terms of modes presented in previous sections, the fields satisfy the boundary conditions at the transverse boundaries of the waveguides. By enforcing the boundary conditions across the interface at $z = 0$ – the tangential components of the electric and magnetic fields must be continuous across the interface – a matrix formulation for this step discontinuity may be derived, as shown in the remainder of this section.

6.4.1 Boundary Conditions Across the Interface

The continuity of the tangential fields across the interface at $z = 0$, written for the transverse Cartesian components of the electric and magnetic fields, is specified by

$$\lim_{z\uparrow 0} E_x^I = \lim_{z\downarrow 0} E_x^{II}, \tag{6.97}$$

$$\lim_{z\uparrow 0} E_y^I = \lim_{z\downarrow 0} E_y^{II}, \tag{6.98}$$

$$\lim_{z\uparrow 0} H_x^I = \lim_{z\downarrow 0} H_x^{II}, \tag{6.99}$$

$$\lim_{z\uparrow 0} H_y^I = \lim_{z\downarrow 0} H_y^{II}, \tag{6.100}$$

where the superscripts I and II relate to the field components in waveguides I and II, respectively.

Using equations (6.38)–(6.41), equations (6.42)–(6.45), equations (6.50)–(6.53) and

$$e^-_{x\mathrm{TE}mn} = e^+_{x\mathrm{TE}mn}, \tag{6.101}$$

$$e^-_{y\mathrm{TE}mn} = e^+_{y\mathrm{TE}mn}, \tag{6.102}$$

$$e^-_{x\mathrm{TM}mn} = -e^+_{x\mathrm{TM}mn}, \tag{6.103}$$

$$e^-_{y\mathrm{TM}mn} = -e^+_{y\mathrm{TM}mn}, \tag{6.104}$$

within the boundaries $-a/2 \leq x \leq a/2$, $-b/2 \leq y \leq b/2$, results in

$$\sum_{m=0}^{M}\sum_{n=0}^{N}\left\{\left[A_{mn}^{I+} + A_{mn}^{I-}\right]e^{I+}_{x\mathrm{TE}mn} + \left[A_{mn}^{\prime I+} - A_{mn}^{\prime I-}\right]e^{I+}_{x\mathrm{TM}mn}\right\}$$
$$= \begin{cases} \displaystyle\sum_{l=0}^{L}\sum_{w=0}^{W}\left\{\left[A_{lw}^{II+} + A_{lw}^{II-}\right]e^{II+}_{x\mathrm{TE}lw} + \left[A_{lw}^{\prime II+} - A_{lw}^{\prime II-}\right]e^{II+}_{x\mathrm{TM}lw}\right\}, \\ \quad -\dfrac{a'}{2} + x_c \leq x \leq \dfrac{a'}{2} + x_c \wedge -\dfrac{b'}{2} + y_c \leq y \leq \dfrac{b'}{2} + y_c, \\ 0, \quad \text{otherwise}, \end{cases} \tag{6.105}$$

$$\sum_{m=0}^{M}\sum_{n=0}^{N}\left\{\left[A_{mn}^{I+}+A_{mn}^{I-}\right]e_{y\mathrm{TE}_{mn}}^{I+}+\left[A_{mn}^{\prime I+}-A_{mn}^{\prime I-}\right]e_{y\mathrm{TM}_{mn}}^{I+}\right\}$$

$$=\begin{cases}\displaystyle\sum_{l=0}^{L}\sum_{w=0}^{W}\left\{\left[A_{lw}^{II+}+A_{lw}^{II-}\right]e_{y\mathrm{TE}_{lw}}^{II+}+\left[A_{lw}^{\prime II+}-A_{lw}^{\prime II-}\right]e_{y\mathrm{TM}_{lw}}^{II+}\right\}, \\ \quad -\frac{a'}{2}+x_c\le x\le\frac{a'}{2}+x_c\wedge-\frac{b'}{2}+y_c\le y\le\frac{b'}{2}+y_c, \\ 0, \quad \text{otherwise,}\end{cases} \tag{6.106}$$

$$\sum_{m=0}^{M}\sum_{n=0}^{N}\left\{\left[-A_{mn}^{I+}+A_{mn}^{I-}\right]Y_{\mathrm{TE}_{mn}}^{I}e_{y\mathrm{TE}_{mn}}^{I+}+\left[-A_{mn}^{\prime I+}-A_{mn}^{\prime I-}\right]Y_{\mathrm{TM}_{mn}}^{I}e_{y\mathrm{TM}_{mn}}^{I+}\right\}$$

$$=\sum_{l=0}^{L}\sum_{w=0}^{W}\left\{\left[-A_{lw}^{\prime I+}+A_{lw}^{\prime I-}\right]Y_{\mathrm{TE}_{lw}}^{II}e_{y\mathrm{TE}_{lw}}^{II+}+\left[-A_{lw}^{\prime II+}-A_{lw}^{\prime II-}\right]Y_{\mathrm{TM}_{lw}}^{II}e_{y\mathrm{TM}_{lw}}^{II+}\right\},$$

$$-\frac{a'}{2}+x_c\le x\le\frac{a'}{2}+x_c\wedge-\frac{b'}{2}+y_c\le y\le\frac{b'}{2}+y_c, \tag{6.107}$$

$$\sum_{m=0}^{M}\sum_{n=0}^{N}\left\{\left[A_{mn}^{I+}-A_{mn}^{I-}\right]Y_{\mathrm{TE}_{mn}}^{I}e_{x\mathrm{TE}_{mn}}^{I+}+\left[A_{mn}^{\prime I+}+A_{mn}^{\prime I-}\right]Y_{\mathrm{TM}_{mn}}^{I}e_{x\mathrm{TM}_{mn}}^{I+}\right\}$$

$$=\sum_{l=0}^{L}\sum_{w=0}^{W}\left\{\left[A_{lw}^{\prime I+}-A_{lw}^{\prime I-}\right]Y_{\mathrm{TE}_{lw}}^{II}e_{x\mathrm{TE}_{lw}}^{II+}+\left[A_{lw}^{\prime II+}+A_{lw}^{\prime II-}\right]Y_{\mathrm{TM}_{lw}}^{II}e_{x\mathrm{TM}_{lw}}^{II+}\right\},$$

$$-\frac{a'}{2}+x_c\le x\le\frac{a'}{2}+x_c\wedge-\frac{b'}{2}+y_c\le y\le\frac{b'}{2}+y_c. \tag{6.108}$$

In the above equations, A_{mn}^{I+} and A_{mn}^{I-} are the mode amplitude coefficients in waveguide I for the TE_{mn} mode traveling in the increasing and the decreasing z direction, respectively. $A_{mn}^{\prime I+}$ and $A_{mn}^{\prime I-}$ are the mode amplitude coefficients in waveguide I for the TM_{mn} mode traveling in the increasing and the decreasing z direction, respectively. A_{lw}^{II+} and A_{lw}^{II-} are the mode amplitude coefficients in waveguide II for the TE_{lw} mode traveling in the increasing and the decreasing z direction, respectively, and $A_{lw}^{\prime II+}$ and $A_{lw}^{\prime II-}$ are the mode amplitude coefficients in this waveguide for the TM_{lw} mode traveling in the increasing and the decreasing z direction, respectively.

The mode functions (solutions of the two-dimensional Helmholtz equation) and mode admittances are superscripted with a 'I' or a 'II', depending on which waveguide is being considered. For waveguide I, the mode functions are given by equations (6.46)–(6.49); for waveguide II, the same equations are used, but a' is substituted for a, b' is substituted for b, l is substituted for m and w is substituted for n. The TE and TM admittance functions for waveguide I are given by equations (6.27) and (6.35). For waveguide II, these functions are obtained by the same equations upon substitution of l for m and w for n.

Equations (6.105)–(6.108) are exact for $M=\infty$, $N=\infty$ and $L=\infty$, $W=\infty$. For the exact equations, the matching procedure leads to an infinite system of linear equations with $A_{mn}^{I\pm}$, $A_{mn}^{\prime I\pm}$, $A_{mn}^{II\pm}$ and $A_{mn}^{\prime II\pm}$ as unknowns. Fortunately, it turns out that not all modes contribute equally strongly to the GSM characterization of a waveguide junction. In fact, only the fundamental waveguide modes and a limited set of higher-order modes suffice for an accurate characterization. In the next subsection, the creation of a finite system of linear equations will be discussed.

6.4.2 Creation of a Finite System of Linear Equations

We start by truncating the infinite summations to finite ones, as has been done in equations (6.105)–(6.108). We now have a finite system of linear equations. However, these equations must be satisfied simultaneously at all positions (x, y). Therefore, the system cannot be solved at this point. To get a system of K ($K \in \mathbb{N}$) linear, independent equations with K unknowns that can be solved, we need to establish tangential-field equalities in a weak form. To that end, we must, integrate the tangential electric and magnetic fields over the cross section. By applying an appropriate weighting function, we can arrive at the required finite system of linear equations.

The appropriate weighting functions are provided by the transverse mode functions of the rectangular waveguide, owing to their orthogonality properties [9]. By utilizing these orthogonality properties (see appendix 6.10), we are able to reduce a double summation in the integration to a single term only, as will be shown in the remainder of this section.

We start by taking the vector product of both sides of the electric-field continuity equations (equations (6.97) and (6.98)) with the complex conjugate of the magnetic-field mode vector of waveguide I. Then a dot product is formed WITH the unit vector in the direction of propagation and, finally, the resulting equation is integrated over the cross section of the interface. Remembering that S^I is the cross section of waveguide I, S^{II} is the cross section of waveguide II and S is the cross-sectional area of waveguide I, excluding the cross-sectional area of waveguide II, we obtain

$$\iint_S \left(\boldsymbol{E}_\mathrm{T}^I \times \boldsymbol{h}_{\mathrm{T}_{ij}}^{\mathrm{I}*}\right) \cdot \hat{\mathbf{i}}_z \, dS = 0, \quad i, j \in \mathbb{N}, \tag{6.109}$$

$$\iint_{S^{II}} \left(\boldsymbol{E}_\mathrm{T}^I \times \boldsymbol{h}_{\mathrm{T}_{ij}}^{I*}\right) \cdot \hat{\mathbf{i}}_z \, dS = \iint_{S^{II}} \left(\boldsymbol{E}_\mathrm{T}^{II} \times \boldsymbol{h}_{\mathrm{T}_{ij}}^{I*}\right) \cdot \hat{\mathbf{i}}_z \, dS, \quad i, j \in \mathbb{N}, \tag{6.110}$$

where

$$\boldsymbol{E}_\mathrm{T}^I = E_x^I \hat{\boldsymbol{i}}_x + E_y^I \hat{\boldsymbol{i}}_y, \tag{6.111}$$

$$\boldsymbol{E}_\mathrm{T}^{II} = E_x^{II} \hat{\boldsymbol{i}}_x + E_y^{II} \hat{\boldsymbol{i}}_y \tag{6.112}$$

and

$$\boldsymbol{h}_{\mathrm{T}_{ij}}^I = h_{x_{ij}}^I \hat{\boldsymbol{i}}_x + h_{y_{ij}}^I \hat{\boldsymbol{i}}_y, \quad i, j \in \mathbb{N}. \tag{6.113}$$

The superscript asterisk denotes the complex conjugate.[9]

[9]Equation (6.109) states in effect that the integral of the kernel over the cross-sectional surface of waveguide I is equal to the integral of the same kernel over the cross-sectional surface of waveguide II. This does not mean, however, that the choice of the modal weighting function is arbitrary. We have to choose the modal weighting function as that of waveguide I and understand that equation (6.109) is implicitly incorporated into equation (6.110). If we had taken the modal weighting function as that of waveguide II, than we would only have enforced the continuity of the electric field over the common aperture, and not over the complete discontinuity as we have done now [38].

Next, we take the vector product of both sides of the magnetic-field continuity equation (equations (6.99) and (6.100)) with the complex conjugate of the electric-field mode vector of waveguide II. Then a dot product is formed with the unit vector in the direction of propagation and the resulting equation is integrated over the cross section of the interface:

$$\iint_{S^{II}} \left(\boldsymbol{H}_{\mathrm{T}}^{I} \times \boldsymbol{e}_{\mathrm{T}_{ij}}^{II*} \right) \cdot \hat{\boldsymbol{i}}_z \, dS = \iint_{S^{II}} \left(\boldsymbol{H}_{\mathrm{T}}^{II} \times \boldsymbol{e}_{\mathrm{T}_{ij}}^{II*} \right) \cdot \hat{\boldsymbol{i}}_z \, dS, \quad i, j \in \mathbb{N}, \tag{6.114}$$

where

$$\boldsymbol{H}_{\mathrm{T}}^{I} = H_x^{I} \hat{\boldsymbol{i}}_x + H_y^{I} \hat{\boldsymbol{i}}_y, \tag{6.115}$$

$$\boldsymbol{H}_{\mathrm{T}}^{II} = H_x^{II} \hat{\boldsymbol{i}}_x + H_y^{II} \hat{\boldsymbol{i}}_y \tag{6.116}$$

and

$$\boldsymbol{e}_{\mathrm{T}_{ij}}^{II} = e_{x_{ij}}^{II} \hat{\boldsymbol{i}}_x + e_{y_{ij}}^{II} \hat{\boldsymbol{i}}_y, \quad i, j \in \mathbb{N}. \tag{6.117}$$

Substitution of equations (6.42)–(6.45) into equations (6.110) and (6.114) leads to the following set of scalar equations:[10]

$$\iint_{S^{I}} \left(E_x^{I} e_{x\mathrm{TE}_{ij}}^{I+^*} + E_y^{I} e_{y\mathrm{TE}_{ij}}^{I+^*} \right) dS = \iint_{S^{II}} \left(E_x^{II} e_{x\mathrm{TE}_{ij}}^{I+^*} + E_y^{II} e_{y\mathrm{TE}_{ij}}^{I+^*} \right) dS, \tag{6.118}$$

$$\iint_{S^{I}} \left(E_x^{I} e_{x\mathrm{TM}_{ij}}^{I+^*} + E_y^{I} e_{y\mathrm{TM}_{ij}}^{I+^*} \right) dS = \iint_{S^{II}} \left(E_x^{II} e_{x\mathrm{TM}_{ij}}^{I+^*} + E_y^{II} e_{y\mathrm{TM}_{ij}}^{I+^*} \right) dS, \tag{6.119}$$

$$\iint_{S^{II}} \left(H_x^{I} e_{y\mathrm{TE}_{ij}}^{II+^*} - H_y^{I} e_{x\mathrm{TE}_{ij}}^{II+^*} \right) dS = \iint_{S^{II}} \left(H_x^{II} e_{y\mathrm{TE}_{ij}}^{II+^*} - H_y^{II} e_{x\mathrm{TE}_{ij}}^{II+^*} \right) dS, \tag{6.120}$$

$$\iint_{S^{II}} \left(H_x^{I} e_{y\mathrm{TM}_{ij}}^{II+^*} - H_y^{I} e_{x\mathrm{TM}_{ij}}^{II+^*} \right) dS = \iint_{S^{II}} \left(H_x^{II} e_{y\mathrm{TM}_{ij}}^{II+^*} - H_y^{II} e_{x\mathrm{TM}_{ij}}^{II+^*} \right) dS, \tag{6.121}$$

where $i, j \in \mathbb{N}$ and where, in equations (6.118) and (6.119), use has been made of equation (6.109).

Upon expanding the transverse field components in the above equations into mode contributions, the orthogonality characteristics of the modes (see Appendix 6.10) show that non-trivial solutions exist only for $i = m$ and $j = n$ in domain I and for $i = l$ and $j = w$ in domain II.

[10]Equations (6.110) and (6.114) ensure conservation of complex power across the discontinuity. Another implementation of the mode-matching method that is employed does not perform weighting with complex conjugate mode functions, but with the mode functions themselves. In this implementation, self-reaction across the discontinuity is ensured. The two formulations are equivalent as long as the structure is lossless, because the mode functions are then real-valued [38].

Use of equations (6.38)–(6.41), (6.42)–(6.45), (6.46)–(6.49) and (6.50)–(6.53) results in

$$\left[A_{mn}^{I+} + A_{mn}^{I-}\right] = \sum_{l=0}^{L}\sum_{w=0}^{W}\left\{\frac{[H1_{mnlw} + H3_{mnlw}][A_{lw}^{II+} + A_{lw}^{II-}] + [H2_{mnlw} + H4_{mnlw}][A_{lw}^{\prime II+} - A_{lw}^{\prime II-}]}{[O_1(m,n,a,b) + O_5(m,n,a,b)]}\right\}, \tag{6.122}$$

$$\left[A_{mn}^{\prime I+} - A_{mn}^{\prime I-}\right] = \sum_{l=0}^{L}\sum_{w=0}^{W}\left\{\frac{[H5_{mnlw} + H7_{mnlw}][A_{lw}^{II+} + A_{lw}^{II-}] + [H6_{mnlw} + H8_{mnlw}][A_{lw}^{\prime II+} - A_{lw}^{\prime II-}]}{[O_4(m,n,a,b) + O_8(m,n,a,b)]}\right\}, \tag{6.123}$$

where $m = 0, 1, 2, \ldots, M$ and $n = 0, 1, 2, \ldots, N$, and

$$\left[A_{lw}^{II+} - A_{lw}^{II-}\right] = \sum_{m=0}^{M}\sum_{n=0}^{N}\left\{\frac{Y_{\mathrm{TE}_{mn}}^{I}[H9_{lwmn} + H11_{lwmn}][A_{mn}^{I+} - A_{mn}^{I-}]}{Y_{\mathrm{TE}_{lw}}^{II}[O_1(l,w,a',b') + O_5(l,w,a',b')]} + \frac{Y_{\mathrm{TM}_{mn}}^{I}[H10_{lwmn} + H12_{lwmn}][A_{mn}^{\prime I+} + A_{mn}^{\prime I-}]}{Y_{\mathrm{TE}_{lw}}^{II}[O_1(l,w,a',b') + O_5(l,w,a',b')]}\right\}, \tag{6.124}$$

$$\left[A_{lw}^{\prime II+} + A_{lw}^{\prime II-}\right] = \sum_{m=0}^{M}\sum_{n=0}^{N}\left\{\frac{Y_{\mathrm{TE}_{mn}}^{I}[H13_{lwmn} + H15_{lwmn}][A_{mn}^{I+} - A_{mn}^{I-}]}{Y_{\mathrm{TM}_{lw}}^{II}[O_4(l,w,a',b') + O_8(l,w,a',b')]} + \frac{Y_{\mathrm{TM}_{mn}}^{I}[H14_{lwmn} + H16_{lwmn}][A_{mn}^{\prime I+} + A_{mn}^{\prime I-}]}{Y_{\mathrm{TM}_{lw}}^{II}[O_4(l,w,a',b') + O_8(l,w,a',b')]}\right\}, \tag{6.125}$$

where $l = 0, 1, 2, \ldots, L$ and $w = 0, 1, 2, \ldots, W$.

The functions $H_{i_{mnlw}}$, $i = 1, 2, \ldots, 8$, and $H_{j_{lwmn}}$, $j = 9, 10, \ldots, 16$, are the mode-coupling integrals

$$H1_{mnlw} = \int_{x=-a'/2+x_c}^{a'/2+x_c}\int_{y=-b'/2+y_c}^{b'/2+y_c} e_{x\mathrm{TE}_{mn}}^{I+^*} e_{x\mathrm{TE}_{lw}}^{II+}\,dx\,dy, \tag{6.126}$$

$$H2_{mnlw} = \int_{x=-a'/2+x_c}^{a'/2+x_c}\int_{y=-b'/2+y_c}^{b'/2+y_c} e_{x\mathrm{TE}_{mn}}^{I+^*} e_{x\mathrm{TM}_{lw}}^{II+}\,dx\,dy, \tag{6.127}$$

$$H3_{mnlw} = \int_{x=-a'/2+x_c}^{a'/2+x_c}\int_{y=-b'/2+y_c}^{b'/2+y_c} e_{y\mathrm{TE}_{mn}}^{I+^*} e_{y\mathrm{TE}_{lw}}^{II+}\,dx\,dy, \tag{6.128}$$

$$H4_{mnlw} = \int_{x=-a'/2+x_c}^{a'/2+x_c}\int_{y=-b'/2+y_c}^{b'/2+y_c} e_{y\mathrm{TE}_{mn}}^{I+^*} e_{y\mathrm{TM}_{lw}}^{II+}\,dx\,dy, \tag{6.129}$$

$$H5_{mnlw} = \int_{x=-a'/2+x_c}^{a'/2+x_c} \int_{y=-b'/2+y_c}^{b'/2+y_c} e^{I^{+*}}_{x\mathrm{TM}_{mn}} e^{II^{+}}_{x\mathrm{TE}_{lw}} \, dx\, dy, \tag{6.130}$$

$$H6_{mnlw} = \int_{x=-a'/2+x_c}^{a'/2+x_c} \int_{y=-b'/2+y_c}^{b'/2+y_c} e^{I^{+*}}_{x\mathrm{TM}_{mn}} e^{II^{+}}_{x\mathrm{TM}_{lw}} \, dx\, dy, \tag{6.131}$$

$$H7_{mnlw} = \int_{x=-a'/2+x_c}^{a'/2+x_c} \int_{y=-b'/2+y_c}^{b'/2+y_c} e^{I^{+*}}_{y\mathrm{TM}_{mn}} e^{II^{+}}_{y\mathrm{TE}_{lw}} \, dx\, dy, \tag{6.132}$$

$$H8_{mnlw} = \int_{x=-a'/2+x_c}^{a'/2+x_c} \int_{y=-b'/2+y_c}^{b'/2+y_c} e^{I^{+*}}_{y\mathrm{TM}_{mn}} e^{II^{+}}_{y\mathrm{TM}_{lw}} \, dx\, dy, \tag{6.133}$$

$$H9_{lwmn} = \int_{x=-a'/2+x_c}^{a'/2+x_c} \int_{y=-b'/2+y_c}^{b'/2+y_c} e^{II^{+*}}_{y\mathrm{TE}_{lw}} e^{I^{+}}_{y\mathrm{TE}_{mn}} \, dx\, dy, \tag{6.134}$$

$$H10_{lwmn} = \int_{x=-a'/2+x_c}^{a'/2+x_c} \int_{y=-b'/2+y_c}^{b'/2+y_c} e^{II^{+*}}_{y\mathrm{TE}_{lw}} e^{I^{+}}_{y\mathrm{TM}_{mn}} \, dx\, dy, \tag{6.135}$$

$$H11_{lwmn} = \int_{x=-a'/2+x_c}^{a'/2+x_c} \int_{y=-b'/2+y_c}^{b'/2+y_c} e^{II^{+*}}_{x\mathrm{TE}_{lw}} e^{I^{+}}_{x\mathrm{TE}_{mn}} \, dx\, dy, \tag{6.136}$$

$$H12_{lwmn} = \int_{x=-a'/2+x_c}^{a'/2+x_c} \int_{y=-b'/2+y_c}^{b'/2+y_c} e^{II^{+*}}_{x\mathrm{TE}_{lw}} e^{I^{+}}_{x\mathrm{TM}_{mn}} \, dx\, dy, \tag{6.137}$$

$$H13_{lwmn} = \int_{x=-a'/2+x_c}^{a'/2+x_c} \int_{y=-b'/2+y_c}^{b'/2+y_c} e^{II^{+*}}_{y\mathrm{TM}_{lw}} e^{I^{+}}_{y\mathrm{TE}_{mn}} \, dx\, dy, \tag{6.138}$$

$$H14_{lwmn} = \int_{x=-a'/2+x_c}^{a'/2+x_c} \int_{y=-b'/2+y_c}^{b'/2+y_c} e^{II^{+*}}_{y\mathrm{TM}_{lw}} e^{I^{+}}_{y\mathrm{TM}_{mn}} \, dx\, dy, \tag{6.139}$$

$$H15_{lwmn} = \int_{x=-a'/2+x_c}^{a'/2+x_c} \int_{y=-b'/2+y_c}^{b'/2+y_c} e^{II^{+*}}_{x\mathrm{TM}_{lw}} e^{I^{+}}_{x\mathrm{TE}_{mn}} \, dx\, dy, \tag{6.140}$$

$$H16_{lwmn} = \int_{x=-a'/2+x_c}^{a'/2+x_c} \int_{y=-b'/2+y_c}^{b'/2+y_c} e^{II^{+*}}_{x\mathrm{TM}_{lw}} e^{I^{+}}_{x\mathrm{TM}_{mn}} \, dx\, dy. \tag{6.141}$$

The coupling integrals are calculated in Appendix 6.A. The normalization constants O_1, O_4, O_5 and O_8 are defined in Appendix 6.A.

Before we can form the generalized scattering matrix of the waveguide junction, we first cast equations (6.122)–(6.125) into a more convenient form:

$$\left[A^{I^+}_{mn} + A^{I^-}_{mn}\right] = \sum_{l=0}^{L} \sum_{w=0}^{W} \left\{ H_{a_{mnlw}} \left[A^{II^+}_{lw} + A^{II^-}_{lw}\right] + H_{b_{mnlw}} \left[A'^{II^+}_{lw} - A'^{II^-}_{lw}\right] \right\},$$

$$m = 0, 1, \ldots, M, \quad n = 0, 1, \ldots, N, \tag{6.142}$$

$$\left[A'^{I^+}_{mn} - A'^{I^-}_{mn}\right] = \sum_{l=0}^{L} \sum_{w=0}^{W} \left\{ H_{c_{mnlw}} \left[A^{II^+}_{lw} + A^{II^-}_{lw}\right] + H_{d_{mnlw}} \left[A'^{II^+}_{lw} - A'^{II^-}_{lw}\right] \right\},$$

$$m = 0, 1, \ldots, M, \quad n = 0, 1, \ldots, N, \tag{6.143}$$

$$\left[A_{lw}^{II+} - A_{lw}^{II-}\right] = \sum_{m=0}^{M} \sum_{n=0}^{N} \left\{ H_{e_{lwmn}} \left[A_{mn}^{I+} - A_{mn}^{I-}\right] + H_{f_{lwmn}} \left[A_{mn}^{\prime I+} + A_{mn}^{\prime I-}\right] \right\},$$
$$l = 0, 1, \ldots, L, \quad w = 0, 1, \ldots, W, \tag{6.144}$$

$$\left[A_{lw}^{\prime II+} + A_{lw}^{\prime II-}\right] = \sum_{m=0}^{M} \sum_{n=0}^{N} \left\{ H_{g_{lwmn}} \left[A_{mn}^{I+} - A_{mn}^{I-}\right] + H_{h_{lwmn}} \left[A_{mn}^{\prime I+} + A_{mn}^{\prime I-}\right] \right\},$$
$$l = 0, 1, \ldots, L, \quad w = 0, 1, \ldots, W, \tag{6.145}$$

where

$$H_{a_{mnlw}} = \frac{H1_{mnlw} + H3_{mnlw}}{O_1(m,n,a,b) + O_5(m,n,a,b)}, \tag{6.146}$$

$$H_{b_{mnlw}} = \frac{H2_{mnlw} + H4_{mnlw}}{O_1(m,n,a,b) + O_5(m,n,a,b)}, \tag{6.147}$$

$$H_{c_{mnlw}} = \frac{H5_{mnlw} + H7_{mnlw}}{O_4(m,n,a,b) + O_8(m,n,a,b)}, \tag{6.148}$$

$$H_{d_{mnlw}} = \frac{H6_{mnlw} + H8_{mnlw}}{O_4(m,n,a,b) + O_8(m,n,a,b)}, \tag{6.149}$$

$$H_{e_{lwmn}} = \frac{Y_{\mathrm{TE}_{mn}}^{I}}{Y_{\mathrm{TE}_{lw}}^{II}} \frac{H9_{lwmn} + H11_{lwmn}}{O_1(l,w,a',b') + O_4(l,w,a',b')}, \tag{6.150}$$

$$H_{f_{lwmn}} = \frac{Y_{\mathrm{TM}_{mn}}^{I}}{Y_{\mathrm{TE}_{lw}}^{II}} \frac{H10_{lwmn} + H12_{lwmn}}{O_1(l,w,a',b') + O_4(l,w,a',b')}, \tag{6.151}$$

$$H_{g_{lwmn}} = \frac{Y_{\mathrm{TE}_{mn}}^{I}}{Y_{\mathrm{TM}_{lw}}^{II}} \frac{H13_{lwmn} + H15_{lwmn}}{O_4(l,w,a',b') + O_8(l,w,a',b')}, \tag{6.152}$$

$$H_{h_{lwmn}} = \frac{Y_{\mathrm{TM}_{mn}}^{I}}{Y_{\mathrm{TM}_{lw}}^{II}} \frac{H14_{lwmn} + H16_{lwmn}}{O_4(l,w,a',b') + O_8(l,w,a',b')}. \tag{6.153}$$

The auxiliary functions defined by equations (6.146)–(6.153) are calculated in Appendix 6.A.

6.4.3 Matrix Formulation and GSM Derivation

Equations (6.142)–(6.145) can be written in matrix form as follows:

$$[A^I]_{\mathrm{TE}}^+ + [A^I]_{\mathrm{TE}}^- = [H_a]\{[A^{II}]_{\mathrm{TE}}^+ + [A^{II}]_{\mathrm{TE}}^-\} + [H_b]\{[A^{II}]_{\mathrm{TM}}^+ - [A^{II}]_{\mathrm{TM}}^-\}, \tag{6.154}$$

$$[A^I]_{\mathrm{TM}}^+ - [A^I]_{\mathrm{TM}}^- = [H_c]\{[A^{II}]_{\mathrm{TE}}^+ + [A^{II}]_{\mathrm{TE}}^-\} + [H_d]\{[A^{II}]_{\mathrm{TM}}^+ - [A^{II}]_{\mathrm{TM}}^-\}, \tag{6.155}$$

$$[A^{II}]_{\mathrm{TE}}^+ - [A^{II}]_{\mathrm{TE}}^- = [H_e]\{[A^I]_{\mathrm{TE}}^+ - [A^I]_{\mathrm{TE}}^-\} + [H_f]\{[A^I]_{\mathrm{TM}}^+ + [A^I]_{\mathrm{TM}}^-\}, \tag{6.156}$$

$$[A^{II}]_{\mathrm{TM}}^+ + [A^{II}]_{\mathrm{TM}}^- = [H_g]\{[A^I]_{\mathrm{TE}}^+ - [A^I]_{\mathrm{TE}}^-\} + [H_h]\{[A^I]_{\mathrm{TM}}^+ + [A^I]_{\mathrm{TM}}^-\}. \tag{6.157}$$

In the above equations, $[A^I]_{\mathrm{TE}}^{\pm}$ and $[A^{II}]_{\mathrm{TE}}^{\pm}$ are column matrices (vectors) containing the (m, n) mode amplitude coefficients for TE waves traveling in the increasing and the decreasing z direction for waveguide I and waveguide II, respectively. Similarly, $[A^I]_{\mathrm{TM}}^{\pm}$

and $[A^{II}]^{\pm}_{\text{TM}}$ are column matrices containing the (m, n) mode amplitude coefficients for TM waves traveling in the increasing and the decreasing z direction for waveguide I and waveguide II, respectively.

$[H_a]$, $[H_b]$, $[H_c]$ and $[H_d]$ are matrices containing the TE_{mn}-to-TE_{lw}, TE_{mn}-to-TM_{lw}, TM_{mn}-to-TE_{lw} and TM_{mn}-to-TM_{lw} mode coupling coefficients, respectively, going from waveguide I to waveguide II. So, in these matrices, the (m, n) variations are in the columns and the (l, w) variations are in the rows. $[H_e]$, $[H_f]$, $[H_g]$ and $[H_h]$ are matrices containing the TE_{lw}-to-TE_{mn}, TE_{lw}-to-TM_{mn}, TM_{lw}-to-TE_{mn} and TM_{lw}-to-TM_{mn} mode-coupling coefficients, respectively, going from waveguide II to waveguide I. So, in these matrices, the (l, w) variations are in the columns and the (m, n) variations are in the rows.

Using the composite matrices

$$[A^I]^{\pm} = \begin{bmatrix} [A^I]^{\pm}_{\text{TE}} \\ [A^I]^{\pm}_{\text{TM}} \end{bmatrix},$$

$$[A^{II}]^{\pm} = \begin{bmatrix} [A^{II}]^{\pm}_{\text{TE}} \\ [A^{II}]^{\pm}_{\text{TM}} \end{bmatrix},$$

equations (6.154)–(6.157) can be written as

$$[A^I]^+ + [U^I][A^I]^- = [G_1][A^{II}]^+ + [G_1][U^I][A^{II}]^-, \tag{6.158}$$

$$[A^{II}]^+ - [U^{II}][A^{II}]^- = [G_3][A^I]^+ - [G_3][U^{II}][A^I]^-, \tag{6.159}$$

where

$$[G_1] = \begin{bmatrix} [H_a] & [H_b] \\ [H_c] & [H_d] \end{bmatrix}, [G_3] = \begin{bmatrix} [H_e] & [H_f] \\ [H_g] & [H_h] \end{bmatrix},$$

and $[U^I]$ and $[U^{II}]$ are modified unit matrices for waveguide I and waveguide II, respectively:

$$[U^{I,II}] = \begin{bmatrix} [I] & [0] \\ [0] & -[I] \end{bmatrix}.$$

The difference between $[U^I]$ and $[U^{II}]$ is in the dimensions of the matrices. The minus sign before the bottom right identity submatrix is due to propagation to the left.

The submatrices of the generalized scattering matrix for a reduction in a rectangular waveguide are related to the mode amplitude coefficients through

$$[A^I]^- = [S_{11}]^{\text{r}}[A^I]^+ + [S_{12}]^{\text{r}}[A^{II}]^-, \tag{6.160}$$

$$[A^{II}]^+ = [S_{21}]^{\text{r}}[A^I]^+ + [S_{22}]^{\text{r}}[A^{II}]^-, \tag{6.161}$$

where

$$[S]^{\text{r}} = \begin{bmatrix} [S_{11}] & [S_{12}] \\ [S_{21}] & [S_{22}] \end{bmatrix}.$$

The substitution of equations (6.160) and (6.161) into equations (6.158) and (6.159) finally yields

$$[S_{11}]^{\mathrm{r}} = [U^I]([G_1][G_3] + [I])^{-1}([G_1][G_3] - [I]), \tag{6.162}$$

$$[S_{12}]^{\mathrm{r}} = 2[U^I]([G_1][G_3] + [I])^{-1}[G_1][U^{II}], \tag{6.163}$$

$$[S_{21}]^{\mathrm{r}} = [G_3]([I] - [U^I][S_{11}]^{\mathrm{r}}), \tag{6.164}$$

$$[S_{22}]^{\mathrm{r}} = [U^{II}] - [G_3][U^I][S_{12}]^{\mathrm{r}}, \tag{6.165}$$

where $[I]$ is the unit matrix.

The submatrices of the GSM for a waveguide cross-sectional enlargement, $[S]^{\mathrm{e}}$, are constructed from the above submatrices by virtue of reciprocity:

$$[S_{11}]^{\mathrm{e}} = [S_{22}]^{\mathrm{r}}, \tag{6.166}$$

$$[S_{12}]^{\mathrm{e}} = [S_{21}]^{\mathrm{r}}, \tag{6.167}$$

$$[S_{21}]^{\mathrm{e}} = [S_{12}]^{\mathrm{r}}, \tag{6.168}$$

$$[S_{22}]^{\mathrm{e}} = [S_{11}]^{\mathrm{r}}. \tag{6.169}$$

6.5 JUNCTION BETWEEN A RECTANGULAR WAVEGUIDE AND A UNIT CELL

In this section, we shall derive the generalized scattering matrix for a transition from one rectangular waveguide (out of the infinite number of rectangular waveguides that make up an infinite array antenna) to a unit cell, i.e. to a homogeneous half-space. The waveguide apertures of the array antenna are assumed to be positioned in an electrically conducting ground plane. The rectangular waveguide, with metallic walls, is assumed to be positioned centrally in a rectangular unit cell (which is a rectangular waveguide with phase shift walls) as shown in Figure 6.4. The transition is shown again in Figure 6.6.

The input is the rectangular waveguide (with metallic walls), designated as waveguide I ($z < 0$). The cross-sectional dimensions of this waveguide are a and b. The output is the rectangular unit cell (with phase shift walls), designated as waveguide II ($z > 0$). The cross-sectional dimensions of the unit cell are s and $t\sin(\Omega)$. The cross-sectional dimensions of the two waveguides are such that $s \geq a$ and $t\sin(\Omega) \geq b$. The junction of the two waveguides is positioned at $z = 0$ in a rectangular coordinate system (Figure 6.6). Furthermore, S^I is the area of the cross section of waveguide $I(-a/2 \leq x \leq a/2, -b/2 \leq y \leq b/2)$, S^{II} is the area of the cross section of waveguide $II(-s/2 \leq x \leq s/2, -t/2\sin(\Omega) \leq y \leq t/2\sin(\Omega))$ and S is the area of the cross section of waveguide II, excluding the cross section of waveguide I. S is assumed to be perfectly electrically conducting.

The derivation of the generalized scattering matrix for the rectangular-waveguide-to-unit-cell junction proceeds along the same lines as explained in detail in section 6.4 for a rectangular-waveguide-to-rectangular-waveguide junction. The only difference in comparison with that situation is in the modes in waveguide II. Therefore, in this section, we shall only outline the major steps in the derivation of the GSM.

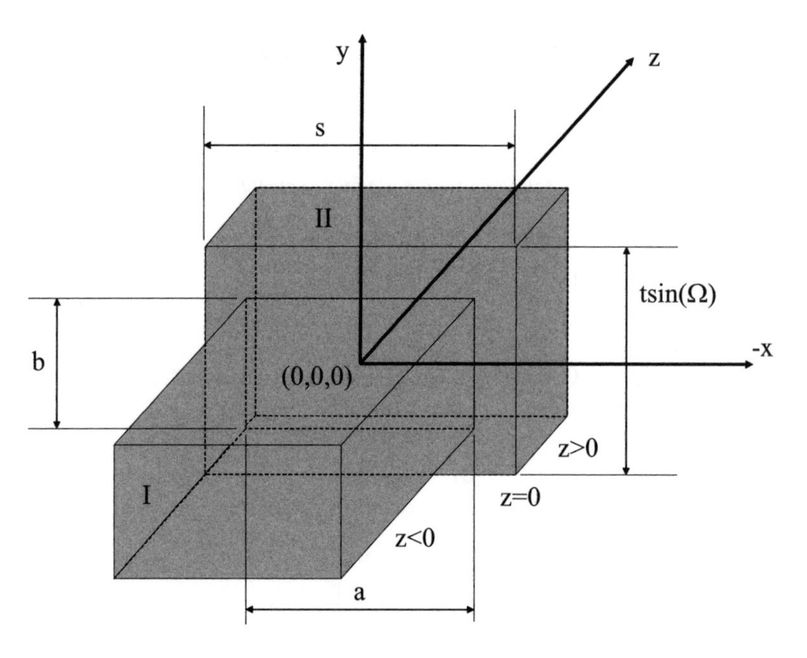

Figure 6.6 Rectangular-waveguide-to-unit-cell junction in the plane $z = 0$.

6.5.1 GSM Derivation

Establishing tangential-field continuity in a weak form across the waveguide aperture at $z = 0$ leads to

$$\left[A_{pq}^{II^+} + A_{pq}^{II^-}\right] = \sum_{m=0}^{M}\sum_{n=0}^{N}\left\{W_{a_{pqmn}}\left[A_{mn}^{I^+} + A_{mn}^{I^-}\right] + W_{b_{pqmn}}\left[A_{mn}^{\prime I^+} - A_{mn}^{\prime I^-}\right]\right\},$$
$$p = -P, \ldots, -1, 0, 1, \ldots, P, \quad q = -Q, \ldots, -1, 0, 1, \ldots, Q, \tag{6.170}$$

$$\left[A_{pq}^{\prime II^+} - A_{pq}^{\prime II^-}\right] = \sum_{m=0}^{M}\sum_{n=0}^{N}\left\{W_{c_{pqmn}}\left[A_{mn}^{I^+} + A_{mn}^{I^-}\right] + W_{d_{pqmn}}\left[A_{mn}^{\prime I^+} - A_{mn}^{\prime I^-}\right]\right\},$$
$$p = -P, \ldots, -1, 0, 1, \ldots, P, \quad q = -Q, \ldots, -1, 0, 1, \ldots, Q, \tag{6.171}$$

$$\left[A_{mn}^{I^+} - A_{mn}^{I^-}\right] = \sum_{p=-P}^{P}\sum_{q=-Q}^{Q}\left\{W_{e_{mnpq}}\left[A_{pq}^{II^+} - A_{pq}^{II^-}\right] + W_{f_{mnpq}}\left[A_{pq}^{\prime II^+} + A_{pq}^{\prime II^-}\right]\right\},$$
$$m = 0, 1, 2, \ldots, M, \quad n = 0, 1, 2, \ldots, N, \tag{6.172}$$

$$\left[A_{mn}^{\prime I^+} + A_{mn}^{\prime I^-}\right] = \sum_{p=-P}^{P}\sum_{q=-Q}^{Q}\left\{W_{g_{mnpq}}\left[A_{pq}^{II^+} - A_{pq}^{II^-}\right] + W_{h_{mnpq}}\left[A_{pq}^{\prime II^+} + A_{pq}^{\prime II^-}\right]\right\},$$
$$m = 0, 1, 2, \ldots, M, \quad n = 0, 1, 2, \ldots, N. \tag{6.173}$$

In the above equations, which are exact for $M, N, P, Q = \infty$, A_{mn}^{I+} and A_{mn}^{I-} are the mode amplitude coefficients in waveguide I for the TE_{mn} mode traveling in the increasing and the decreasing z direction, respectively. $A_{mn}^{\prime I+}$ and $A_{mn}^{\prime I-}$ are the mode amplitude coefficients in waveguide I for the TM_{mn} mode traveling in the increasing and the decreasing z direction, respectively. A_{pq}^{II+} and A_{pq}^{II-} are the (Floquet) mode amplitude coefficients in waveguide II for the TE_{pqn} mode traveling in the increasing and the decreasing z direction, respectively, and $A_{pq}^{\prime II+}$ and $A_{pq}^{\prime II-}$ are the (Floquet) mode amplitude coefficients in this waveguide for the TM_{pq} mode traveling in the increasing and the decreasing z direction, respectively. The auxiliary functions $W_{a_{pqmn}}$, $W_{b_{pqmn}}$, $W_{c_{pqmn}}$, $W_{d_{pqmn}}$, $W_{e_{mnpq}}$, $W_{f_{mnpq}}$, $W_{g_{mnpq}}$ and $W_{h_{mnpq}}$ are defined and calculated in Appendix 6.D, using Appendix 6.C.

These equations may be written in matrix form as

$$[A^{II}]_{\mathrm{TE}}^{+} + [A^{II}]_{\mathrm{TE}}^{-} = [W_a]\left\{[A^{I}]_{\mathrm{TE}}^{+} + [A^{I}]_{\mathrm{TE}}^{-}\right\} + [W_b]\left\{[A^{I}]_{\mathrm{TM}}^{+} - [A^{I}]_{\mathrm{TM}}^{-}\right\}, \tag{6.174}$$

$$[A^{II}]_{\mathrm{TM}}^{+} - [A^{II}]_{\mathrm{TM}}^{-} = [W_c]\left\{[A^{I}]_{\mathrm{TE}}^{+} + [A^{I}]_{\mathrm{TE}}^{-}\right\} + [W_d]\left\{[A^{I}]_{\mathrm{TM}}^{+} - [A^{I}]_{\mathrm{TM}}^{-}\right\}, \tag{6.175}$$

$$[A^{I}]_{\mathrm{TE}}^{+} - [A^{I}]_{\mathrm{TE}}^{-} = [W_e]\left\{[A^{II}]_{\mathrm{TE}}^{+} - [A^{II}]_{\mathrm{TE}}^{-}\right\} + [W_f]\left\{[A^{II}]_{\mathrm{TM}}^{+} + [A^{II}]_{\mathrm{TM}}^{-}\right\}, \tag{6.176}$$

$$[A^{I}]_{\mathrm{TM}}^{+} + [A^{I}]_{\mathrm{TM}}^{-} = [W_g]\left\{[A^{II}]_{\mathrm{TE}}^{+} - [A^{II}]_{\mathrm{TE}}^{-}\right\} + [W_h]\left\{[A^{II}]_{\mathrm{TM}}^{+} + [A^{II}]_{\mathrm{TM}}^{-}\right\}, \tag{6.177}$$

where $[A^{I}]_{\mathrm{TE}}^{\pm}$ and $[A^{II}]_{\mathrm{TE}}^{\pm}$ are column matrices (vectors) containing the (m, n) and (p, q) mode amplitude coefficients for TE waves traveling in the increasing and the decreasing z directions in waveguide I and waveguide II, respectively. Similarly, $[A^{I}]_{\mathrm{TM}}^{\pm}$ and $[A^{II}]_{\mathrm{TM}}^{\pm}$ contain the (m, n) and (p, q) mode amplitude coefficients for TM waves traveling in the increasing and the decreasing z directions in waveguide I and waveguide II, respectively.

$[W_a]$, $[W_b]$, $[W_c]$ and $[W_d]$ are matrices containing the TE_{mn}-to-TE_{pq}, TE_{mn}-to-TM_{pq}, TM_{mn}-to-TE_{pq} and TM_{mn}-to-TM_{pq} mode-coupling coefficients, respectively, going from waveguide I to waveguide II. $[W_e]$, $[W_f]$, $[W_g]$ and $[W_h]$ are matrices containing the TE_{pq}-to-TE_{mn}, TE_{pq}-to-TM_{mn}, TM_{pq}-to-TE_{mn} and TM_{pq}-to-TM_{mn} mode coupling coefficients, respectively, going from waveguide II to waveguide I.

Using the composite matrices

$$[A^{I}]^{\pm} = \begin{bmatrix} [A^{I}]_{\mathrm{TE}}^{\pm} \\ [A^{I}]_{\mathrm{TM}}^{\pm} \end{bmatrix}, \tag{6.178}$$

$$[A^{II}]^{\pm} = \begin{bmatrix} [A^{II}]_{\mathrm{TE}}^{\pm} \\ [A^{II}]_{\mathrm{TM}}^{\pm} \end{bmatrix}, \tag{6.179}$$

equations (6.174)–(6.177) can be written as

$$[A^{II}]^{+} + [U^{II}][A^{II}]^{-} = [V_1][A^{I}]^{+} + [V_1][U^{I}][A^{I}]^{-}, \tag{6.180}$$

$$[A^{I}]^{+} - [U^{I}][A^{I}]^{-} = [V_3][A^{II}]^{+} - [V_3][U^{II}][A^{II}]^{-}, \tag{6.181}$$

where

$$[V_1] = \begin{bmatrix} [W_a] & [W_b] \\ [W_c] & [W_d] \end{bmatrix}, \tag{6.182}$$

$$[V_3] = \begin{bmatrix} [W_e] & [W_f] \\ [W_g] & [W_h] \end{bmatrix}, \tag{6.183}$$

and $[U^I]$ and $[U^{II}]$ are modified unit matrices for waveguide I and waveguide II, respectively:

$$[U^{I,II}] = \begin{bmatrix} [I] & [0] \\ [0] & -[I] \end{bmatrix}. \tag{6.184}$$

The submatrices of the generalized scattering matrix for a rectangular-waveguide-to-unit-cell junction are related to the mode amplitude coefficients through

$$[A^I]^- = [S_{11}]^{\mathrm{u}}[A^I]^+ + [S_{12}]^{\mathrm{u}}[A^{II}]^-, \tag{6.185}$$

$$[A^{II}]^+ = [S_{21}]^{\mathrm{u}}[A^I]^+ + [S_{22}]^{\mathrm{u}}[A^{II}]^-. \tag{6.186}$$

Combining equations (6.180) and (6.181) with equations (6.185) and (6.186) finally results in

$$[S_{11}]^{\mathrm{u}} = [U^I]([I] + [V_3][V_1])^{-1}([I] - [V_3][V_1]), \tag{6.187}$$

$$[S_{12}]^{\mathrm{u}} = 2[U^I]([I] + [V_3][V_1])^{-1}[V_3][U^{II}], \tag{6.188}$$

$$[S_{21}]^{\mathrm{u}} = [V_1]([I] + [U^I][S_{11}]^{\mathrm{u}}), \tag{6.189}$$

$$[S_{22}]^{\mathrm{u}} = [V_1][U^I][S_{12}]^{\mathrm{u}} - [U^{II}], \tag{6.190}$$

where $[I]$ is the unit matrix.

The submatrices of the GSM for a unit-cell-to-rectangular-waveguide transition, $[S]^{\mathrm{w}}$, are constructed from the above submatrices by virtue of reciprocity:

$$[S_{11}]^{\mathrm{w}} = [S_{22}]^{\mathrm{u}}, \tag{6.191}$$

$$[S_{12}]^{\mathrm{w}} = [S_{21}]^{\mathrm{u}}, \tag{6.192}$$

$$[S_{21}]^{\mathrm{w}} = [S_{12}]^{\mathrm{u}}, \tag{6.193}$$

$$[S_{22}]^{\mathrm{w}} = [S_{11}]^{\mathrm{u}}. \tag{6.194}$$

6.6 DIELECTRIC STEP IN A UNIT CELL

Only the GSM of a dielectric step in a unit cell will be treated explicitly. For a dielectric step in a metallic-wall rectangular waveguide, we can use the GSM of a cross-sectional enlargement or reduction in a waveguide. Since there is no such thing for a phase-shift-wall waveguide – the unit cell dimensions are uniquely related to the array lattice – we have to treat a step in the dielectric explicitly. A dielectric step in a unit cell will in practice be a step in the permittivity, but a step in the permeability or both is also allowed. The geometry of the configuration under consideration is shown in Figure 6.7.

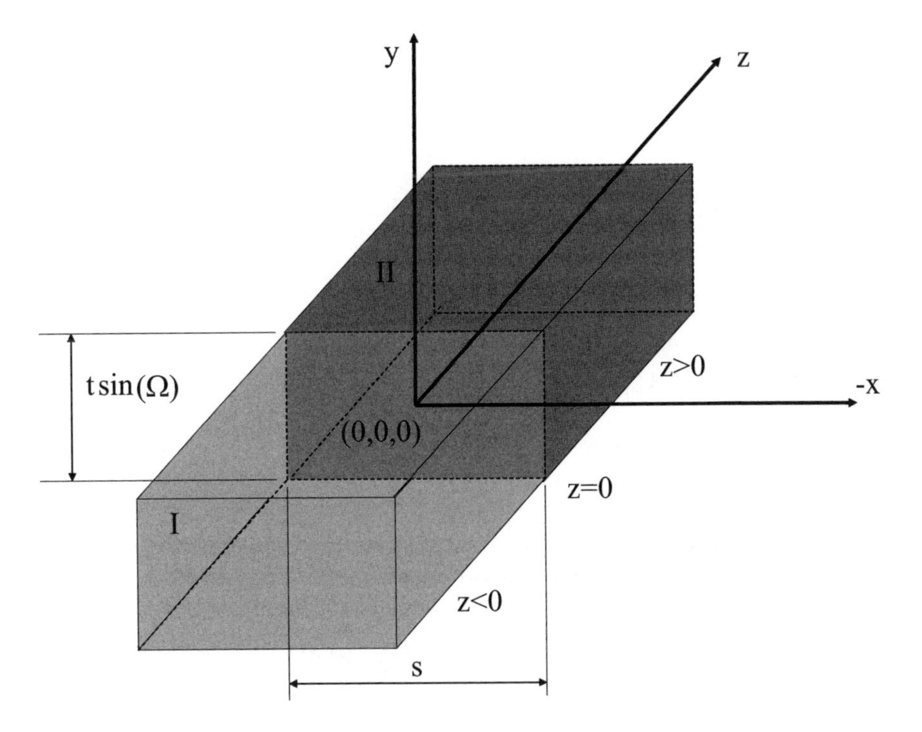

Figure 6.7 Dielectric step in rectangular unit cell.

The cross-sectional dimensions of the unit cell are s and $t\sin(\Omega)$. A rectangular coordinate system is assumed, centrally positioned at the step interface in the substrate. In domain I, the relative permittivity and permeability of the substrate are denoted by $\varepsilon_{\mathrm{r}}^{I}$ and μ_{r}^{I}, respectively. The corresponding quantities in domain II are $\varepsilon_{\mathrm{r}}^{II}$ and μ_{r}^{II}.

The derivation of the GSM follows the by know well-known mode-matching procedure. It leads to a diagonal GSM matrix, the elements of which may be stored in a vector.

6.6.1 GSM Derivation

Establishing tangential-field continuity in a weak form over the discontinuity leads to

$$\left[A_{pq}^{II+} + A_{pq}^{II-}\right] = \frac{\mu_{\mathrm{r}}^{I}}{\mu_{\mathrm{r}}^{II}}\left[A_{pq}^{I+} + A_{pq}^{I-}\right],$$

$$p = -P, \ldots, -1, 0, 1, \ldots, P, \quad q = -Q, \ldots, -1, 0, 1, \ldots, Q, \tag{6.195}$$

$$\left[A'^{II+}_{pq} - A'^{II-}_{pq}\right] = \frac{\gamma^{I}_{pq}}{\gamma^{II}_{pq}}\left[A'^{I+}_{pq} - A'^{I-}_{pq}\right],$$
$$p = -P, \ldots, -1, 0, 1, \ldots, P, \quad q = -Q, \ldots, -1, 0, 1, \ldots, Q, \tag{6.196}$$

$$\left[A^{I+}_{pq} - A^{I-}_{pq}\right] = \frac{\gamma^{II}_{pq}}{\gamma^{I}_{pq}}\left[A^{II+}_{pq} - A^{II-}_{pq}\right],$$
$$p = -P, \ldots, -1, 0, 1, \ldots, P, \quad q = -Q, \ldots, -1, 0, 1, \ldots, Q, \tag{6.197}$$

$$\left[A'^{I+}_{pq} + A'^{I-}_{pq}\right] = \frac{\varepsilon^{II}_{\mathrm{r}}}{\varepsilon^{I}_{\mathrm{r}}}\left[A'^{II+}_{pq} + A'^{II-}_{pq}\right],$$
$$p = -P, \ldots, -1, 0, 1, \ldots, P, \quad q = -Q, \ldots, -1, 0, 1, \ldots, Q, \tag{6.198}$$

where the mode amplitude coefficients $A^{I,II\pm}_{pq}$ are as defined in previous sections.

These equations may be written in matrix form as

$$[A^{II}]^{+}_{\mathrm{TE}} + [A^{II}]^{-}_{\mathrm{TE}} = \frac{\mu^{I}_{\mathrm{r}}}{\mu^{II}_{\mathrm{r}}}\{[A^{I}]^{+}_{\mathrm{TE}} + [A^{I}]^{-}_{\mathrm{TE}}\}, \tag{6.199}$$

$$[A^{II}]^{+}_{\mathrm{TM}} - [A^{II}]^{-}_{\mathrm{TM}} = [J_1]\{[A^{I}]^{+}_{\mathrm{TM}} - [A^{I}]^{-}_{\mathrm{TM}}\}, \tag{6.200}$$

$$[A^{I}]^{+}_{\mathrm{TE}} - [A^{I}]^{-}_{\mathrm{TE}} = [J_1]^{-1}\{[A^{II}]^{+}_{\mathrm{TE}} - [A^{II}]^{-}_{\mathrm{TE}}\}, \tag{6.201}$$

$$[A^{I}]^{+}_{\mathrm{TM}} + [A^{I}]^{-}_{\mathrm{TM}} = \frac{\varepsilon^{II}_{\mathrm{r}}}{\varepsilon^{I}_{\mathrm{r}}}\{[A^{II}]^{+}_{\mathrm{TM}} + [A^{II}]^{-}_{\mathrm{TM}}\}, \tag{6.202}$$

where the column matrices (vectors) $[A^{I,II}]^{\pm}_{\mathrm{TE,TM}}$ have their usual meaning, explained in previous sections, and $[J_1]$ is a diagonal matrix containing the TM-to-TM coupling coefficients going from domain II to domain I.

Using the usual composite column matrices, these equations may be written as

$$[A^{II}]^{+} + [U][A^{II}]^{-} = [R_1][A^{I}]^{+} + [R_1][U][A^{I}]^{-}, \tag{6.203}$$

$$[A^{I}]^{+} - [U][A^{I}]^{-} = [R_3][A^{II}]^{+} - [U][R_3][A^{II}]^{-}, \tag{6.204}$$

where

$$[R_1] = \begin{bmatrix} (\mu^{I}_{\mathrm{r}}/\mu^{II}_{\mathrm{r}})[I] & [0] \\ [0] & [J_1] \end{bmatrix}, \tag{6.205}$$

$$[R_3] = \begin{bmatrix} [J_1]^{-1} & [0] \\ [0] & (\varepsilon^{II}_{\mathrm{r}}/\varepsilon^{I}_{\mathrm{r}})[I] \end{bmatrix}, \tag{6.206}$$

and where

$$[U] = \begin{bmatrix} [I] & [0] \\ [0] & -[I] \end{bmatrix}. \tag{6.207}$$

The submatrices of the generalized scattering matrix for a dielectric step in a unit cell are related to the mode amplitude coefficients through

$$[A^I]^- = [S_{11}]^s[A^I]^+ + [S_{12}]^s[A^{II}]^-, \quad (6.208)$$

$$[A^{II}]^+ = [S_{21}]^s[A^I]^+ + [S_{22}]^s[A^{II}]^-. \quad (6.209)$$

Combining equations (6.203) and (6.204) with equations (6.208) and (6.209) finally results in[11]

$$[S_{11}]^s = [U]([I] + [R_3][R_1])^{-1}([I] - [R_3][R_1]), \quad (6.210)$$

$$[S_{12}]^s = 2[U]([I] + [R_3][R_1])^{-1}[R_3][U], \quad (6.211)$$

$$[S_{21}]^s = [R_1]([I] + [U][S_{11}]^s), \quad (6.212)$$

$$[S_{22}]^s = [R_1][U][S_{12}]^s - [U]. \quad (6.213)$$

6.7 FINITE-LENGTH TRANSMISSION LINE

The last building block will be a finite-length transmission line. This transmission line can be either a piece of rectangular waveguide with metallic walls or a piece of rectangular waveguide with phase shift walls (a unit cell). The geometry is shown in Figure 6.8.

The transmission line has cross-sectional dimensions a and b if a rectangular-waveguide transmission line is considered and cross-sectional dimensions s and $t\sin(\Omega)$ if a 'unit cell transmission line' is considered. A rectangular coordinate system as shown in Figure 6.8 is assumed, where the origin is positioned at the center of the cross section of the transmission line at $z = 0$. The transmission line has a length d. The beginning of the transmission line ($z = 0$) is designated as domain I, and the end of the transmission line ($z = d$) is designated as domain II.

In the following, we shall restrict ourselves to the derivation of the GSM for a rectangular-waveguide transmission line with metallic walls. The GSM for a 'unit cell transmission line' can then be obtained by inspection from the GSM for a waveguide transmission line, because of the similarity.

The field components in domain II are obtained from those in domain I upon substitution of a phase factor $e^{-j\gamma^I_{mn}d}$ in equations (6.38)–(6.41). To apply a mode-matching procedure in the plane $z = 0$, a coordinate transformation needs to be applied first to the field components in domain II. This coordinate transformation is implemented by replacing z by $z - d$ in the descriptions of the fields for domain II.

[11] Since the matrices $[U]$, $[R_1]$ and $[R_3]$ are all diagonal matrices, the scattering-matrix elements may be calculated directly (and stored in a vector) when implemented in software.

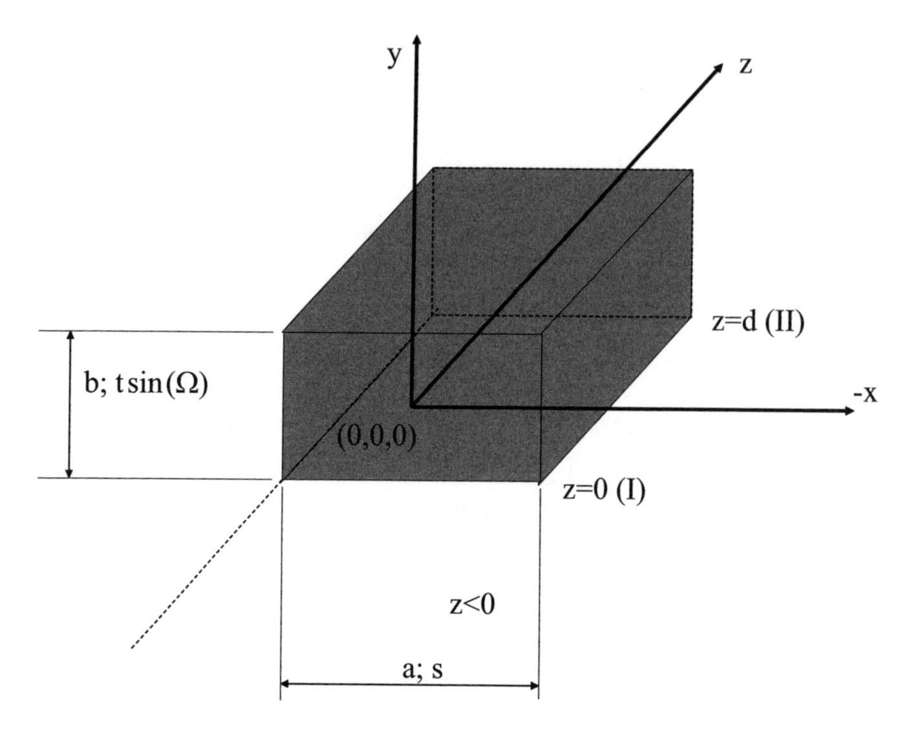

Figure 6.8 Finite-length transmission line.

6.7.1 GSM Derivation

Establishing tangential-field continuity in a weak form across the interface leads to

$$\left[A_{mn}^{I+} + A_{mn}^{I-}\right] = \left[A_{mn}^{II+} e^{j\gamma_{mn}^{I} d} + A_{mn}^{II-} e^{-j\gamma_{mn}^{I} d}\right],$$
$$m = 0, 1, 2, \ldots, M, \quad n = 0, 1, 2, \ldots, N, \tag{6.214}$$

$$\left[A_{mn}^{\prime I+} - A_{mn}^{\prime I-}\right] = \left[A_{mn}^{\prime II+} e^{j\gamma_{mn}^{I} d} - A_{mn}^{\prime II-} e^{-j\gamma_{mn}^{I} d}\right],$$
$$m = 0, 1, 2, \ldots, M, \quad n = 0, 1, 2, \ldots, N, \tag{6.215}$$

$$\left[A_{mn}^{I+} - A_{mn}^{I-}\right] = \left[A_{mn}^{II+} e^{j\gamma_{mn}^{I} d} - A_{mn}^{II-} e^{-j\gamma_{mn}^{I} d}\right],$$
$$m = 0, 1, 2, \ldots, M, \quad n = 0, 1, 2, \ldots, N, \tag{6.216}$$

$$\left[A_{mn}^{\prime I+} + A_{mn}^{\prime I-}\right] = \left[A_{mn}^{\prime II+} e^{j\gamma_{mn}^{I} d} + A_{mn}^{\prime II-} e^{-j\gamma_{mn}^{I} d}\right],$$
$$m = 0, 1, 2, \ldots, M, \quad n = 0, 1, 2, \ldots, N. \tag{6.217}$$

These equations may be written in matrix form as

$$[A^{I}]_{\mathrm{TE}}^{+} + [A^{I}]_{\mathrm{TE}}^{-} = [D_1][A^{II}]_{\mathrm{TE}}^{+} + [D_2][A^{II}]_{\mathrm{TE}}^{-}, \tag{6.218}$$

$$[A^I]^+_{\text{TE}} - [A^I]^-_{\text{TE}} = [D_1][A^{II}]^+_{\text{TE}} - [D_2][A^{II}]^-_{\text{TE}}, \tag{6.219}$$

$$[A^I]^+_{\text{TM}} + [A^I]^-_{\text{TM}} = [D_3][A^{II}]^+_{\text{TM}} + [D_4][A^{II}]^-_{\text{TM}}, \tag{6.220}$$

$$[A^I]^+_{\text{TM}} - [A^I]^-_{\text{TM}} = [D_3][A^{II}]^+_{\text{TM}} - [D_4][A^{II}]^-_{\text{TM}}, \tag{6.221}$$

where the column matrices $[A^{I,II}]^{\pm}_{\text{TE,TM}}$ have their usual meaning, explained in previous sections, and $[D_1]$ and $[D_2]$ are diagonal matrices containing the TE-to-TE coupling coefficients. The nontrivial elements of $[D_1]$ are $e^{j\gamma^I_{mn}d}$, and the non-trivial elements of $[D_2]$ are $e^{-j\gamma^I_{mn}d}$, for $m = 0, 1, 2, \ldots, M$ and $n = 0, 1, 2, \ldots, N$. $[D_3]$ and $[D_4]$ are diagonal matrices containing the TM-to-TM coupling coefficients. The nontrivial elements of $[D_3]$ are $e^{j\gamma^I_{mn}d}$, and the non-trivial elements of $[D_4]$ are $e^{-j\gamma^I_{mn}d}$, for $m = 0, 1, 2, \ldots, M$ and $n = 0, 1, 2, \ldots, N$. $[D_1]$, and $[D_3]$, and similarly $[D_2]$ and $[D_4]$, differ only in size.

Using the usual composite column matrices, the above equations may be written as

$$[A^I]^+ + [A^I]^- = [H_1][A^{II}]^+ + [H_2][A^{II}]^-, \tag{6.222}$$

$$[A^I]^+ - [A^I]^- = [H_1][A^{II}]^+ - [H_2][A^{II}]^-, \tag{6.223}$$

where

$$[H_1] = \begin{bmatrix} [D_1] & [0] \\ [0] & [D_3] \end{bmatrix} \tag{6.224}$$

and

$$[H_2] = \begin{bmatrix} [D_2] & [0] \\ [0] & [D_4] \end{bmatrix}. \tag{6.225}$$

The submatrices of the generalized scattering matrix for a finite-length rectangular-waveguide transmission line are related to the mode amplitude coefficients through

$$[A^I]^- = [S_{11}]^{\text{w}}[A^I]^+ + [S_{12}]^{\text{w}}[A^{II}]^-, \tag{6.226}$$

$$[A^{II}]^+ = [S_{21}]^{\text{w}}[A^I]^+ + [S_{22}]^{\text{w}}[A^{II}]^-. \tag{6.227}$$

Combining equations (6.222) and (6.223) with equations (6.226) and (6.227) finally results in

$$[S_{11}]^{\text{w}} = [S_{22}]^{\text{w}} = [0], \tag{6.228}$$

$$[S_{12}]^{\text{w}} = [S_{21}]^{\text{w}} = [H_2]. \tag{6.229}$$

The derivation of the GSM of a finite-length 'unit cell transmission line', $[S]^d$, is similar to the derivation of the GSM of a finite-length rectangular-waveguide transmission line. Therefore, for a 'unit cell transmission line' of length d, the results can be obtained by inspection from equations (6.228) and (6.229):

$$[S_{11}]^d = [S_{22}]^d = [0], \tag{6.230}$$

$$[S_{12}]^d = [S_{21}]^d = [L], \tag{6.231}$$

where $[L]$ is a diagonal matrix with nontrivial elements $e^{-j\gamma^I_{pq}d}$ for $p = -P, \ldots, -1, 0, 1, \ldots, P$ and $q = -Q, \ldots, -1, 0, 1, \ldots, Q$.

Since the submatrices are diagonal matrices, the nontrivial elements may be stored in a single vector.

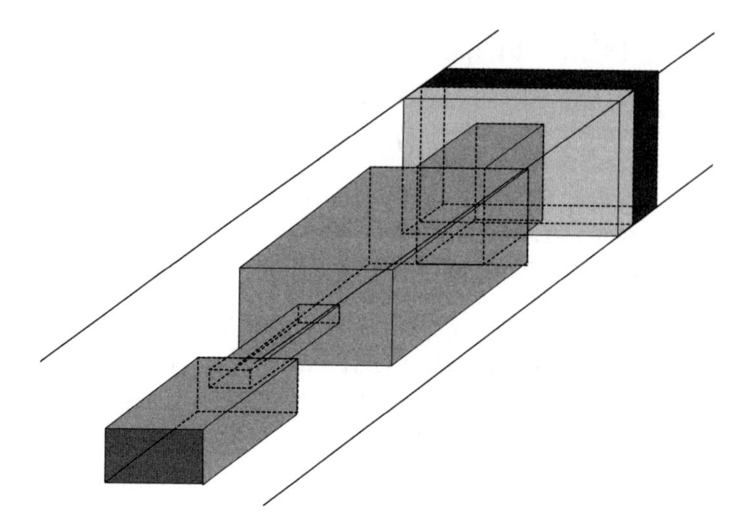

Figure 6.9 Waveguide structure consisting of cascaded rectangular-waveguide discontinuities and transmission lines.

6.8 OVERALL GSM OF A CASCADED RECTANGULAR-WAVEGUIDE STRUCTURE

Now that we have learned how to calculate the generalized scattering matrices of discontinuities in rectangular waveguides and unit cells, as well as GSMs for waveguide and unit cell transmission lines, it is necessary to develop a method for obtaining the GSM of a cascade of these transmission lines and discontinuities. An example of such a cascade is given in Figure 6.9.

The example shown in the figure consists of a rectangular waveguide connected to an asymmetrically positioned smaller rectangular waveguide, which is connected symmetrically to a larger waveguide. This waveguide terminates into a symmetrical inductive iris [9] of finite thickness, which radiates into a unit cell that is filled with two different layers of dielectric material, of finite thickness. The last junction consists of a dielectric step to free space. The cascaded structure consists of six transmission lines (four waveguide transmission lines and two 'unit cell transmission lines') and six discontinuities.

Of the several possible ways to combine the local GSMs, the most obvious one is to 'directly' cascade these GSM's, i.e. to repeatedly construct a combined GSM from two local GSM's. We have chosen this method, since alternative methods – aimed at improving the cascading process by reducing the number of matrix inversion operations – impose restrictions on the length of the transmission line for the applicability of the cascading process.

Mansour and Macphie [28], for example, proposed a transmission matrix formulation to reduce the number of matrix inversion operations. Van Schaik [8, 33] also used transmission matrices. The major drawback of using a transmission matrix formulation is that diagonal matrices will evolve that have nontrivial elements of the form $e^{j\gamma d}$, where d is the distance

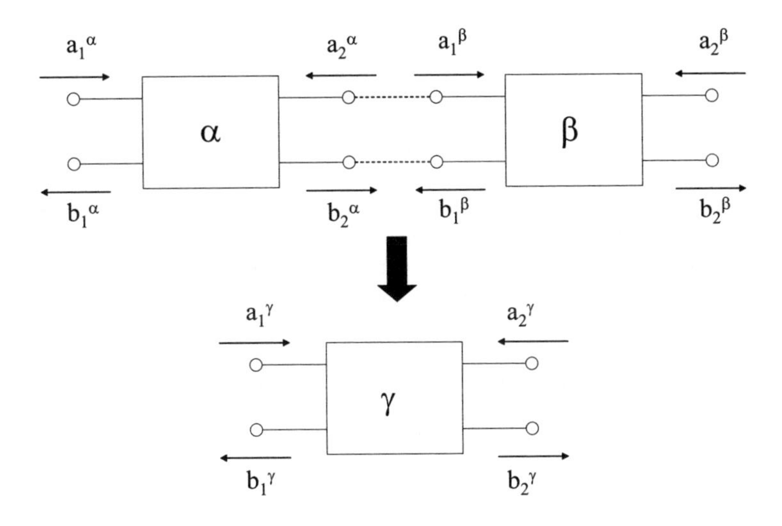

Figure 6.10 Cascading of two 2-ports.

between two discontinuities. If this distance becomes too large, it may give rise to exploding exponentials. If GSMs are cascaded 'directly', only diagonal matrices that have nontrivial elements of the form $e^{-j\gamma d}$ (see section 6.7) will evolve. These matrices will not give rise to numerical instabilities with increasing distance d. Besides numerical stability, a further advantage of cascading 'directly' is that it is not necessary to have an equal number of modes in the waveguides on both sides of a discontinuity [28, 39].

The admittance matrix formulation, as proposed by Alessandri *et al.* [40], is restricted to structures where the cascaded discontinuities are first of the enlargement type and second of the reduction type, thus not providing the degree of freedom desired for analyzing structures such as those shown in Figure 6.9.

So, although the method of 'directly' cascading the local GSMs has the drawback of needing one matrix inversion operation per discontinuity, it guarantees numerical stability with respect to exponentials and general applicability. Truncation effects will be discussed in the next section.

To visualize the cascading process, we consider two 2-ports to be cascaded, as shown in Figure 6.10. The 2-ports α and β are cascaded, resulting in the 2-port γ. The scattering equations for the three 2-ports in Figure 6.10 are, considering outgoing waves to be a result of ingoing waves,

$$b_1^\alpha = S_{11}^\alpha a_1^\alpha + S_{12}^\alpha a_2^\alpha, \tag{6.232}$$

$$b_2^\alpha = S_{21}^\alpha a_1^\alpha + S_{22}^\alpha a_2^\alpha, \tag{6.233}$$

$$b_1^\beta = S_{11}^\beta a_1^\beta + S_{12}^\beta a_2^\beta, \tag{6.234}$$

$$b_2^\alpha = S_{21}^\beta a_1^\beta + S_{22}^\alpha a_2^\beta \tag{6.235}$$

and

$$b_1^\gamma = S_{11}^\gamma a_1^\gamma + S_{12}^\gamma a_2^\gamma, \tag{6.236}$$
$$b_2^\gamma = S_{21}^\gamma a_1^\gamma + S_{22}^\gamma a_2^\gamma. \tag{6.237}$$

Combining equations (6.232)–(6.237) leads to the scattering parameters of the 2-port γ, expressed in terms of the (known) scattering parameters of the 2-ports α and β. Owing to the direct relation between 2-port scattering parameters and the scattering submatrices of an N-port [41], we find the following for the overall GSM of γ, which is the result of cascading the two local GSM's of α and β:

$$[S_{11}]^\gamma = [S_{11}]^\alpha + [S_{12}]^\alpha([I] - [S_{11}]^\beta[S_{22}]^\alpha)^{-1}[S_{11}]^\beta[S_{21}]^\alpha, \tag{6.238}$$
$$[S_{12}]^\gamma = [S_{12}]^\alpha([I] - [S_{11}]^\beta[S_{22}]^\alpha)^{-1}[S_{12}]^\beta, \tag{6.239}$$
$$[S_{21}]^\gamma = [S_{21}]^\beta([I] + [S_{22}]^\alpha([I] - [S_{11}]^\beta[S_{22}]^\alpha)^{-1}[S_{11}]^\beta)[S_{21}]^\alpha, \tag{6.240}$$
$$[S_{22}]^\gamma = [S_{21}]^\beta[S_{22}]^\alpha([I] - [S_{11}]^\beta[S_{22}]^\alpha)^{-1}[S_{12}]^\beta + [S_{22}]^\beta. \tag{6.241}$$

The above equations show that every cascading of two GSM's requires one matrix inversion. The overall GSM of a complete waveguide structure (see for example Figure 6.9) is obtained by repeatedly using the cascading operation, reducing the number of local GSM's in every iteration. From the overall GSM, we may find the reflection coefficients for all incident waves. The element GSM_{11}, for example, is the reflection coefficient for an incident TE_{10} wave.

6.9 VALIDATION

The theory discussed in the previous sections has been implemented in a computer code. To validate the theory and its implementation in code, analysis results will be compared, for several different structures, with simulation results from the literature and simulation results obtained with an independently developed GSM mode-matching code [31].

The validation will be obtained in two steps. First, waveguide structures without open boundaries will be analyzed. We shall refer to these structures as *filter structures*. Next, waveguide structures with open boundaries in the form of unit cells will be analyzed. These structures will be referred to as *array antenna structures*.

Before we start with the validation, we need to discuss the choice of modes and the number of modes. To be able to perform analyses at all, we need to truncate the various infinite series. The way in which we truncate the series at both sides of a waveguide junction cannot be chosen arbitrarily. A phenomenon called *relative convergence* (RC) has to be taken into account.

6.9.1 Initial Choice of Modes

In general, a waveguide structure is designed for operation in the fundamental (TE_{10}) input waveguide mode. When the structure is excited with this mode, the geometrical features of the cross-sectional steps in the waveguide will dictate which modes should be taken into account

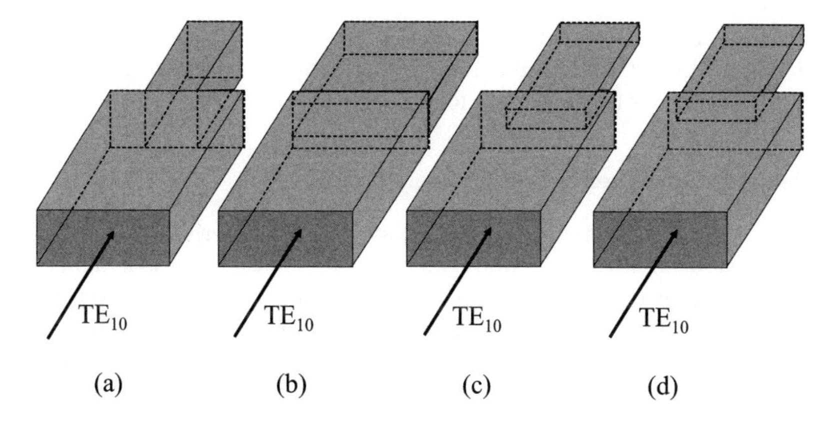

Figure 6.11 Cross-sectional discontinuities in a rectangular waveguide. (a) Symmetric inductive step. (b) Symmetric capacitive step. (c) Symmetric dual step. (d) Asymmetric dual step.

Table 6.1 Mode preselection criteria for rectangular-waveguide discontinuities subject to an incident TE_{10} mode.

Waveguide step	Excited modes	Restrictions
Symmetric inductive	TE_{m0}	m odd
Symmetric capacitive	TE_{1n}, TM_{1n}	None
Symmetric dual	TE_{mn}, TM_{mn}	m odd, n even
Asymmetric dual	TE_{mn}, TM_{mn}	None

in an analysis of the discontinuity. Using only these modes in the analysis of the junction will enhance the efficiency of the computer code.

Several possible cross-sectional discontinuities in a rectangular waveguide are shown in Figure 6.11: a symmetric inductive step, a symmetric capacitive step, a symmetric dual step and an asymmetric dual step.

Through rigorous but straightforward analysis of the TE-to-TE, TE-to-TM, TM-to-TE and TM-to-TM coupling coefficients given in equations (6.154)–(6.157), using the coupling integrals in Appendix 6.10, the modes excited at the discontinuities shown in Figure 6.11 may be found. For a TE_{10} mode incident into a rectangular waveguide that terminates in a symmetric inductive step (Figure 6.11(a)), only TE_{m0} modes will be excited, where $m \in N$ and m is odd. For a TE_{10} mode incident into a rectangular waveguide that terminates in a symmetric capacitive step (Figure 6.11(b)), TE_{1n} and TM_{1n} modes will be excited, where $n \in N$. A TE_{10} mode incident into a rectangular waveguide that terminates in a symmetric dual step (Figure 6.11(c)) gives rise to TE_{mn} and TM_{mn} modes, where $m, n \in N$ and m is odd and n is even. When the rectangular waveguide terminates in an asymmetrical dual step, all TE and TM modes are excited.

Use of these mode preselection criteria, which are tabulated in Table 6.1 for convenience, will help in developing an efficient computer code for analyzing waveguide structures, avoiding the use of unnecessary matrix elements and thus memory.

What remains to be discussed now is how to appropriately truncate the infinite series encountered in the theory and how the truncated series at the two sides of a cross-sectional discontinuity (step) in a waveguide relate to one another.

6.9.2 Relative Convergence and Choice of Modes

As explained in the foregoing, we have to truncate the infinite series to obtain a numerical solution to the problem of cascaded waveguide discontinuities. Because a digital computer will nearly always give a solution, even if the problem to be solved is not well posed or does not have a unique solution, we have to be careful about applying these truncations. It is only permitted to truncate an infinite series if the series is convergent. To validate the convergence, we have to look at the desired analysis parameters, and we consider the series at hand to be convergent – and thus the truncation to be appropriate – if the change in the parameters is smaller than certain criteria when the number of modes applied in the truncation is increased. This convergence criterion in itself, however, is not enough to ensure correct solutions for the analysis method that we are using!

In the analysis of waveguide discontinuities by mode matching, as described in previous sections, we have to truncate two or more infinite series on both sides of the discontinuity simultaneously. It is known that when mode matching is applied, the numerical results may converge to different values depending on the way the series are truncated [42]. This phenomenon of relative convergence was first studied in [15] and was believed to originate from violation of field distributions at the edge of a conductor at a boundary, an explanation also found in [42]. In [43], it was explained that using the 'cure' for the RC problem stated in [15] indeed results in a bounded solution, but that it is not necessary to satisfy an explicit edge condition. It was shown that the origin of the RC phenomenon is in the behavior of the linear system to be solved. It was proved that the numerical results converge to the exact solution if the linear system is well conditioned. The 'cure' described in [15], which consists of taking the ratio of the numbers of modes on the two sides of a cross-sectional discontinuity to be approximately equal to the ratio of the waveguide dimensions on the two sides, ensures a well-conditioned linear system.

Thus

$$\frac{M_1}{M_2} \begin{cases} \geq \dfrac{a_1}{a_2} & \text{if } \dfrac{a_1}{a_2} \geq 1 \\ \leq \dfrac{a_1}{a_2} & \text{if } \dfrac{a_1}{a_2} \leq 1 \end{cases} \tag{6.242}$$

and

$$\frac{N_1}{N_2} \begin{cases} \geq \dfrac{b_1}{b_2} & \text{if } \dfrac{b_1}{b_2} \geq 1 \\ \leq \dfrac{b_1}{b_2} & \text{if } \dfrac{b_1}{b_2} \leq 1, \end{cases} \tag{6.243}$$

where the inequality sign applies to the most appropriate choice. In the above equations, M_1 and M_2 are the maximum mode indices m in waveguide 1 and waveguide 2, respectively, and N_1 and N_2 are the maximum mode indices n in waveguide 1 and waveguide 2, respectively,

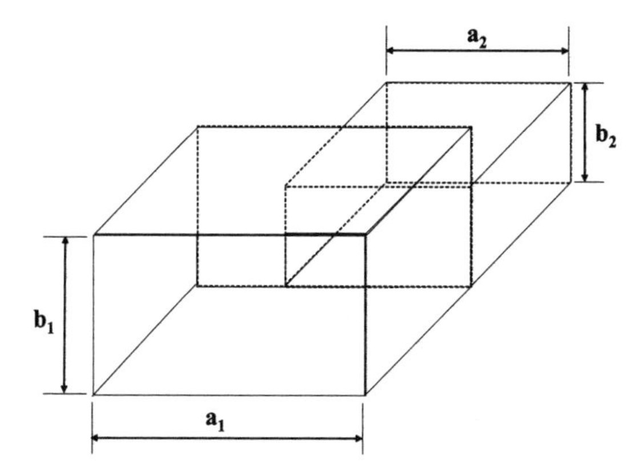

Figure 6.12 Asymmetric junction between two rectangular waveguides.

for TE_{mn} and TM_{mn} modes, where $m, n \in \mathbb{N}$. The horizontal and vertical dimensions of waveguides 1 and 2 are a_1, a_2 and b_1, b_2, respectively.

The 'RC condition' stated in equations (6.242) and (6.243) ensures that the highest cutoff wave numbers on the two sides of the discontinuity are (almost) equal (see also equation (6.23)). Therefore the fastest transversely fluctuating signal at either side of the waveguide cross-sectional discontinuity is supported by the signals at the other side, which is not true when the RC condition is violated. So, the spectral content in the Fourier representations is equal for the two sides. When the equality sign cannot be realized, we choose to have more modes in the larger waveguide, thus ensuring that the fastest fluctuation with the lowest mode number, on both sides of the junction, is accounted for.

To visualize the RC phenomenon, we have analyzed an asymmetric junction between two rectangular waveguides where both transverse dimensions are different (Figure 6.12). The dimensions of the larger rectangular waveguide were $a_1 = 15.8$ mm and $b_1 = 7.9$ mm; those of the smaller waveguide were $a_1 = 11.85$ mm and $b_1 = 6.0$ mm.

If we choose the maximum wave numbers in the larger guide, M_1 and N_1, to be $(M_1, N_1) = (1, 1)$[12] and increase the maximum wave numbers in the smaller guide, M_2 and N_2, we seem to reach convergence in the fundamental-mode reflection coefficient of the large waveguide as a function of frequency after only a few iterations (Figure 6.13).

Although convergence is reached *for our choice of input waveguide modes*, the reflection coefficient results are not correct. We have clearly violated the RC condition, so that rapidly fluctuating transverse fields in the smaller waveguide are not supported by the fields in the larger one. Taking the RC condition into account, i.e. choosing the maximum mode numbers in accordance with the ratio of transverse waveguide dimensions, leads to convergence by increasing the maximum mode numbers at both sides of the junction (Figure 6.14). We observe that the reflection coefficient as a function of frequency now converges to different (correct) values.

[12]This means that we allow the TE_{01}, TE_{10}, TE_{11} and TM_{11} modes to be present in the waveguide with the largest cross section.

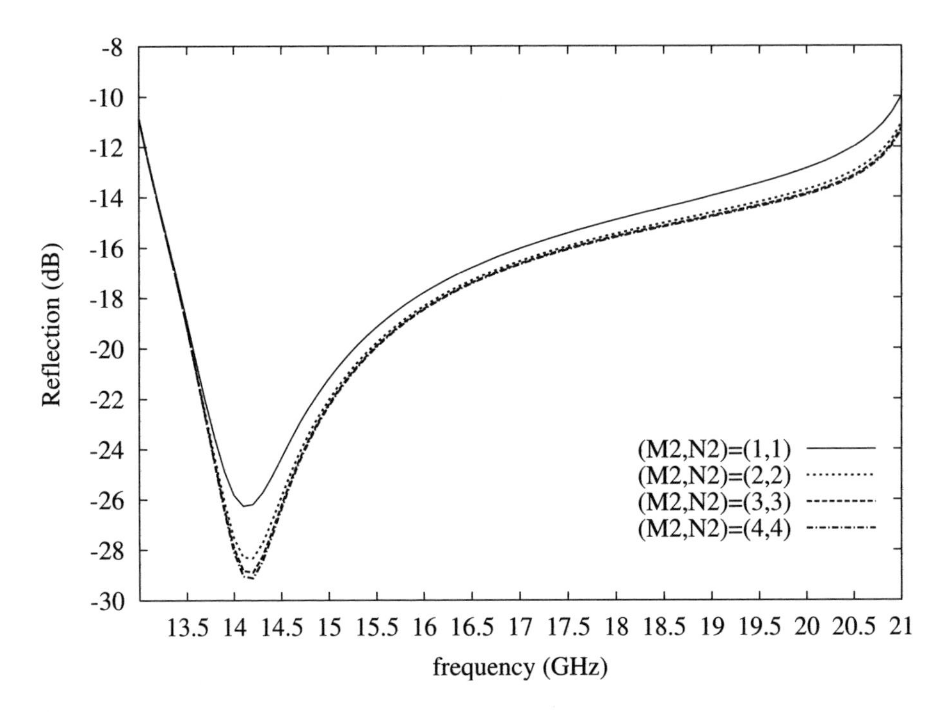

Figure 6.13 TE_{10}-mode reflection coefficient of the larger-cross-sectional waveguide versus frequency as a function of the maximum mode numbers in the smaller guide. $(M_1, N_1) = (1, 1)$.

As long as the system is well conditioned, increasing the number of modes in the input and output waveguides, regardless of the RC condition, eventually leads to convergence to the correct solution. However, choosing the modes according to equations (6.242) and (6.243) leads to the fastest convergence and prevents an overflow of memory. So, in all situations, the RC condition should be met.

By meeting the RC condition, we ensure that the highest cutoff wave numbers on the two sides of the junction are identical, as can be concluded from inspection of equations (6.23), (6.32), (6.242) and (6.243), thus ensuring that the fastest-fluctuating field along the transverse directions in either of the two guides is supported by the field in the other guide. It is this observation that will help us in selecting the Floquet modes when dealing with a junction formed by a metallic-wall waveguide radiating into a free-space unit cell, for which situation no such relations as stated in equations (6.242) and (6.243) can be easily derived.

From the above observation, a guideline can be found for selecting the Floquet modes necessary for ensuring a well-posed problem, thus eliminating the need for numerical experiments. First, the waveguide modes needed to represent the waveguide aperture are selected. Next, the cutoff wave numbers of the Floquet modes for a given scan angle (ϑ_0, φ_0) are computed, and assembled in increasing order. All Floquet modes that have a cut-off wave number less than or equal to that of the cut-off wave number of the highest waveguide mode are used for the expansion in the external half-space.

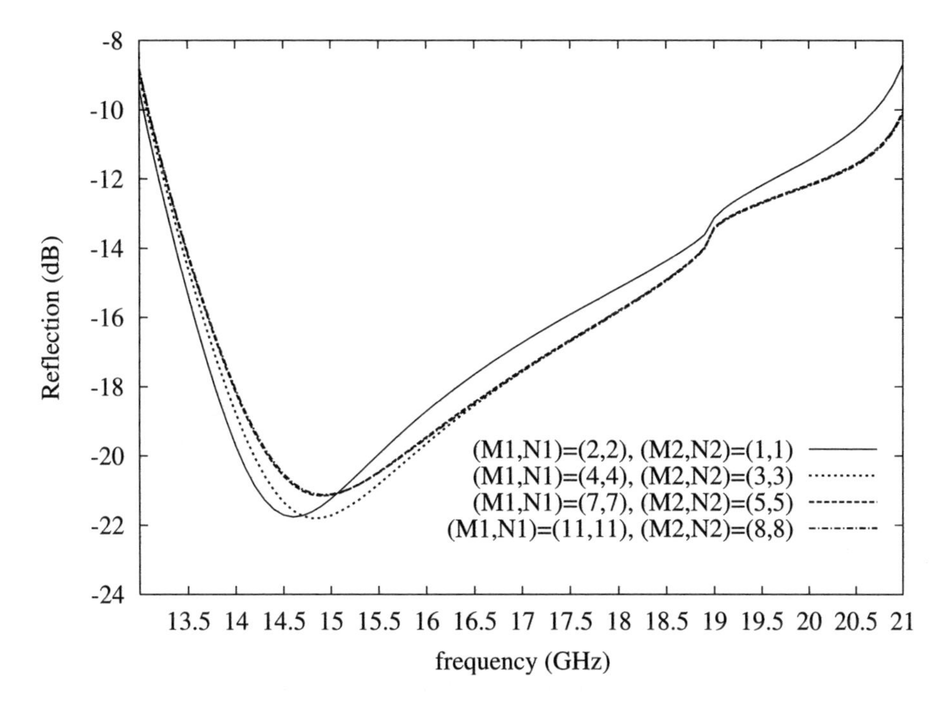

Figure 6.14 TE_{10}-mode reflection coefficient of the larger-cross-sectional waveguide versus frequency as a function of the maximum mode numbers in both waveguides, chosen according to the RC condition.

We have now achieved a situation, by using the RC condition and the Floquet mode selection guideline, where ensuring the validity of an analysis of a waveguide structure is restricted to a convergence check only. We start by selecting a set of modes in an input rectangular waveguide. Next, the modes in the subsequent waveguide sections are chosen according to the RC condition, utilizing only the ratios of the transverse dimensions of the waveguide sections. Floquet modes in phase-shift-wall waveguides are chosen according to the described selection guideline described above. Next, we increase the set of modes in the input waveguide and change the modes in the subsequent waveguide sections and in free space, with use of the RC condition and the Floquet mode selection guideline, until the change in the parameter(s) obtained from the analysis is smaller than some chosen criterion, which depends on the required accuracy.

By using the waveguide mode preselection criteria described in section 6.9.1, we may speed up the numerical analysis. Note that using the preselection criteria influences only the computational efficiency; it does not influence the accuracy of the analysis. The computational efficiency may be improved further by leaving out the highest-order modes in relatively long (with respect to wavelength) waveguide sections. Since these modes will decay rapidly with distance, their contribution may become negligible for a long enough waveguide section.

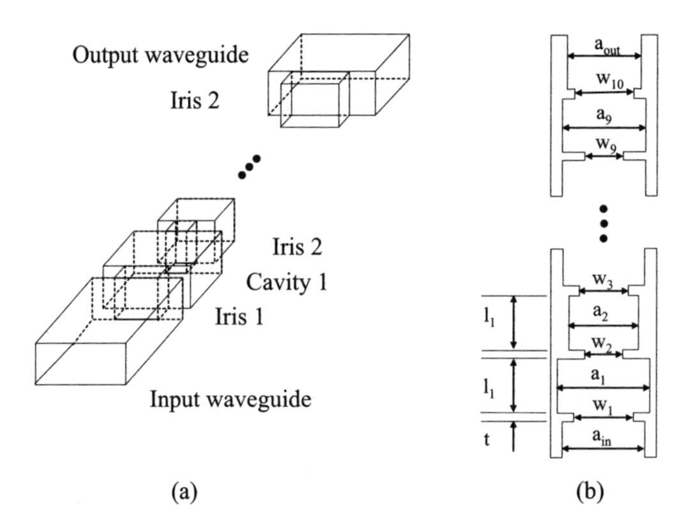

Figure 6.15 Nine-pole, H-step rectangular-waveguide filter. (a) Layout. (b) Top view and dimensions.

With the mode preselection scheme, the convergence scheme described above and the additional measures to improve computational efficiency, we may now analyze waveguide structures. We shall start by demonstrating the validity of the method for filter structures. Then, we expand the cascaded metallic-wall waveguide sections with a phase-shift-wall waveguide, thus demonstrating the validity of the method for infinite array antennas.

6.9.3 Filter Structures

For all of the metal-wall waveguide structures for which analysis results will be shown in this section, convergence checks as discussed above were performed. Therefore, in general, the numbers of modes employed in the analyses were slightly more than necessary.

We start with rectangular-waveguide filter structures employing symmetric inductive irises of finite thickness (i.e. length). The filter structures analyzed were derived from a nine-pole waveguide filter described in [44]. The layout of the filter is shown in Figure 6.15, and the dimensions are stated in Table 6.2.

In Figure 6.16, the scattering parameter S_{21} is shown as a function of frequency for the first cavity of the nine-pole filter. The measurement results obtained from [44] are also shown in this figure. Since only symmetric inductive irises are present in the structure, only TE_{m0} modes, $m \in \mathbb{N}$, were used in the analysis; 45 modes were used in the input waveguide. Figure 6.17 shows the analysis results for the first two cavities of the filter. For this analysis, 90 modes were used in the input waveguide.

Both figures show fair to good agreement between the simulation and measurement results (the difference is less than 0.2 db). The differences are believed to be largely due to extracting the measurement results from the graphs in [44] and to waveguide wall losses that were not incorporated into the analysis.

Table 6.2 Dimensions of the nine-pole filter shown in Figure 6.15. The height of the waveguide sections was 4.318 mm (WR-42 waveguide), and the thickness of the irises t was 2.0 mm.

Cavity widths (mm)	Iris widths (mm)	Cavity lengths (mm)
$a_{\text{in}} = 10.68$		
$a_1 = 7.50$	$w_1 = 5.364$	$l_1 = 9.818$
$a_2 = 8.00$	$w_2 = 3.675$	$l_2 = 9.663$
$a_3 = 8.50$	$w_3 = 3.183$	$l_3 = 8.998$
$a_4 = 9.00$	$w_4 = 3.010$	$l_4 = 8.478$
$a_5 = 9.50$	$w_5 = 2.924$	$l_5 = 8.094$
$a_6 = 10.00$	$w_6 = 2.879$	$l_6 = 7.800$
$a_7 = 10.20$	$w_7 = 2.872$	$l_7 = 7.689$
$a_8 = 10.40$	$w_8 = 2.919$	$l_8 = 7.523$
$a_9 = 10.60$	$w_9 = 3.181$	$l_9 = 6.678$
$a_{\text{out}} = 10.68$	$w_{10} = 4.964$	

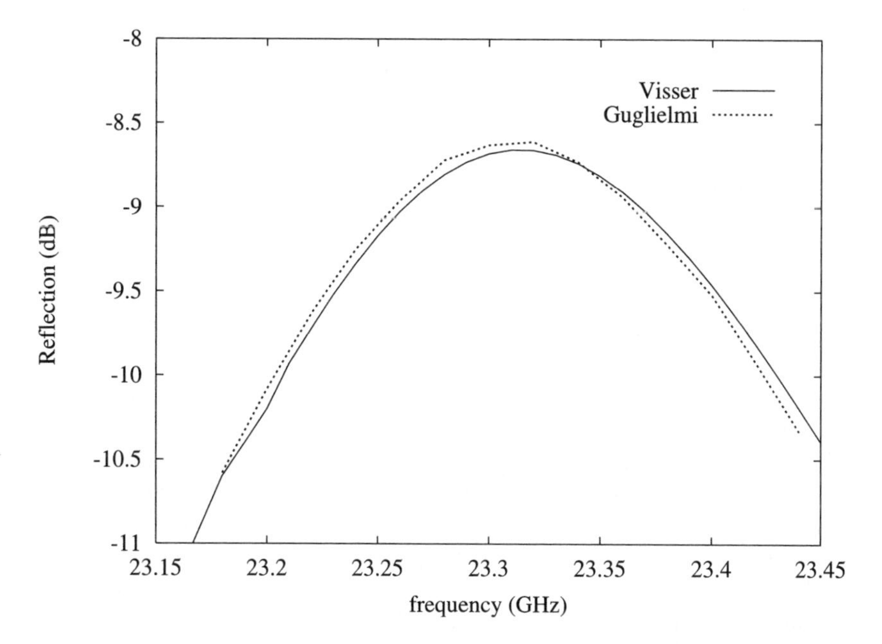

Figure 6.16 Simulation and measurement results for the first cavity of the nine-pole filter of Figure 6.15.

The analysis results for the complete nine-pole waveguide filter are shown in Figure 6.18. The agreement between the simulation and measurement results is fair to good, considering the low reflection levels and the rather coarse frequency resolution ($\Delta f = 5.0$ MHz).

Next, we take a look at symmetrical dual steps. In Figure 6.19, the TE_{10}-mode reflection coefficient is shown as a function of frequency for a Ku-band-to-X-band concentric step. Measurement results obtained from [26] are also shown in this figure. For this simulation, we did not use mode pre-selection, but we did choose maximum mode numbers such as to meet

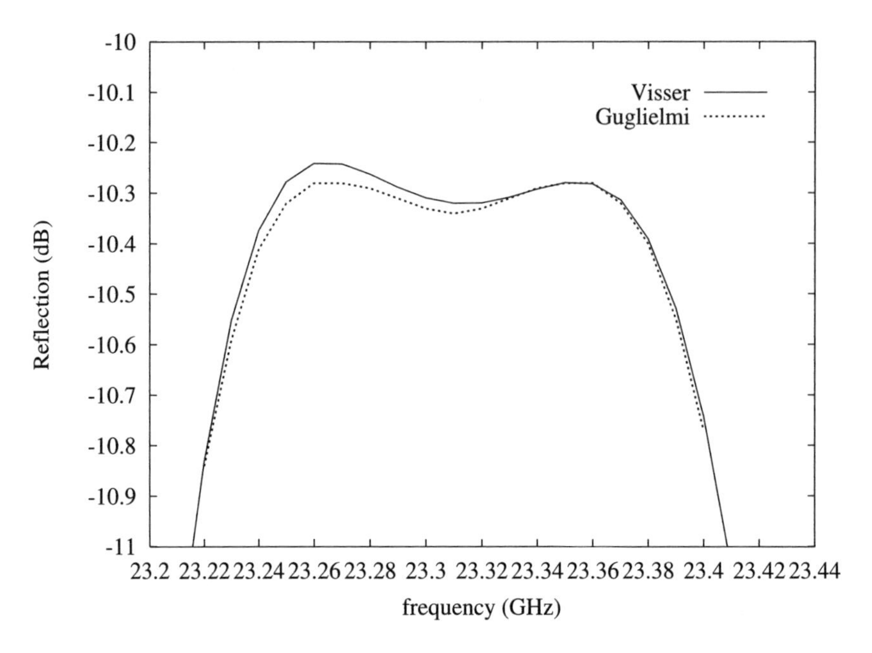

Figure 6.17 Simulation and measurement results for the first two cavities of the nine-pole filter of Figure 6.15.

the RC condition. For the Ku-band waveguide (15.8 mm × 7.9 mm), we chose $(M_1, N_1) = (9, 6)$, and for the X-band waveguide (22.86 mm × 10.16 mm,) we chose $(M_2, N_2) = (13, 9)$.

We observe a good match between the simulations and measurements from 12.5 GHz onwards (the difference is less than 0.2 db). The large difference at 10 GHz between the measurement and simulations is also present in [26], which casts suspicion on the measurement value. The other differences are believed to originate from extracting the measurement results from the graphs in [26]. This assumption is strengthened by the good agreement between the simulation and measurement results (from [26]) over the whole frequency band for a resonant iris, shown in Figure 6.20. Again, we did not use mode preselection. For the waveguide (15.8 mm × 7.9 mm), we chose $(M_1, N_1) = (13, 9)$ and for the iris (11.17 mm × 5.59 mm, thickness 2 mm),we chose $(M_1, N_1) = (9, 6)$.

To check the validity of the calculated TE_{10}-mode reflection coefficient in terms of both amplitude and phase, we looked at an H-plane step discontinuity as analyzed in [45]. Simulation results, together with the simulation results from [45], are shown in Figure 6.21. The width of the larger waveguide was 22.86 mm, and that of the smaller waveguide was 15 mm. The height of both waveguides was 10.16 mm. In the larger waveguide, seven modes were used (TE_{10}, TE_{20}, TE_{30}, TE_{40}, TE_{50}, TE_{60} and TE_{70}), and in the smaller waveguide, five modes were used (TE_{10}, TE_{20}, TE_{30}, TE_{40} and TE_{50}), meeting the RC condition. The results match very well: the difference in amplitude is less than 0.2 dB, and the difference in phase is less than 2°.

As a last validation check for filter structures, we looked at an asymmetric iris (11.85 mm × 6.0 mm) of finite thickness (3.0 mm), placed in a Ku-band waveguide

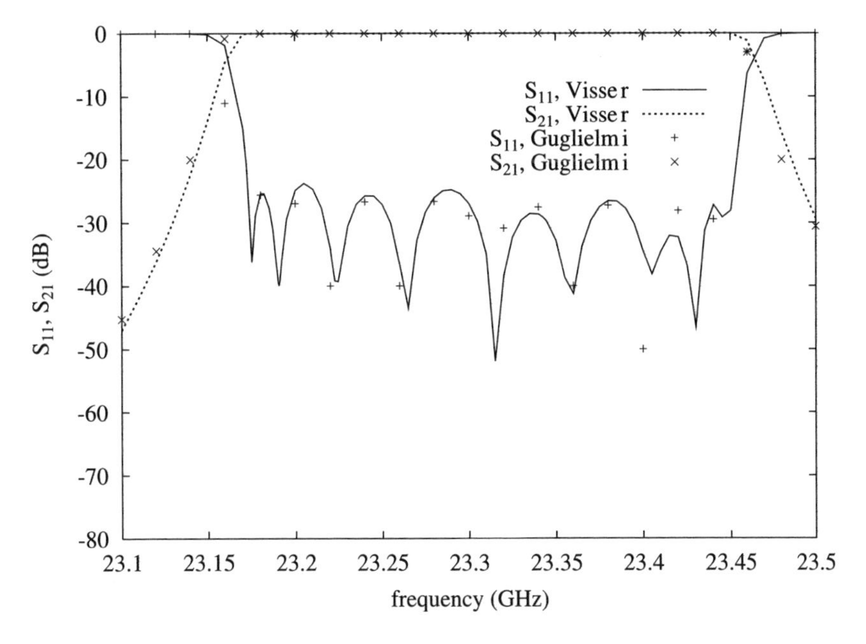

Figure 6.18 Simulation and measurement results for the complete nine-pole filter of Figure 6.15.

(15.8 mm × 7.9 mm) (Figure 6.22). The analysis results for the TE_{10}-mode reflection coefficient are compared with measurement results from [26] in Figure 6.22.

With the exception of the values below −25 db, the simulation and measurement data agree reasonably well (the differences are less than 1.5 db). Here also, errors were introduced by extracting measurement data from the graphs in [26]. Since the simulated results in [26] for levels below −25 db differ in the same way from the measurement results as in Figure 6.22, we may attribute these differences to measurement errors around the resonance frequency.

Now that the method and its software implementation have been thoroughly validated for filter structures, we have to demonstrate validation when radiation into a unit cell is incorporated.

6.9.4 Array Antenna Structures

For the evaluation of large, uniformly excited phased array antennas consisting of open-ended waveguide radiators, it suffices to analyze the reflection coefficient of the fundamental mode of the feeding waveguide. First of all, the reflection coefficient tells us directly about the impedance matching and, secondly, through the *scan element pattern*, we can learn directly about the radiation pattern of the phased array.

6.9.4.1 Scan Element Pattern As explained in [37, 46] for a linear array, the gain of a fully, uniformly excited array scanned to the angular position (ϑ_0, φ_0), in that direction,

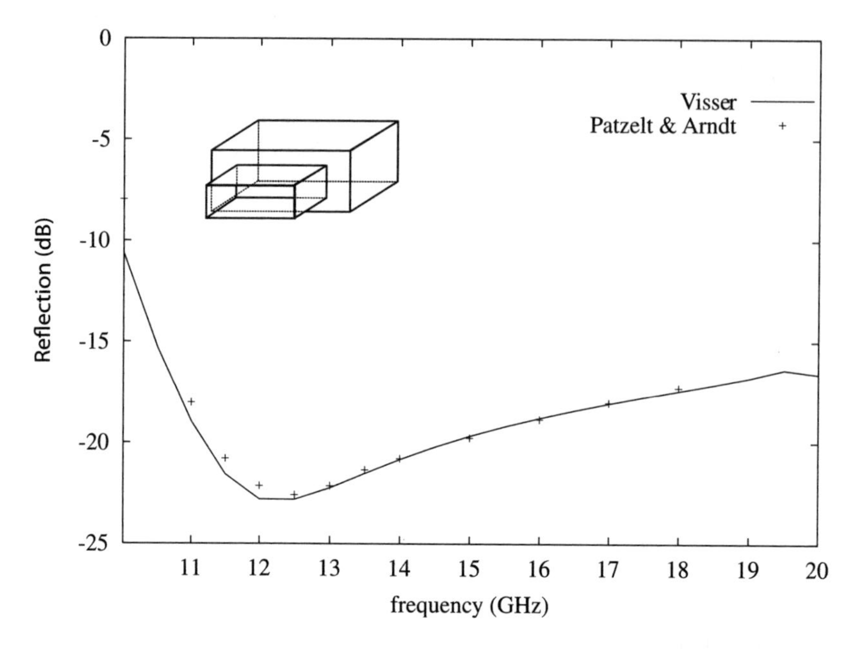

Figure 6.19 Simulated and measured reflection coefficient versus frequency for a concentric Ku-to-X-band step.

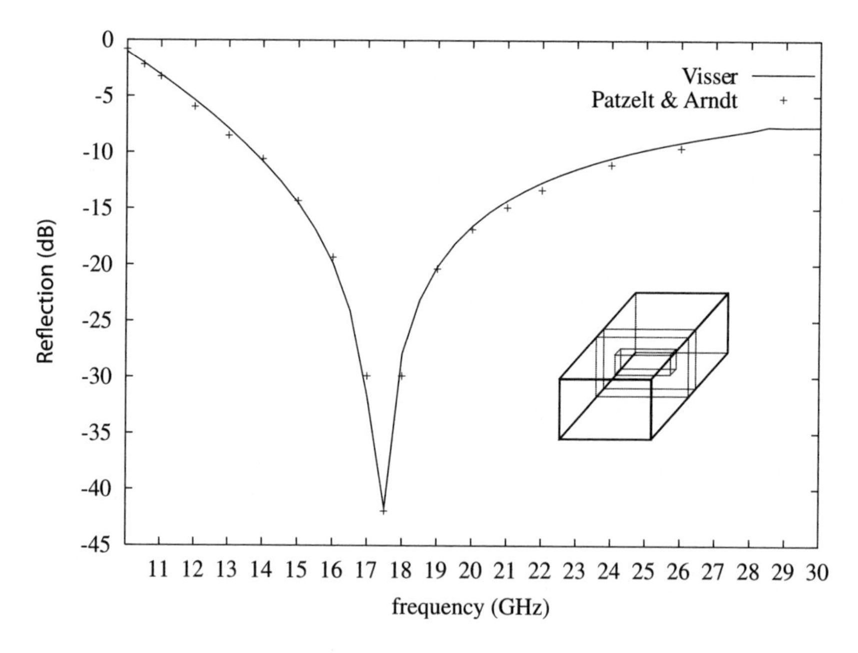

Figure 6.20 Simulated and measured reflection coefficient versus frequency for a resonant iris.

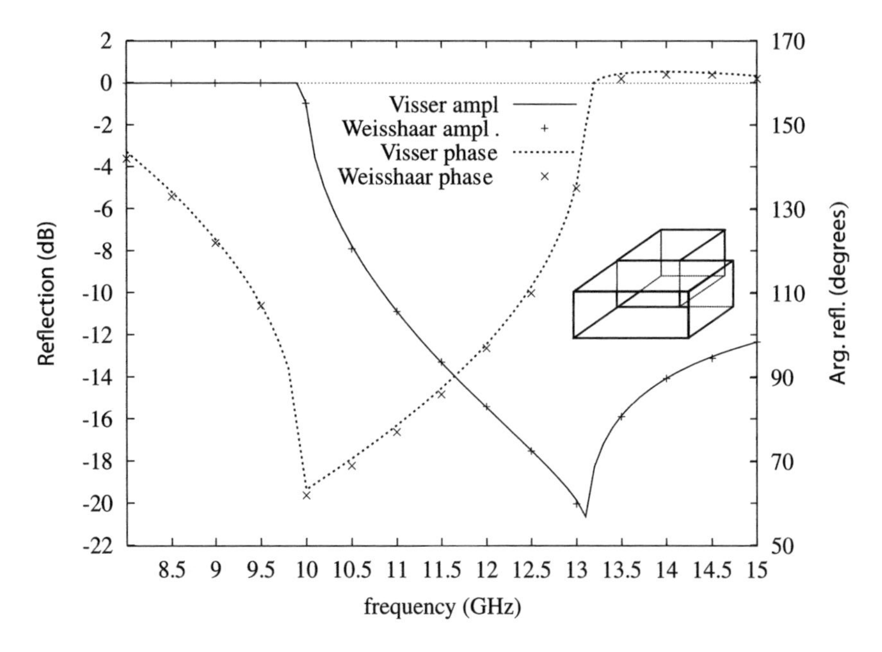

Figure 6.21 Simulated and measured reflection coefficient versus frequency for an H-plane step discontinuity.

$G_a(\vartheta_0, \varphi_0)$, is given by

$$G_a(\vartheta_0, \varphi_0) \approx KG_k^s(\vartheta_0, \varphi_0), \tag{6.244}$$

where K is the number of elements and $G_k^s(\vartheta_0, \varphi_0)$ is the scan element pattern for the angular position (ϑ_0, φ_0).

The *scan element pattern* is the radiation pattern of a single excited element in its array environment, when all other elements are terminated into matched loads. The scan element pattern is given by

$$G_k^s(\vartheta, \varphi) = G_e(\vartheta, \varphi)[1 - |\Gamma(\vartheta, \varphi)|^2], \tag{6.245}$$

where $\Gamma(\vartheta, \varphi)$ is the *scan reflection coefficient* and $G_e(\vartheta, \varphi)$ is the gain of an isolated element.

Since all coupling effects are contained in the scan reflection coefficient, the pattern of the phased array antenna may now be constructed by multiplying the array factor by the scan element pattern. All anomalies in the scan reflection coefficient as a function of scan angle – such as 'dips' or 'nulls' – will appear in the scan element pattern and thus in the pattern of the phased array antenna.

6.9.4.2 Infinite Open-Ended Waveguide Arrays To validate the theory and the computer code that we developed with results from the open literature,[13] we refer to the array layout

[13]Most large, open-ended waveguide phased array antennas have been designed and realized for military radar and therefore only a limited amount of publicly available analysis and measurement data exists.

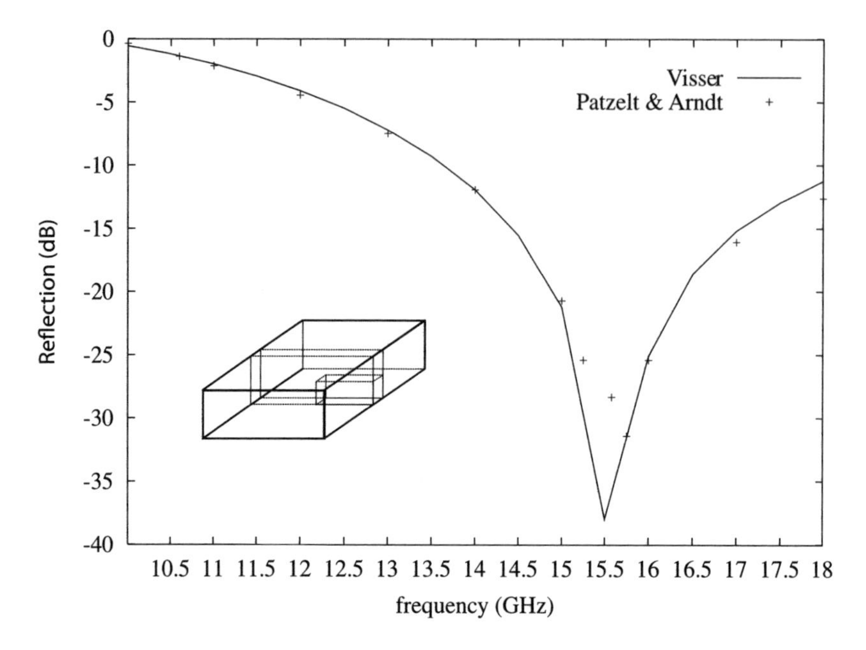

Figure 6.22 Simulated and measured reflection coefficient versus frequency for a finite-thickness asymmetric iris placed in a waveguide.

shown in Figure 6.23. The grid here is triangular in general, the rectangular grid ($\Omega = \pi/2$) being a special case. The dielectric sheets, combined with the aperture iris, serve the purpose of wide-angle impedance matching.

We shall make our first comparisons with analysis results for the CAISSA antenna [8,33], for which simulation and measurement data is publicly available. For the CAISSA antenna (consisting of 849 iris-loaded rectangular-waveguide elements and an external matching sheet), $s = 0.942\lambda_0$, $t = 0.5444\lambda_0$, $\Omega = 30.097°$, $a = 0.659\lambda_0$, $b = d = 0.217\lambda_0$, $c = 0.650\lambda_0$, $t = 0.000$, $\tau_1 = \tau_2 = 0.0942\lambda_0$, $\varepsilon_1 = 1.000$ and $\varepsilon_2 = 2.300$, where λ_0 is the free-space wavelength.

The calculated reflection coefficients for a number of scan positions in the H plane ($\varphi_0 = 0$), D plane ($\varphi_0 = 45°$) and E plane ($\varphi_0 = 90°$)[14] are shown in Table 6.3, as calculated with the theory of [47], the theory of [17], the theory of [8,33] and the current theory, for an infinite antenna without a matching structure, i.e. without irises and a matching sheet. In Table 6.4, results are shown for the CAISSA antenna without a matching sheet, for different iris sizes, with the beam scanned to $(\varphi_0, \vartheta_0) = (45°, 60°)$. (M_1, N_1) give the maximum mode numbers in the feeding waveguide, and (M_2, N_2) give the maximum mode numbers in the aperture. Finally, in Table 6.5, results for the CAISSA antenna with its internal and external matching structures are given.

[14]The x axis is parallel to the long side of the waveguides, and the y axis is parallel to the short side.

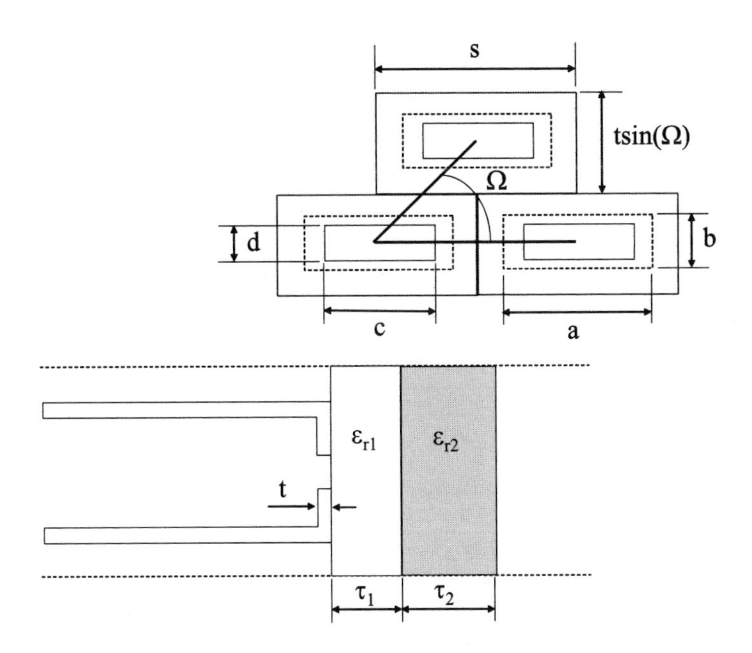

Figure 6.23 Infinite array of open-ended waveguide radiators, having a finite-thickness iris in the waveguide aperture and two dielectric sheets in front of the array aperture.

Table 6.3 Calculated reflection coefficients for the CAISSA antenna with $c = b$ and $\varepsilon_2 = 1.000$.

φ_0	ϑ_0	Γ [47]	Γ [17]	Γ [8]	Γ
0	0	$0.19\angle 7.29°$	$0.18\angle 9.63°$	$0.18\angle 13.21°$	$0.18\angle 10.79°$
0	20	$0.27\angle -6.49°$	$0.27\angle -6.16°$	$0.26\angle -4.53°$	$0.27\angle -5.26°$
0	40	$0.48\angle -23.67°$	$0.48\angle -24.72°$	$0.48\angle -24.36°$	$0.48\angle -24.12°$
0	60	$0.70\angle -33.98°$	$0.71\angle -35.11°$	$0.71\angle -34.79°$	$0.71\angle -36.64°$
45	20	$0.21\angle -1.99°$	$0.21\angle -0.81°$	$0.20\angle 1.46°$	$0.21\angle 0.18°$
45	40	$0.28\angle -23.80°$	$0.28\angle -24.22°$	$0.27\angle -23.74°$	$0.27\angle -23.64°$
45	60	$0.31\angle -53.93°$	$0.32\angle -54.81°$	$0.31\angle -55.52°$	$0.31\angle -54.92°$
90	40	$0.06\angle -3.85°$	$0.06\angle 2.73°$	$0.04\angle 6.97°$	$0.05\angle 4.01°$
90	60	$0.15\angle -162.90°$	$0.15\angle -165.26°$	$0.17\angle -163.96°$	$0.16\angle -164.78°$

Looking at Table 6.4, we see a very poor match, especially in the phase. The reason must be due to the infinitely-thin-iris model employed in [8, 17]. However, in Tables 6.3 and 6.5, we observe a much better correspondence between the results of the various calculations. In general, we can say that our mode-matching method appears to produce correct results, but at the same time we see that it is very difficult to make this judgment on a point-to-point basis. It is better to make the comparisons over a range of scan angles for a geometrically fixed array antenna system. In the remainder, the comparisons that we shall make will be of this type.

To start with, we analyze a square grid array and compare the results with analysis results and measurement results for a 19×19 element array reported in [19]. The relevant array

Table 6.4 Calculated reflection coefficients for the CAISSA antenna without sheet, scanned to (45°, 60°).

c/a	(M_1, N_1)	(M_2, N_2)	Γ [17]	Γ [8]	Γ
1.0	(7,2)	(7,2)	$0.32\angle 304.86°$	$0.31\angle 304.35°$	$0.31\angle 10.79°$
0.9	(7,2)	(6,2)	$0.29\angle 315.29°$	$0.29\angle 308.53°$	$0.27\angle 354.74°$
0.8	(8,2)	(6,2)	$0.25\angle 343.46°$	$0.26\angle 324.20°$	$0.23\angle 335.88°$
0.7	(10,2)	(7,2)	$0.27\angle 25.80°$	$0.23\angle 2.62°$	$0.26\angle 323.36°$
0.6	(10,2)	(6,2)	$0.47\angle 81.08°$	$0.37\angle 63.33°$	$0.50\angle 0.18°$
0.5	(10,2)	(5,2)	$0.86\angle 136.80°$	$0.59\angle 98.50°$	$0.72\angle 336.36°$
0.4	(10,2)	(4,2)	$0.91\angle 145.05°$	$0.87\angle 137.51°$	$0.93\angle 305.08°$
0.3	(10,2)	(3,2)	$0.89\angle 148.06°$	$0.84\angle 140.08°$	$0.97\angle 4.010°$

Table 6.5 Calculated reflection coefficients for the CAISSA antenna with its matching structures.

φ_0	ϑ_0	Γ [17]	Γ [8]	Γ
0	0	$0.27\angle 102.27°$	$0.24\angle 105.50°$	$0.30\angle 103.42°$
0	20	$0.24\angle 90.72°$	$0.21\angle 90.25°$	$0.27\angle 92.69°$
0	40	$0.18\angle 51.16°$	$0.17\angle 39.90°$	$0.21\angle 59.38°$
0	60	$0.19\angle 334.41°$	$0.21\angle 329.60°$	$0.18\angle 337.53°$
45	20	$0.23\angle 98.79°$	$0.19\angle 102.21°$	$0.26\angle 100.20°$
45	40	$0.08\angle 66.58°$	$0.04\angle 56.79°$	$0.11\angle 76.78°$
45	60	$0.18\angle 287.71°$	$0.21\angle 279.87°$	$0.14\angle 294.48°$
90	20	$0.22\angle 108.03°$	$0.19\angle 115.31°$	$0.25\angle 108.71°$
90	40	$0.10\angle 129.35°$	$0.10\angle 158.96°$	$0.12\angle 124.51°$
90	60	$0.05\angle 142.14°$	$0.08\angle 194.71°$	$0.07\angle 131.33°$

dimensions, referring to Figure 6.23, are $a = b = c = d = 0.5354\lambda_0$, $s = t = 0.5714\lambda_0$, $\Omega = 90°$ and $\varepsilon_{r1} = \varepsilon_{r2} = 1.000$, where λ_0 is the free-space wavelength.

Figure 6.24 shows the amplitude of the H-plane scan reflection coefficient as calculated in [19], as calculated with the theory described in this chapter and as measured in [19]. The phase, as calculated in [19] and as calculated with the mode-matching method of this chapter, is shown in Figure 6.25. The maximum number of modes applied in the waveguide aperture is given by $(M, N) = (5, 5)$.

The figures show fair agreement up to 48° in the amplitude (the difference is less than 1.5 db) for both simulation methods. The measurements fail to register the deep 'null' around 48°. Most likely the null is (partly) filled owing to coupling effects in the finite array. These couplings are also visible in the fluctuations of the measured amplitude of the reflection coefficient as a function of the scan angle around the values for the reflection coefficient of the infinite array. The relatively high reflection coefficients measured for scan angles beyond 48° are probably caused by the same effect that has filled up the null. Obviously, the 19×19 array is not 'infinite enough' to realize the low reflection coefficient amplitudes predicted by the infinite-array antenna model.

The phase as a function of the scan angle calculated with the current method agrees well with the value calculated in [19]; the difference over all scan angles is less than 5°.

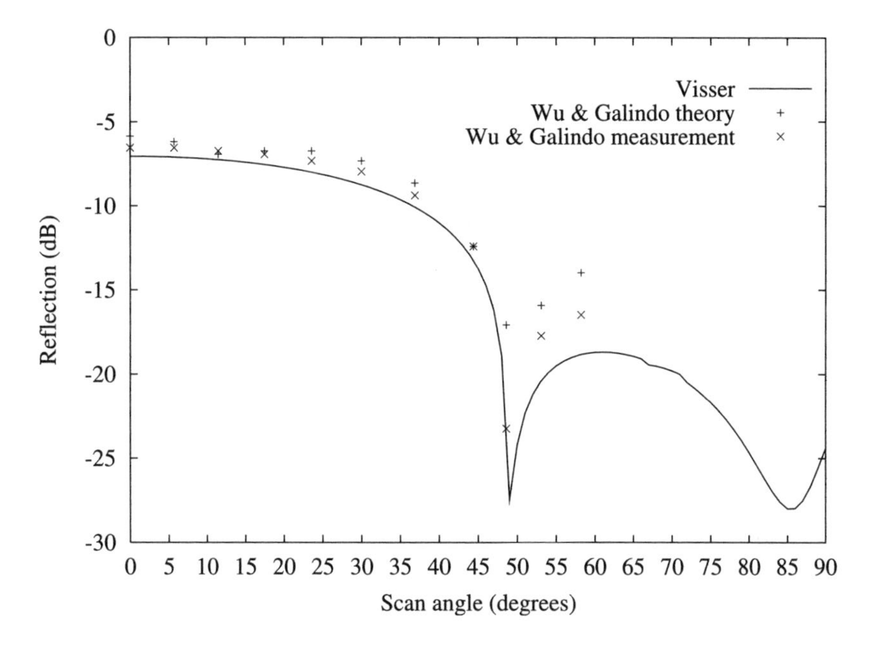

Figure 6.24 Amplitude of reflection coefficient for a square-grid, open-ended waveguide array, as calculated for an infinite array and measured for a 19 × 19 element array.

To compare the analysis results for open-ended waveguide array antennas with *thick* irises in the apertures ($t \neq 0$ in Figure 6.23), we have to resort to published analysis results [48].

We start with an infinite array of open-ended waveguide radiators without a matching structure. The relevant parameters for this geometry (Figure 6.23) are $a = c = 0.660\lambda_0$, $b = d = 0.159\lambda_0$, $s = 0.856\lambda_0$, $t = 0.495\lambda_0$, $\Omega = 32.440°$ and $\varepsilon_{r1} = \varepsilon_{r2} = 1.000$, where λ_0 is the free-space wavelength. The amplitude of the reflection coefficient for an H-plane scan ($\varphi_0 = 0$) and for an E-plane scan ($\varphi_0 = 90°$) is shown in Figure 6.26, as calculated in [48] and as calculated by the mode-matching method described in this chapter. The number of waveguide modes was set to 42, with $(M, N) = (8, 2)$, identical to the number of modes used in [48]. The phase as a function of the scan angle is shown in Figure 6.27.

Considering the difficulty of extracting the analysis data from the graphs in [48], the results of our analysis match those of [31] well. The difference in amplitude is less than 1.7 dB, except around resonance, and the difference in phase is less than 4.5°, excluding the last E-plane analysis result from [48].

Finally, we shall compare analysis results for an infinite array of open-ended waveguides, arranged in a triangular lattice, having a thick iris in the waveguide aperture. The relevant dimensions are $a = 0.660\lambda_0$, $b = d = 0.159\lambda_0$, $c = 0.500\lambda_0$, $s = 0.856\lambda_0$, $t = 0.495\lambda_0$, $\Omega = 32.440°$, $t = 0.015\lambda_0$ and $\varepsilon_{r1} = \varepsilon_{r2} = 1.000$, where λ_0 is the free-space wavelength. The array was analyzed at $f = 1.17 f_0$, where f_0 is the frequency corresponding to the wavelength λ_0.

The amplitude of the reflection coefficient for an H-plane scan ($\varphi_0 = 0$) and for an E-plane scan ($\varphi_0 = 90°$) is shown in Figure 6.28, as calculated in [48] and as calculated

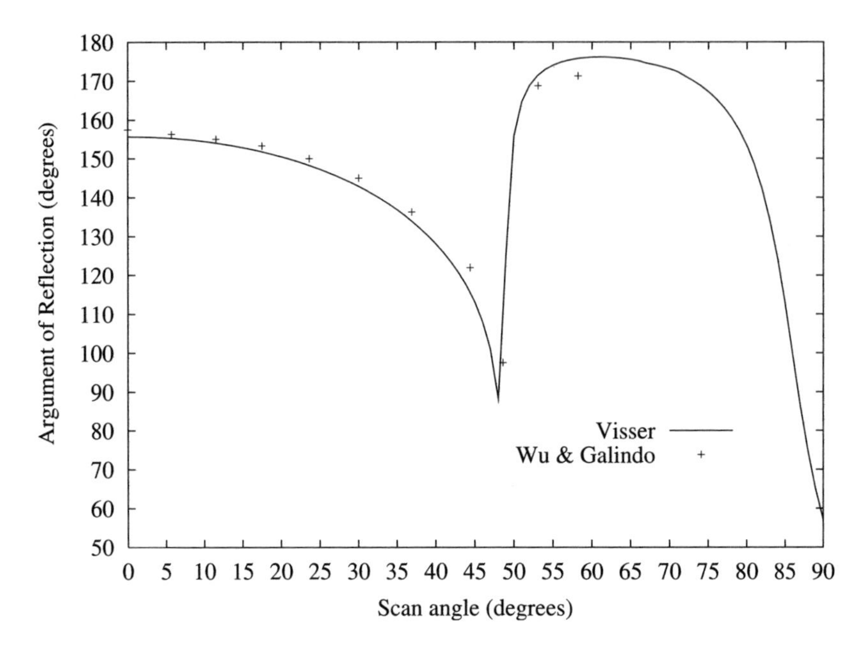

Figure 6.25 Phase of reflection coefficient for a square-grid, open-ended waveguide array, as calculated for an infinite array.

by the mode-matching method described in this chapter, for $f = 1.17 f_0$. The number of waveguide aperture modes was set to 42, with $(M, N) = (8, 2)$, identical to the number of modes used in [48]. The number of waveguide modes in the feeding waveguide then met the RC condition, with $(M, N) = (11, 2)$. The phase as a function of the scan angle is shown in Figure 6.29.

Again considering the difficulty of extracting the analysis data from the graphs in [48], the results of our analysis match those of [31] well. The difference in amplitude is less than 2.5 db over all scan angles, and the difference in phase is less than 5.0°, excluding the last E-plane analysis result from [48].

Our method and its software implementation have now also been thoroughly validated for planar, infinite, open-ended waveguide array antennas.

6.10 CONCLUSIONS

A versatile analysis method has been developed for waveguide structures and infinite waveguide phased array antennas. The method is based on mode matching, which is used to construct generalized scattering matrices for waveguide-to-waveguide and waveguide-to-unit-cell junctions. The code developed has been thoroughly validated by comparing analysis results with analysis and measurement results from the open literature. A mode preselection scheme, a convergence strategy and a Floquet mode selection scheme have been developed and have been applied to the structures analyzed. The convergence strategy, which avoids the

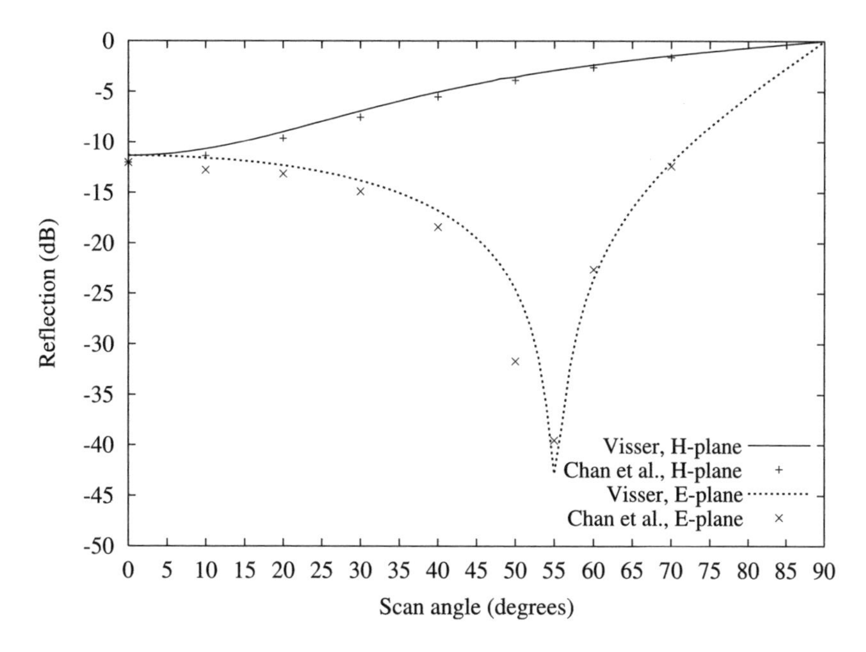

Figure 6.26 Amplitude of reflection coefficient for an infinite array of open-ended, rectangular waveguides in a triangular grid.

occurrence of relative convergence, together with the Floquet mode selection scheme based on the maximum wave number in the waveguide aperture, makes it possible for the code to be used by nonspecialists.

The theory presented in this chapter may be regarded as a 'classic' mode-matching approach for analyzing large arrays of open-ended waveguides, stemming from work done in the mid 1990s. Although valuable in itself, as demonstrated in this chapter, it may also serve in understanding the progress that has been made since then in analyzing this kind of antennas, amongst others, the realization of uniform convergence by virtue of the concept of the multimode equivalent network [2–7].

APPENDIX 6.A. WAVEGUIDE MODE ORTHOGONALITY AND NORMALIZATION FUNCTIONS

In section 6.2, we decomposed the electric and magnetic fields in a rectangular waveguide into modes. The mode vectors, $\boldsymbol{e}_{\mathrm{T}_{\mathrm{TE}_{mn}}}$ for the electric field of a TE mode and $\boldsymbol{e}_{\mathrm{T}_{\mathrm{TM}_{mn}}}$ for the electric field of a TM mode, possess orthogonality characteristics, i.e. each mode vector $\boldsymbol{e}_{\mathrm{T}_{\mathrm{TE}_{mn}}}$ or $\boldsymbol{e}_{\mathrm{T}_{\mathrm{TM}_{mn}}}$ is orthogonal to all other mode vectors [9, 36]:

$$\iint \boldsymbol{e}_{\mathrm{T}_{\mathrm{TE}_{mn}}} \cdot \boldsymbol{e}_{\mathrm{T}_{\mathrm{TE}_{ij}}} dS = \iint \boldsymbol{e}_{\mathrm{T}_{\mathrm{TM}_{mn}}} \cdot \boldsymbol{e}_{\mathrm{T}_{\mathrm{TM}_{ij}}} dS = \begin{cases} 1 & \text{for } m = i \wedge n = j \\ 0 & \text{for } m \neq i \vee n \neq j \end{cases}$$

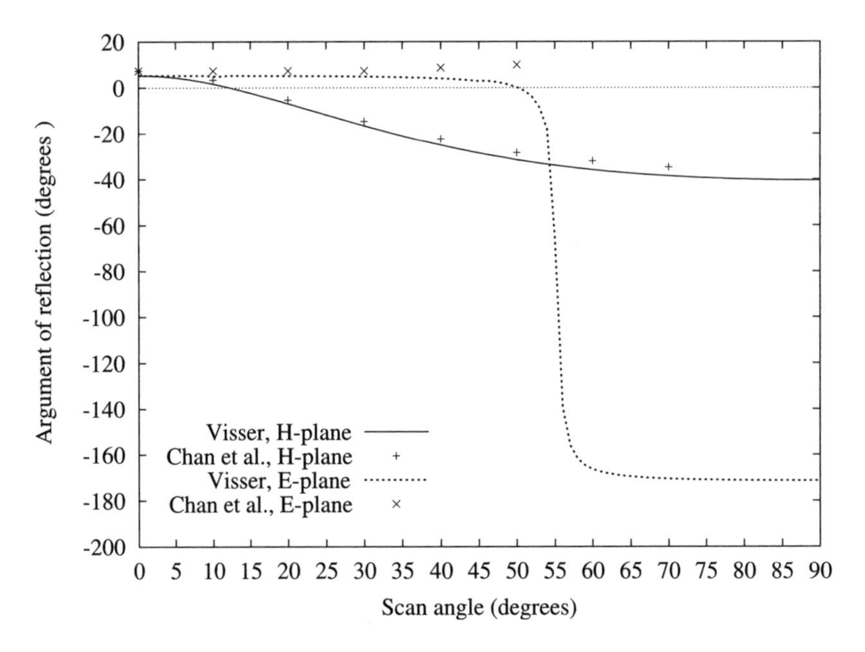

Figure 6.27 Phase of reflection coefficient for an infinite array of open-ended, rectangular waveguides in a triangular grid.

and

$$\iint \boldsymbol{e}_{\mathrm{T}_{\mathrm{TE}_{mn}}} \cdot \boldsymbol{e}_{\mathrm{T}_{\mathrm{TM}_{ij}}} dS = 0. \tag{6.A.1}$$

The fields of the magnetic modes, $\boldsymbol{h}_{\mathrm{T}_{\mathrm{TE}_{mn}}}$ and $\boldsymbol{h}_{\mathrm{T}_{\mathrm{TM}_{mn}}}$, follow from the fields of the electric modes through [9]

$$\boldsymbol{h}_{\mathrm{T}_{\mathrm{TE}_{mn}}} = \hat{\boldsymbol{i}}_z \times \boldsymbol{e}_{\mathrm{T}_{\mathrm{TE}_{mn}}} \tag{6.A.2}$$

and

$$\boldsymbol{h}_{\mathrm{T}_{\mathrm{TM}_{mn}}} = \hat{\boldsymbol{i}}_z \times \boldsymbol{e}_{\mathrm{T}_{\mathrm{TM}_{mn}}}. \tag{6.A.3}$$

Taking the mode functions of section 6.2, equations (6.45)–(6.49), into account, these orthogonality relations can be written in terms of the following pair of integrals,

$$NC(\xi, \eta) = \int_{-1/2}^{1/2} C_\xi(\kappa) C_\eta(\kappa)\, d\kappa, \tag{6.A.4}$$

$$NS(\xi, \eta) = \int_{-1/2}^{1/2} S_\xi(\kappa) S_\eta(\kappa)\, d\kappa, \tag{6.A.5}$$

where the cosine function $C_\xi(\kappa)$ and sine function $S_\xi(\kappa)$ are defined by equations (6.54) and (6.55), respectively:

$$C_\xi(\kappa) = \cos\left[\xi\pi\left(\kappa + \frac{1}{2}\right)\right] \tag{6.A.6}$$

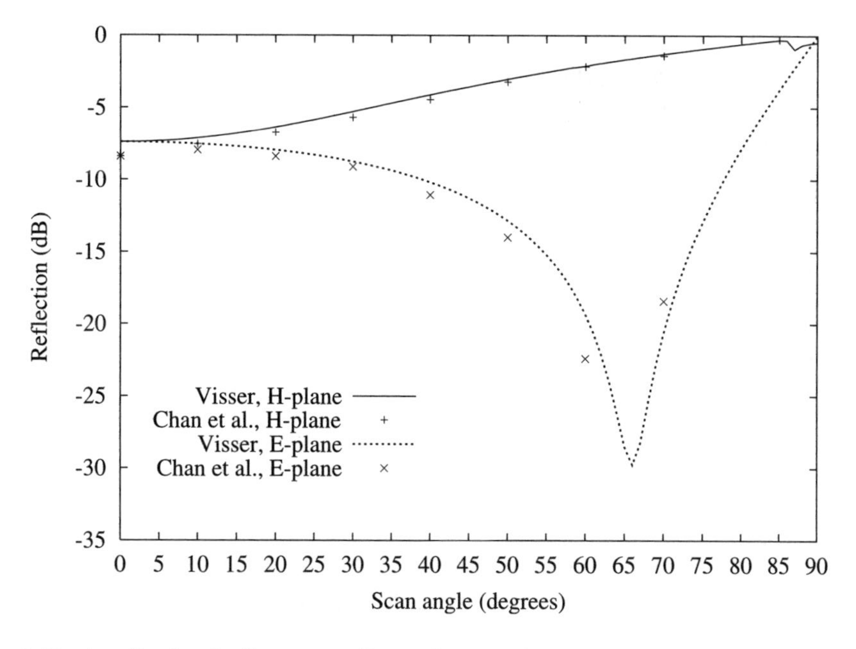

Figure 6.28 Amplitude of reflection coefficient for an infinite array of open-ended, rectangular waveguides with a thick aperture iris in a triangular grid, evaluated at $f = 1.17 f_0$.

and

$$S_\xi(\kappa) = \sin\left[\xi\pi\left(\kappa + \frac{1}{2}\right)\right]. \tag{6.A.7}$$

Calculation of the integrals $NC(\xi, \eta)$ and $NS(\xi, \eta)$ yields

$$NC(\xi, \eta) = \begin{cases} 1 & \text{if } \xi = \eta = 0 \\ \frac{1}{2} & \text{if } \xi = \eta \neq 0 \\ 0 & \text{if } \xi \neq \eta, \end{cases} \tag{6.A.8}$$

$$NS(\xi, \eta) = \begin{cases} 1 & \text{if } \xi = \eta = 0 \\ \frac{1}{2} & \text{if } \xi = \eta \neq 0 \\ 0 & \text{if } \xi \neq \eta. \end{cases} \tag{6.A.9}$$

Using the above, we can calculate the normalization functions $O_i(\xi, \eta, \alpha, \beta)$, $i = 1, 2, \ldots, 8$, $\xi, \eta \in \mathbb{N}$, $\alpha, \beta \in \mathbb{R}$, as

$$\begin{aligned} O_1(\xi, \eta, \alpha, \beta) &= \int_{x=-\alpha/2}^{\alpha/2} \int_{y=-\beta/2}^{\beta/2} e^{+*}_{x\mathrm{TE}_{ij}} e^{+}_{x\mathrm{TE}_{\xi\eta}} \, dx \, dy \\ &= \frac{\alpha\beta}{k^4_{\mathrm{T}_{\xi\eta}}} (\omega\mu_0\mu_\mathrm{r})^2 \left(\frac{\eta\pi}{\beta}\right)^2 NC(i, \xi) NS(j, \eta), \end{aligned} \tag{6.A.10}$$

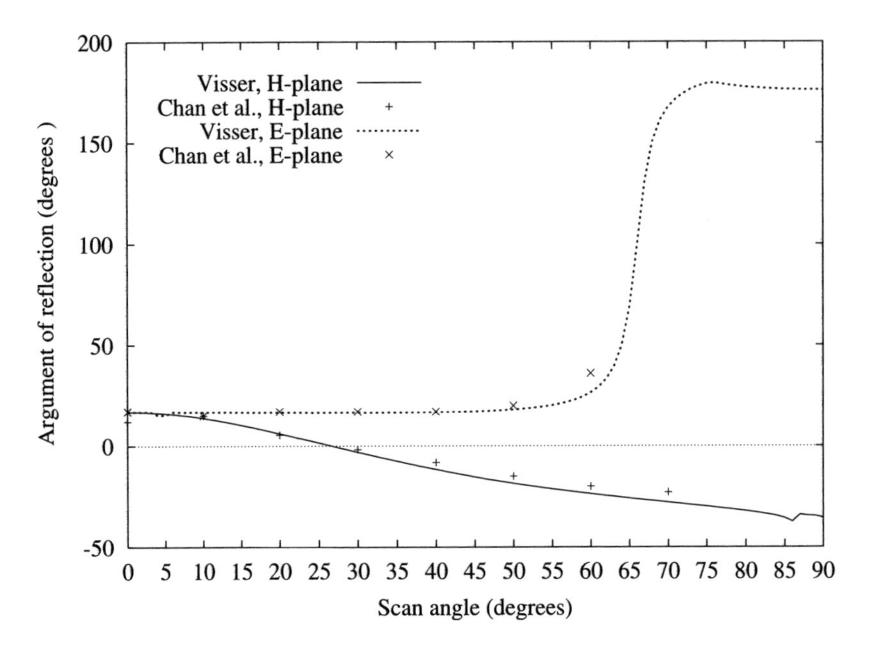

Figure 6.29 Phase of reflection coefficient for an infinite array of open-ended, rectangular waveguides with a thick aperture iris in a triangular grid, evaluated at $f = 1.17f_0$.

$$\begin{aligned} O_2(\xi,\eta,\alpha,\beta) &= \int_{x=-\alpha/2}^{\alpha/2}\int_{y=-\beta/2}^{\beta/2} e^{+^*}_{x\mathrm{TE}_{ij}} e^{+}_{x\mathrm{TM}_{\xi\eta}}\,dx\,dy \\ &= -\frac{\alpha\beta}{k^4_{\mathrm{T}_{\xi\eta}}}(\omega\mu_0\mu_\mathrm{r})\gamma_{\xi\eta}\left(\frac{\xi\pi}{\alpha}\right)\left(\frac{\eta\pi}{\beta}\right)NC(i,\xi)NS(j,\eta), \end{aligned} \tag{6.A.11}$$

$$\begin{aligned} O_3(\xi,\eta,\alpha,\beta) &= \int_{x=-\alpha/2}^{\alpha/2}\int_{y=-\beta/2}^{\beta/2} e^{+^*}_{x\mathrm{TM}_{ij}} e^{+}_{x\mathrm{TE}_{\xi\eta}}\,dx\,dy \\ &= -\frac{\alpha\beta}{k^4_{\mathrm{T}_{\xi\eta}}}(\omega\mu_0\mu_\mathrm{r})\gamma^*_{\xi\eta}\left(\frac{\xi\pi}{\alpha}\right)\left(\frac{\eta\pi}{\beta}\right)NC(i,\xi)NS(j,\eta), \end{aligned} \tag{6.A.12}$$

$$\begin{aligned} O_4(\xi,\eta,\alpha,\beta) &= \int_{x=-\alpha/2}^{\alpha/2}\int_{y=-\beta/2}^{\beta/2} e^{+^*}_{x\mathrm{TM}_{ij}} e^{+}_{x\mathrm{TM}_{\xi\eta}}\,dx\,dy \\ &= \frac{\alpha\beta}{k^4_{\mathrm{T}_{\xi\eta}}}|\gamma^*_{\xi\eta}|^2\left(\frac{\xi\pi}{\alpha}\right)^2 NC(i,\xi)NS(j,\eta), \end{aligned} \tag{6.A.13}$$

$$\begin{aligned} O_5(\xi,\eta,\alpha,\beta) &= \int_{x=-\alpha/2}^{\alpha/2}\int_{y=-\beta/2}^{\beta/2} e^{+^*}_{y\mathrm{TE}_{ij}} e^{+}_{y\mathrm{TE}_{\xi\eta}}\,dx\,dy \\ &= -\frac{\alpha\beta}{k^4_{\mathrm{T}_{\xi\eta}}}(\omega\mu_0\mu_\mathrm{r})^2\left(\frac{\xi\pi}{\alpha}\right)^2 NS(i,\xi)NC(j,\eta), \end{aligned} \tag{6.A.14}$$

$$O_6(\xi,\eta,\alpha,\beta) = \int_{x=-\alpha/2}^{\alpha/2}\int_{y=-\beta/2}^{\beta/2} e^{+*}_{y\mathrm{TE}_{ij}} e^{+}_{y\mathrm{TM}_{\xi\eta}}\,dx\,dy$$
$$= \frac{\alpha\beta}{k^4_{\mathrm{T}_{\xi\eta}}}(\omega\mu_0\mu_\mathrm{r})\gamma_{\xi\eta}\left(\frac{\xi\pi}{\alpha}\right)\left(\frac{\eta\pi}{\beta}\right) NS(i,\xi)NC(j,\eta), \tag{6.A.15}$$

$$O_7(\xi,\eta,\alpha,\beta) = \int_{x=-\alpha/2}^{\alpha/2}\int_{y=-\beta/2}^{\beta/2} e^{+*}_{y\mathrm{TM}_{ij}} e^{+}_{y\mathrm{TE}_{\xi\eta}}\,dx\,dy$$
$$= \frac{\alpha\beta}{k^4_{\mathrm{T}_{\xi\eta}}}(\omega\mu_0\mu_\mathrm{r})\gamma^*_{\xi\eta}\left(\frac{\xi\pi}{\alpha}\right)\left(\frac{\eta\pi}{\beta}\right) NS(i,\xi)NC(j,\eta), \tag{6.A.16}$$

$$O_8(\xi,\eta,\alpha,\beta) = \int_{x=-\alpha/2}^{\alpha/2}\int_{y=-\beta/2}^{\beta/2} e^{+*}_{y\mathrm{TM}_{ij}} e^{+}_{y\mathrm{TM}_{\xi\eta}}\,dx\,dy$$
$$= \frac{\alpha\beta}{k^4_{\mathrm{T}_{\xi\eta}}}|\gamma^*_{\xi\eta}|^2\left(\frac{\eta\pi}{\beta}\right)^2 NS(i,\xi)NC(j,\eta). \tag{6.A.17}$$

In the above, $i, j \in \mathbb{N}$. Solutions exist only for $i = \xi$ and $j = \eta$. Note that $O_2(\xi,\eta,\alpha,\beta) + O_6(\xi,\eta,\alpha,\beta) = 0$ and $O_3(\xi,\eta,\alpha,\beta) + O_7(\xi,\eta,\alpha,\beta) = 0$.

APPENDIX 6.B. MODE-COUPLING INTEGRALS FOR WAVEGUIDE-TO-WAVEGUIDE JUNCTION

In the mode-matching procedure for a rectangular-waveguide-to-waveguide junction we encountered, in section 6.4.2, the mode-coupling integrals $H_{i_{mnlw}}$, $i = 1, 2, \ldots, 8$ and $H_{j_{lwmn}}$, $j = 9, 10, \ldots, 16$. These coupling integrals can be expressed in terms of the following set of integrals:

$$CC(\xi,\eta,\alpha,\beta) = \int_{\kappa=-1/2\alpha+\beta}^{1/2\alpha+\beta} C_\xi(\kappa)C_\eta(\alpha[\kappa-\beta])\,d\kappa \tag{6.B.1}$$

and

$$SS(\xi,\eta,\alpha,\beta) = \int_{\kappa=-1/2\alpha+\beta}^{1/2\alpha+\beta} S_\xi(\kappa)S_\eta(\alpha[\kappa-\beta])\,d\kappa. \tag{6.B.2}$$

In the above two equations, $\xi, \eta \in \mathbb{N}$ and $\alpha, \beta \in \mathbb{R}$. The cosine function $C_\xi(\kappa)$ and sine function $S_\xi(\kappa)$ are defined by equations (6.54) and (6.55), respectively:

$$C_\xi(\kappa) = \cos\left[\xi\pi\left(\kappa + \frac{1}{2}\right)\right] \tag{6.B.3}$$

and

$$S_\xi(\kappa) = \sin\left[\xi\pi\left(\kappa + \frac{1}{2}\right)\right]. \tag{6.B.4}$$

The calculation of the integrals $CC(\eta, \xi, \alpha, \beta)$ and $CC(\eta, \xi, \alpha, \beta)$ results in

$$CC(\eta, \xi, \alpha, \beta) = \begin{cases} \dfrac{1}{\alpha} & \text{if } \xi = \eta\alpha = 0 \\ \dfrac{1}{\pi}\dfrac{\xi}{\xi^2 - (\eta\alpha)^2}\left\{(-1)^\eta \sin\left[\xi\pi\dfrac{\alpha+1+2\alpha\beta}{2\alpha}\right] - \sin\left[\xi\pi\dfrac{\alpha-1+2\alpha\beta}{2\alpha}\right]\right\} & \text{if } \xi \neq \eta\alpha \\ \dfrac{1}{2\alpha}\cos\left[\eta\pi\dfrac{\alpha-1+2\alpha\beta}{2}\right] & \text{if } \xi = \eta\alpha \neq 0 \end{cases} \tag{6.B.5}$$

and

$$SS(\eta, \xi, \alpha, \beta) = \begin{cases} 0 & \text{if } \xi = 0 \vee \eta = 0 \\ \dfrac{\eta\alpha}{\xi} CC(\eta, \xi, \alpha, \beta) & \text{if } \xi \neq \eta\alpha \wedge \xi \neq 0 \\ CC(\eta, \xi, \alpha, \beta) & \text{if } \xi = \eta\alpha \neq 0. \end{cases} \tag{6.B.6}$$

The coupling integrals are then found as

$$H1_{mnlw} = \frac{(\omega\mu_0)^2 \mu_r^I \mu_r^{II} (n\pi/b)(w\pi/b')ab}{[(m\pi/a)^2 + (n\pi/b)^2][(l\pi/a')^2 + (w\pi/b')^2]} \times CC\left(m, l, \frac{a}{a'}, \frac{x_c}{a}\right) SS\left(n, w, \frac{b}{b'}, \frac{y_c}{b}\right), \tag{6.B.7}$$

$$H2_{mnlw} = \begin{cases} -Y_{\mathrm{TE}_{lw}}^{II} \dfrac{lb'}{wa'} H1_{mnlw} & \text{if } w \neq 0 \\ 0 & \text{if } w = 0, \end{cases} \tag{6.B.8}$$

$$H3_{mnlw} = \frac{(\omega\mu_0)^2 \mu_r^I \mu_r^{II} (m\pi/a)(l\pi/a')ab}{[(m\pi/a)^2 + (n\pi/b)^2][(l\pi/a')^2 + (w\pi/b')^2]} \times CC\left(n, w, \frac{b}{b'}, \frac{y_c}{b}\right) SS\left(m, l, \frac{a}{a'}, \frac{x_c}{a}\right), \tag{6.B.9}$$

$$H4_{mnlw} = \begin{cases} Y_{\mathrm{TE}_{lw}}^{II} \dfrac{wa'}{lb'} H3_{mnlw} & \text{if } l \neq 0 \\ 0 & \text{if } l = 0, \end{cases} \tag{6.B.10}$$

$$H5_{mnlw} = \begin{cases} -Y_{\mathrm{TE}_{mn}}^{I*} \dfrac{mb}{na} H1_{mnlw} & \text{if } n \neq 0 \\ 0 & \text{if } n = 0, \end{cases} \tag{6.B.11}$$

$$H6_{mnlw} = \begin{cases} Y_{\mathrm{TE}_{mn}}^{I*} Y_{\mathrm{TE}_{lw}}^{II} \dfrac{mlbb'}{nwaa'} H1_{mnlw} & \text{if } n \neq 0 \wedge w \neq 0 \\ 0 & \text{if } n = 0 \vee w = 0, \end{cases} \tag{6.B.12}$$

$$H_{7_{mnlw}} = \begin{cases} Y^{I*}_{\mathrm{TE}_{mn}} \dfrac{na}{mb} H_{3_{mnlw}} & \text{if } m \neq 0 \\ 0 & \text{if } m = 0, \end{cases} \tag{6.B.13}$$

$$H_{8_{mnlw}} = \begin{cases} Y^{I*}_{\mathrm{TE}_{mn}} Y^{II}_{\mathrm{TE}_{lw}} \dfrac{nwaa'}{mlbb'} H_{3_{mnlw}} & \text{if } m \neq 0 \wedge l \neq 0 \\ 0 & \text{if } m = 0 \vee l = 0, \end{cases} \tag{6.B.14}$$

$$H_{9_{lwmn}} = H_{3_{mnlw}}, \tag{6.B.15}$$

$$H_{10_{lwmn}} = \begin{cases} Y^{I}_{\mathrm{TE}_{mn}} \dfrac{na}{mb} H_{3_{mnlw}} & \text{if } m \neq 0 \\ 0 & \text{if } m = 0, \end{cases} \tag{6.B.16}$$

$$H_{11_{lwmn}} = H_{1_{mnlw}}, \tag{6.B.17}$$

$$H_{12_{lwmn}} = \begin{cases} -Y^{I}_{\mathrm{TE}_{mn}} \dfrac{mb}{na} H_{1_{mnlw}} & \text{if } n \neq 0 \\ 0 & \text{if } n = 0, \end{cases} \tag{6.B.18}$$

$$H_{13_{lwmn}} = \begin{cases} Y^{II*}_{\mathrm{TE}_{lw}} \dfrac{wa'}{lb'} H_{3_{mnlw}} & \text{if } l \neq 0 \\ 0 & \text{if } l = 0, \end{cases} \tag{6.B.19}$$

$$sH_{14_{lwmn}} = \begin{cases} Y^{I}_{\mathrm{TE}_{mn}} Y^{II*}_{\mathrm{TE}_{lw}} \dfrac{nwaa'}{mlbb'} H_{3_{mnlw}} & \text{if } m \neq 0 \wedge l \neq 0 \\ 0 & \text{if } m = 0 \vee l = 0, \end{cases} \tag{6.B.20}$$

$$H_{15_{lwmn}} = \begin{cases} -Y^{II*}_{\mathrm{TE}_{lw}} \dfrac{lb'}{wa'} H_{1_{mnlw}} & \text{if } w \neq 0 \\ 0 & \text{if } w = 0, \end{cases} \tag{6.B.21}$$

$$H_{16_{lwmn}} = \begin{cases} Y^{I}_{\mathrm{TE}_{mn}} Y^{II*}_{\mathrm{TE}_{lw}} \dfrac{mlbb'}{nwaa'} H_{1_{mnlw}} & \text{if } n \neq 0 \wedge w \neq 0 \\ 0 & \text{if } n = 0 \vee w = 0. \end{cases} \tag{6.B.22}$$

The coupling integrals, combined and normalized, give rise to the following functions, which are used as matrix elements in section 6.4.2:

$$\begin{aligned} H_{a_{mnlw}} &= \frac{H_{1_{mnlw}} + H_{3_{mnlw}}}{O_1(m,n,a,b) + O_5(m,n,a,b)} \\ &= [H_{1_{mnlw}} + H_{3_{mnlw}}] \frac{2k^4_{\mathrm{T}_{mn}}}{ab(\omega\mu_0\mu^I_{\mathrm{r}})^2} \begin{cases} \left(\dfrac{b}{n\pi}\right)^2 & \text{if } m = 0 \wedge n \neq 0 \\ \left(\dfrac{a}{m\pi}\right)^2 & \text{if } m \neq 0 \wedge n = 0 \\ \dfrac{2}{k^2_{\mathrm{T}_{mn}}} & \text{if } m \neq 0 \wedge n \neq 0, \end{cases} \end{aligned} \tag{6.B.23}$$

$$H_{b_{mnlw}} = \frac{H_{2_{mnlw}} + H_{4_{mnlw}}}{O_1(m,n,a,b) + O_5(m,n,a,b)} = \left[-\frac{lb}{wa'}H_{1_{mnlw}} + \frac{wa'}{lb'}H_{3_{mnlw}}\right]Y^{II}_{\mathrm{TE}_{lw}}$$
$$\times \frac{2k^4_{\mathrm{T}_{mn}}}{ab(\omega\mu_0\mu^I_{\mathrm{r}})^2}\begin{cases} 0 & \text{if } l = 0 \wedge w = 0 \\ \left(\dfrac{b}{n\pi}\right)^2 & \text{if } l \neq 0 \wedge w \neq 0 \wedge m = 0 \wedge n \neq 0 \\ \left(\dfrac{a}{m\pi}\right)^2 & \text{if } l \neq 0 \wedge w \neq 0 \wedge m \neq 0 \wedge n = 0 \\ \dfrac{2}{k^2_{\mathrm{T}_{mn}}} & \text{if } l \neq 0 \wedge w \neq 0 \wedge m \neq 0 \wedge n \neq 0, \end{cases} \tag{6.B.24}$$

$$H_{c_{mnlw}} = \frac{H_{5_{mnlw}} + H_{7_{mnlw}}}{O_4(m,n,a,b) + O_8(m,n,a,b)}$$
$$= \left[-\frac{mb}{na}H_{1_{mnlw}} + \frac{na}{mb}H_{3_{mnlw}}\right]Y^{I*}_{\mathrm{TE}_{mn}}\frac{4k^2_{\mathrm{T}_{mn}}}{ab|\gamma^I_{mn}|^2}\begin{cases} 0 & \text{if } m = 0 \vee n = 0 \\ 1 & \text{if } m \neq 0 \wedge n \neq 0, \end{cases} \tag{6.B.25}$$

$$H_{d_{mnlw}} = \frac{H_{6_{mnlw}} + H_{8_{mnlw}}}{O_4(m,n,a,b) + O_8(m,n,a,b)} = \left[-\frac{mlbb'}{nwaa'}H_{1_{mnlw}} + \frac{nwaa'}{mlbb'}H_{3_{mnlw}}\right]$$
$$\times Y^{I*}_{\mathrm{TE}_{mn}}Y^{II}_{\mathrm{TE}_{lw}}\frac{4k^2_{\mathrm{T}_{mn}}}{ab|\gamma^I_{mn}|^2}\begin{cases} 0 & \text{if } m = 0 \vee n = 0 \vee l = 0 \vee w = 0 \\ 1 & \text{if } m \neq 0 \wedge n \neq 0 \wedge l \neq 0 \wedge w \neq 0, \end{cases} \tag{6.B.26}$$

$$H_{e_{lwmn}} = \frac{Y^I_{\mathrm{TE}_{mn}}}{Y^{II}_{\mathrm{TE}_{lw}}}\frac{H_{9_{lwmn}} + H_{11_{lwmn}}}{O_1(l,w,a',b') + O_5(l,w,a',b')}$$
$$= [H_{1_{mnlw}} + H_{3_{mnlw}}]\frac{2k^4_{\mathrm{T}_{lw}}}{a'b'(\omega\mu_0\mu^{II}_{\mathrm{r}})^2}\frac{Y^I_{\mathrm{TE}_{mn}}}{Y^{II}_{\mathrm{TE}_{lw}}}\begin{cases} \left(\dfrac{b'}{w\pi}\right)^2 & \text{if } l = 0 \wedge w \neq 0 \\ \left(\dfrac{a'}{l\pi}\right)^2 & \text{if } l \neq 0 \wedge w = 0 \\ \dfrac{2}{k^2_{\mathrm{T}_{lw}}} & \text{if } l \neq 0 \wedge w \neq 0, \end{cases} \tag{6.B.27}$$

$$H_{f_{lwmn}} = \frac{Y^I_{\mathrm{TM}_{mn}}}{Y^{II}_{\mathrm{TE}_{lw}}}\frac{H_{10_{lwmn}} + H_{12_{lwmn}}}{O_1(l,w,a',b') + O_5(l,w,a',b')} = \left[-\frac{mb}{na}H_{1_{mnlw}} + \frac{na}{mb}H_{3_{mnlw}}\right]$$
$$\times Y^I_{\mathrm{TE}_{mn}}\frac{Y^I_{\mathrm{TM}_{mn}}}{Y^{II}_{\mathrm{TE}_{lw}}}\frac{2k^4_{\mathrm{T}_{mn}}}{a'b'(\omega\mu_0\mu^{II}_{\mathrm{r}})^2}\begin{cases} 0 & \text{if } m = 0 \wedge n = 0 \\ \left(\dfrac{b'}{w\pi}\right)^2 & \text{if } m \neq 0 \wedge n \neq 0 \wedge l = 0 \wedge w \neq 0 \\ \left(\dfrac{a'}{l\pi}\right)^2 & \text{if } m \neq 0 \wedge n \neq 0 \wedge l \neq 0 \wedge w = 0 \\ \dfrac{2}{k^2_{\mathrm{T}_{lw}}} & \text{if } m \neq 0 \wedge n \neq 0 \wedge l \neq 0 \wedge w \neq 0, \end{cases} \tag{6.B.28}$$

$$H_{g_{lwmn}} = \frac{Y^{I}_{\mathrm{TE}_{mn}}}{Y^{II}_{\mathrm{TM}_{lw}}} \frac{H_{13_{lwmn}} + H_{15_{lwmn}}}{O_4(l, w, a', b') + O_8(l, w, a', b')}$$
$$= \left[-\frac{lb'}{wa'} H_{1_{mnlw}} + \frac{wa'}{lb'} H_{3_{mnlw}} \right] Y^{II*}_{\mathrm{TE}_{lw}} \frac{Y^{I}_{\mathrm{TE}_{mn}}}{Y^{II}_{\mathrm{TM}_{lw}}} \frac{4k^2_{\mathrm{T}_{lw}}}{a'b'|\gamma^{II}_{lw}|^2} \begin{cases} 0 & \text{if } l = 0 \vee w = 0 \\ 1 & \text{if } l \neq 0 \wedge w \neq 0, \end{cases} \tag{6.B.29}$$

$$H_{h_{lwmn}} = \frac{Y^{I}_{\mathrm{TM}_{mn}}}{Y^{II}_{\mathrm{TM}_{lw}}} \frac{H_{14_{lwmn}} + H_{16_{lwmn}}}{O_4(l, w, a', b') + O_8(l, w, a', b')}$$
$$= \left[-\frac{mlbb'}{nwaa'} H_{1_{mnlw}} + \frac{nwaa'}{mlbb'} H_{3_{mnlw}} \right]$$
$$\times Y^{I}_{\mathrm{TE}_{mn}} Y^{II*}_{\mathrm{TE}_{lw}} \frac{Y^{I}_{\mathrm{TM}_{mn}}}{Y^{II}_{\mathrm{TM}_{lw}}} \frac{4k^2_{\mathrm{T}_{lw}}}{a'b'|\gamma^{II}_{lw}|^2} \begin{cases} 0 & \text{if } m = 0 \vee n = 0 \vee l = 0 \vee w = 0 \\ 1 & \text{if } m \neq 0 \wedge n \neq 0 \wedge l \neq 0 \wedge w \neq 0. \end{cases} \tag{6.B.30}$$

APPENDIX 6.C. UNIT CELL MODE ORTHOGONALITY AND NORMALIZATION FUNCTIONS

The mode-matching procedure for a rectangular-waveguide-to-unit-cell junction gives rise to a number of normalization functions, denoted by $U_i(\xi, \eta, \alpha, \beta)$, $i = 1, 2, \ldots, 8$, $\xi, \eta \in \mathbb{N}$, $\alpha, \beta \in \mathbb{R}$. These normalization functions incorporate two normalization integrals:

$$NEU(\alpha, \xi, \zeta) = \int_{\kappa=-\alpha/2}^{\alpha/2} \mathrm{e}^{-\mathrm{j}(2\xi\pi/\alpha - 2\zeta\pi/\alpha)\kappa} \, d\kappa \tag{6.C.1}$$

and

$$NEV(\beta, \xi, \eta, \zeta, \varsigma) = \int_{\kappa=-\beta/2}^{\beta/2} \mathrm{e}^{-\mathrm{j}(2\eta\pi/\beta - 2\varsigma\pi/\beta + 2\beta\zeta\pi - 2\beta\xi\pi)\kappa} \, d\kappa, \tag{6.C.2}$$

where $\zeta, \varsigma \in \mathbb{N}$. These normalization integrals possess the mode orthogonality characteristics and can be calculated as

$$NEU(\alpha, \xi, \zeta) = \begin{cases} \alpha & \text{if } \xi = \zeta \\ 0 & \text{if } \xi \neq \zeta, \end{cases} \tag{6.C.3}$$

$$NEV(\beta, \xi, \eta, \zeta, \varsigma) = \begin{cases} \beta & \text{if } \eta = \varsigma \wedge \xi = \zeta \\ 0 & \text{if } \eta \neq \varsigma \vee \xi \neq \zeta. \end{cases} \tag{6.C.4}$$

The normalization functions are defined and calculated as follows:

$$U_1(\xi, \eta, \alpha, \beta) = \int_{x=-\alpha/2}^{\alpha/2} e^{+*}_{x\mathrm{TE}_{ij}} e^{+}_{x\mathrm{TE}_{\xi\eta}} \, dx \, dy$$
$$= \frac{1}{k^4_{\mathrm{T}_{\xi\eta}}} (\omega\mu_0\mu_\mathrm{r})^2 v^2_{\xi\eta} NEU(\alpha, \xi, i) NEV(\beta, \xi, \eta, i, j), \tag{6.C.5}$$

$$U_2(\xi,\eta,\alpha,\beta) = \int_{x=-\alpha/2}^{\alpha/2} e^{+*}_{x\mathrm{TE}_{ij}} e^{+}_{x\mathrm{TM}_{\xi\eta}}\,dx\,dy$$
$$= \frac{1}{k^4_{\mathrm{T}_{\xi\eta}}}(\omega\mu_0\mu_\mathrm{r})\gamma_{\xi\eta}u_{\xi 0}v_{\xi\eta}NEU(\alpha,\xi,i)NEV(\beta,\xi,\eta,i,j), \tag{6.C.6}$$

$$U_3(\xi,\eta,\alpha,\beta) = \int_{x=-\alpha/2}^{\alpha/2} e^{+*}_{x\mathrm{TM}_{ij}} e^{+}_{x\mathrm{TE}_{\xi\eta}}\,dx\,dy$$
$$= \frac{1}{k^4_{\mathrm{T}_{\xi\eta}}}(\omega\mu_0\mu_\mathrm{r})\gamma^*_{\xi\eta}u_{\xi 0}v_{\xi\eta}NEU(\alpha,\xi,i)NEV(\beta,\xi,\eta,i,j), \tag{6.C.7}$$

$$U_4(\xi,\eta,\alpha,\beta) = \int_{x=-\alpha/2}^{\alpha/2} e^{+*}_{x\mathrm{TM}_{ij}} e^{+}_{x\mathrm{TM}_{\xi\eta}}\,dx\,dy$$
$$= \frac{1}{k^4_{\mathrm{T}_{\xi\eta}}}|\gamma_{\xi\eta}|^2u^2_{\xi 0}NEU(\alpha,\xi,i)NEV(\beta,\xi,\eta,i,j), \tag{6.C.8}$$

$$U_5(\xi,\eta,\alpha,\beta) = \int_{x=-\alpha/2}^{\alpha/2} e^{+*}_{y\mathrm{TE}_{ij}} e^{+}_{y\mathrm{TE}_{\xi\eta}}\,dx\,dy$$
$$= \frac{1}{k^4_{\mathrm{T}_{\xi\eta}}}(\omega\mu_0\mu_\mathrm{r})^2u^2_{\xi 0}NEU(\alpha,\xi,i)NEV(\beta,\xi,\eta,i,j), \tag{6.C.9}$$

$$U_6(\xi,\eta,\alpha,\beta) = \int_{x=-\alpha/2}^{\alpha/2} e^{+*}_{y\mathrm{TE}_{ij}} e^{+}_{y\mathrm{TM}_{\xi\eta}}\,dx\,dy = -U_2(\xi,\eta,\alpha,\beta), \tag{6.C.10}$$

$$U_7(\xi,\eta,\alpha,\beta) = \int_{x=-\alpha/2}^{\alpha/2} e^{+*}_{y\mathrm{TM}_{ij}} e^{+}_{y\mathrm{TE}_{\xi\eta}}\,dx\,dy = -U_3(\xi,\eta,\alpha,\beta), \tag{6.C.11}$$

$$U_8(\xi,\eta,\alpha,\beta) = \int_{x=-\alpha/2}^{\alpha/2} e^{+*}_{y\mathrm{TM}_{ij}} e^{+}_{y\mathrm{TM}_{\xi\eta}}\,dx\,dy$$
$$= \frac{1}{k^4_{\mathrm{T}_{\xi\eta}}}|\gamma_{\xi\eta}|^2v^2_{\xi\eta}NEU(\alpha,\xi,i)NEV(\beta,\xi,\eta,i,j). \tag{6.C.12}$$

In the above, $i, j \in \mathbb{N}$. Non-trivial solutions exist only for $i = \xi$ and $j = \eta$.

APPENDIX 6.D. MODE-COUPLING INTEGRALS FOR RECTANGULAR-WAVEGUIDE-TO-UNIT-CELL JUNCTION

In the mode-matching procedure for a rectangular-waveguide-to-unit-cell junction we encountered, in section 6.4.2, the mode-coupling integrals $W_{i_{pqmn}}$, $i = 1, 2, \ldots, 8$, and $W_{j_{mnpq}}$, $j = 9, 10, \ldots, 16$. These coupling integrals make use of the following set of (Fourier) integrals:

$$FCU(\alpha,\xi,\eta) = \int_{\kappa=-\alpha/2}^{\alpha/2} C_\xi\left(\frac{\kappa}{\alpha}\right)e^{\mathrm{j}u_{\eta 0}\kappa}\,d\kappa, \tag{6.D.1}$$

$$FCUm(\alpha, \xi, \eta) = \int_{\kappa=-\alpha/2}^{\alpha/2} C_\xi\left(\frac{\kappa}{\alpha}\right) \mathrm{e}^{-\mathrm{j}u_{\eta 0}\kappa}\, d\kappa, \tag{6.D.2}$$

$$FSU(\alpha, \xi, \eta) = \int_{\kappa=-\alpha/2}^{\alpha/2} S_\xi\left(\frac{\kappa}{\alpha}\right) \mathrm{e}^{\mathrm{j}u_{\eta 0}\kappa}\, d\kappa, \tag{6.D.3}$$

$$FSUm(\alpha, \xi, \eta) = \int_{\kappa=-\alpha/2}^{\alpha/2} S_\xi\left(\frac{\kappa}{\alpha}\right) \mathrm{e}^{-\mathrm{j}u_{\eta 0}\kappa}\, d\kappa, \tag{6.D.4}$$

$$FCV(\alpha, \xi, \eta, \tau) = \int_{\kappa=-\alpha/2}^{\alpha/2} C_\xi\left(\frac{\kappa}{\alpha}\right) \mathrm{e}^{\mathrm{j}v_{\eta\tau}\kappa}\, d\kappa, \tag{6.D.5}$$

$$FCVm(\alpha, \xi, \eta, \tau) = \int_{\kappa=-\alpha/2}^{\alpha/2} C_\xi\left(\frac{\kappa}{\alpha}\right) \mathrm{e}^{-\mathrm{j}v_{\eta\tau}\kappa}\, d\kappa, \tag{6.D.6}$$

$$FSV(\alpha, \xi, \eta, \tau) = \int_{\kappa=-\alpha/2}^{\alpha/2} S_\xi\left(\frac{\kappa}{\alpha}\right) \mathrm{e}^{\mathrm{j}v_{\eta\tau}\kappa}\, d\kappa, \tag{6.D.7}$$

$$FSVm(\alpha, \xi, \eta, \tau) = \int_{\kappa=-\alpha/2}^{\alpha/2} S_\xi\left(\frac{\kappa}{\alpha}\right) \mathrm{e}^{-\mathrm{j}v_{\eta\tau}\kappa}\, d\kappa, \tag{6.D.8}$$

where $\xi, \eta, \tau \in \mathbb{N}$ and $\alpha \in \mathbb{R}$. The transverse-wave-number components $u_{\eta 0}$ and $v_{\eta\tau}$ are defined by equations (6.64) and (6.65), respectively. The cosine function $C_\xi(\kappa)$ and sine function $S_\xi(\kappa)$ are defined by equations (6.54) and (6.55), respectively. We repeat these expressions here for completeness:

$$C_\xi(\kappa) = \cos\left[\xi\pi\left(\kappa + \frac{1}{2}\right)\right] \tag{6.D.9}$$

and

$$S_\xi(\kappa) = \sin\left[\xi\pi\left(\kappa + \frac{1}{2}\right)\right]. \tag{6.D.10}$$

Calculation of the above integrals gives

$$FCU(\alpha, \xi, \eta) = \begin{cases} \alpha & \text{if } \xi = u_{\eta 0} = 0 \\ \dfrac{\alpha}{2}\mathrm{e}^{-\mathrm{j}\xi\pi/2} & \text{if } \dfrac{\xi\pi}{\alpha} = u_{\eta 0} \\ j\dfrac{u_{\eta 0}}{(\xi\pi/\alpha)^2 - u_{\eta 0}^2}\{(-1)^\xi \mathrm{e}^{\mathrm{j}u_{\eta 0}\alpha/2} - \mathrm{e}^{-\mathrm{j}u_{\eta 0}\alpha/2}\} & \text{if } \dfrac{\xi\pi}{\alpha} \neq u_{\eta 0}, \end{cases} \tag{6.D.11}$$

$$FCUm(\alpha, \xi, \eta) = (-1)^\xi FCU(\alpha, \xi, \eta), \tag{6.D.12}$$

$$FSU(\alpha, \xi, \eta) = \begin{cases} 0 & \text{if } \xi = 0 \\ j\dfrac{\alpha}{2}\mathrm{e}^{-\mathrm{j}\xi\pi/2} & \text{if } \dfrac{\xi\pi}{\alpha} = u_{\eta 0} \neq 0 \\ -\dfrac{(\xi\pi/\alpha)}{(\xi\pi/\alpha)^2 - u_{\eta 0}^2}\{(-1)^\xi \mathrm{e}^{\mathrm{j}u_{\eta 0}\alpha/2} - \mathrm{e}^{-\mathrm{j}u_{\eta 0}\alpha/2}\} & \text{if } \dfrac{\xi\pi}{\alpha} \neq u_{\eta 0}, \end{cases} \tag{6.D.13}$$

$$FSUm(\alpha,\xi,\eta) = -(-1)^{\xi} FSU(\alpha,\xi,\eta), \tag{6.D.14}$$

$$FCV(\alpha,\xi,\eta,\tau) = \begin{cases} \alpha & \text{if } \xi = v_{\eta\tau} = 0 \\ \dfrac{\alpha}{2} e^{-j\xi\pi/2} & \text{if } \dfrac{\xi\pi}{\alpha} = v_{\eta\tau} \\ j\dfrac{v_{\eta\tau}}{(\xi\pi/\alpha)^2 - v_{\eta\tau}^2}\{(-1)^{\xi} e^{jv_{\eta\tau}\alpha/2} - e^{-jv_{\eta\tau}\alpha/2}\} & \text{if } \dfrac{\xi\pi}{\alpha} \neq v_{\eta\tau}, \end{cases} \tag{6.D.15}$$

$$FCVm(\alpha,\xi,\eta,\tau) = (-1)^{\xi} FCV(\alpha,\xi,\eta,\tau), \tag{6.D.16}$$

$$FSV(\alpha,\xi,\eta,\tau) = \begin{cases} 0 & \text{if } \xi = 0 \\ \dfrac{\alpha}{2} e^{-j\xi\pi/2} & \text{if } \dfrac{\xi\pi}{\alpha} = v_{\eta\tau} \neq 0 \\ -\dfrac{(\xi\pi/\alpha)}{(\xi\pi/\alpha)^2 - v_{\eta\tau}^2}\{(-1)^{\xi} e^{jv_{\eta\tau}\alpha/2} - e^{-jv_{\eta\tau}\alpha/2}\} & \text{if } \dfrac{\xi\pi}{\alpha} \neq v_{\eta\tau}, \end{cases} \tag{6.D.17}$$

$$FSVm(\alpha,\xi,\eta,\tau) = -(-1)^{\xi} FSV(\alpha,\xi,\eta,\tau). \tag{6.D.18}$$

The coupling integrals are then found as

$$\begin{aligned} W1_{pqmn} &= \int_{x=-a/2}^{a/2}\int_{y=-b/2}^{b/2} e^{II^{+*}}_{x\mathrm{TE}_{pq}} e^{I^{+}}_{x\mathrm{TE}_{mn}}\,dx\,dy \\ &= -j\frac{(\omega\mu_0)^2\mu_r^I\mu_r^{II}(n\pi/b)v_{pq}}{[(m\pi/a)^2+(n\pi/b)^2][u_{p0}^2+v_{pq}^2]} FCU(a,m,p)FSV(b,n,p,q), \end{aligned} \tag{6.D.19}$$

$$\begin{aligned} W2_{pqmn} &= \int_{x=-a/2}^{a/2}\int_{y=-b/2}^{b/2} e^{II^{+*}}_{x\mathrm{TE}_{pq}} e^{I^{+}}_{x\mathrm{TM}_{mn}}\,dx\,dy \\ &= \begin{cases} -Y^I_{\mathrm{TE}_{mn}}\left(\dfrac{mb}{na}\right) W1_{pqmn} & \text{if } n \neq 0 \\ 0 & \text{if } n = 0, \end{cases} \end{aligned} \tag{6.D.20}$$

$$\begin{aligned} W3_{pqmn} &= \int_{x=-a/2}^{a/2}\int_{y=-b/2}^{b/2} e^{II^{+*}}_{y\mathrm{TE}_{pq}} e^{I^{+}}_{y\mathrm{TE}_{mn}}\,dx\,dy \\ &= -j\frac{(\omega\mu_0)^2\mu_r^I\mu_r^{II}(m\pi/a)u_{p0}}{[(m\pi/a)^2+(n\pi/b)^2][u_{p0}^2+v_{pq}^2]} FSU(a,m,p)FCV(b,n,p,q), \end{aligned} \tag{6.D.21}$$

$$\begin{aligned} W4_{pqmn} &= \int_{x=-a/2}^{a/2}\int_{y=-b/2}^{b/2} e^{II^{+*}}_{y\mathrm{TE}_{pq}} e^{I^{+}}_{y\mathrm{TM}_{mn}}\,dx\,dy \\ &= \begin{cases} Y^I_{\mathrm{TE}_{mn}}\left(\dfrac{na}{mb}\right) W3_{pqmn} & \text{if } m \neq 0 \\ 0 & \text{if } m = 0, \end{cases} \end{aligned} \tag{6.D.22}$$

$$W5_{pqmn} = \int_{x=-a/2}^{a/2} \int_{y=-b/2}^{b/2} e_{x\mathrm{TM}_{pq}}^{II+^*} e_{x\mathrm{TE}_{mn}}^{I+} \, dx \, dy$$
$$= -j \frac{\omega\mu_0\mu_r^I \gamma_{pq}^{II^*} (n\pi/b) u_{p0}}{[(m\pi/a)^2 + (n\pi/b)^2][u_{p0}^2 + v_{pq}^2]} FCU(a, m, p) FSV(b, n, p, q), \quad (6.D.23)$$

$$W6_{pqmn} = \int_{x=-a/2}^{a/2} \int_{y=-b/2}^{b/2} e_{x\mathrm{TM}_{pq}}^{II+^*} e_{x\mathrm{TM}_{mn}}^{I+} \, dx \, dy$$
$$= \begin{cases} -Y_{\mathrm{TE}_{mn}}^{I} \left(\dfrac{mb}{na} \right) W5_{pqmn} & \text{if } n \neq 0 \\ 0 & \text{if } n = 0, \end{cases} \quad (6.D.24)$$

$$W7_{pqmn} = \int_{x=-a/2}^{a/2} \int_{y=-b/2}^{b/2} e_{y\mathrm{TM}_{pq}}^{II+^*} e_{y\mathrm{TE}_{mn}}^{I+} \, dx \, dy$$
$$= j \frac{\omega\mu_0\mu_r^I \gamma_{pq}^{II^*} (m\pi/a) v_{pq}}{[(m\pi/a)^2 + (n\pi/b)^2][u_{p0}^2 + v_{pq}^2]} FSU(a, m, p) FCV(b, n, p, q), \quad (6.D.25)$$

$$W8_{pqmn} = \int_{x=-a/2}^{a/2} \int_{y=-b/2}^{b/2} e_{y\mathrm{TM}_{pq}}^{II+^*} e_{y\mathrm{TM}_{mn}}^{I+} \, dx \, dy$$
$$= \begin{cases} Y_{\mathrm{TE}_{mn}}^{I} \left(\dfrac{na}{mb} \right) W7_{pqmn} & \text{if } m \neq 0 \\ 0 & \text{if } m = 0, \end{cases} \quad (6.D.26)$$

$$W9_{mnpq} = \int_{x=-a/2}^{a/2} \int_{y=-b/2}^{b/2} e_{y\mathrm{TE}_{mn}}^{I+^*} e_{y\mathrm{TE}_{pq}}^{II+} \, dx \, dy = (-1)^{m+n} W3_{pqmn}, \quad (6.D.27)$$

$$W10_{mnpq} = \int_{x=-a/2}^{a/2} \int_{y=-b/2}^{b/2} e_{y\mathrm{TE}_{mn}}^{I+^*} e_{y\mathrm{TM}_{pq}}^{II+} \, dx \, dy = (-1)^{m+n} \left(\frac{\gamma_{pq}^{II}}{\gamma_{pq}^{II^*}} \right) W7_{pqmn}, \quad (6.D.28)$$

$$W11_{mnpq} = \int_{x=-a/2}^{a/2} \int_{y=-b/2}^{b/2} e_{x\mathrm{TE}_{mn}}^{I+^*} e_{x\mathrm{TE}_{pq}}^{II+} \, dx \, dy = (-1)^{m+n} W1_{pqmn}, \quad (6.D.29)$$

$$W12_{mnpq} = \int_{x=-a/2}^{a/2} \int_{y=-b/2}^{b/2} e_{x\mathrm{TE}_{mn}}^{I+^*} e_{x\mathrm{TM}_{pq}}^{II+} \, dx \, dy = (-1)^{m+n} \left(\frac{\gamma_{pq}^{II}}{\gamma_{pq}^{II^*}} \right) W5_{pqmn}, \quad (6.D.30)$$

$$W13_{mnpq} = \int_{x=-a/2}^{a/2} \int_{y=-b/2}^{b/2} e_{y\mathrm{TM}_{mn}}^{I+^*} e_{y\mathrm{TE}_{pq}}^{II+} \, dx \, dy$$
$$= \begin{cases} (-1)^{m+n} Y_{\mathrm{TE}_{mn}}^{I^*} \left(\dfrac{na}{mb} \right) W3_{pqmn} & \text{if } m \neq 0 \\ 0 & \text{if } m = 0, \end{cases} \quad (6.D.31)$$

$$W14_{mnpq} = \int_{x=-a/2}^{a/2} \int_{y=-b/2}^{b/2} e_{y\mathrm{TM}_{mn}}^{I+^*} e_{y\mathrm{TM}_{pq}}^{II+} \, dx \, dy$$
$$= \begin{cases} (-1)^{m+n} \left(\dfrac{\gamma_{pq}^{II}}{\gamma_{pq}^{II^*}} \right) Y_{\mathrm{TE}_{mn}}^{I^*} \left(\dfrac{na}{mb} \right) W7_{pqmn} & \text{if } m \neq 0 \\ 0 & \text{if } m = 0, \end{cases} \quad (6.D.32)$$

$$W_{15_{mnpq}} = \int_{x=-a/2}^{a/2} \int_{y=-b/2}^{b/2} e_{x\mathrm{TM}_{mn}}^{I+^*} e_{x\mathrm{TE}_{pq}}^{II+} \, dx \, dy$$
$$= \begin{cases} (-1)^{m+n} Y_{\mathrm{TE}_{mn}}^{I^*} \left(\dfrac{mb}{na}\right) W_{1_{pqmn}} & \text{if } n \neq 0 \\ 0 & \text{if } n = 0, \end{cases} \tag{6.D.33}$$

$$W_{16_{mnpq}} = \int_{x=-a/2}^{a/2} \int_{y=-b/2}^{b/2} e_{x\mathrm{TM}_{mn}}^{I+^*} e_{x\mathrm{TM}_{pq}}^{II+} \, dx \, dy$$
$$= \begin{cases} -(-1)^{m+n} \left(\dfrac{\gamma_{pq}^{II}}{\gamma_{pq}^{II^*}}\right) Y_{\mathrm{TE}_{mn}}^{I^*} \left(\dfrac{mb}{na}\right) W_{5_{pqmn}} & \text{if } n \neq 0 \\ 0 & \text{if } n = 0. \end{cases} \tag{6.D.34}$$

Combining the coupling integrals gives rise to the following functions, which are used as matrix elements in section 6.5.1:

$$W_{a_{pqmn}} = \frac{W_{1_{pqmn}} + W_{3_{pqmn}}}{U_1(p,q,s,t\sin(\Omega)) + U_5(p,q,s,t\sin(\Omega))}$$
$$= \frac{k_{\mathrm{T}_{pq}}^2}{(\omega\mu_0\mu_{\mathrm{r}}^{II})^2 st \sin(\Omega)} \{W_{1_{pqmn}} + W_{3_{pqmn}}\}, \tag{6.D.35}$$

$$W_{b_{pqmn}} = \frac{W_{2_{pqmn}} + W_{4_{pqmn}}}{U_1(p,q,s,t\sin(\Omega)) + U_5(p,q,s,t\sin(\Omega))}$$
$$= \frac{k_{\mathrm{T}_{pq}}^2 Y_{\mathrm{TE}_{mn}}^I}{(\omega\mu_0\mu_{\mathrm{r}}^{II})^2 st \sin(\Omega)}$$
$$\times \begin{cases} \left[-\left(\dfrac{mb}{na}\right) W_{1_{pqmn}} + \left(\dfrac{na}{mb}\right) W_{3_{pqmn}}\right] & \text{if } m \neq 0 \wedge n \neq 0 \\ 0 & \text{if } m = 0 \vee n = 0, \end{cases} \tag{6.D.36}$$

$$W_{c_{pqmn}} = \frac{W_{5_{pqmn}} + W_{7_{pqmn}}}{U_4(p,q,s,t\sin(\Omega)) + U_8(p,q,s,t\sin(\Omega))}$$
$$= \frac{k_{\mathrm{T}_{pq}}^2}{|\gamma_{pq}^{II}|^2 st \sin(\Omega)} \{W_{5_{pqmn}} + W_{7_{pqmn}}\}, \tag{6.D.37}$$

$$W_{d_{pqmn}} = \frac{W_{6_{pqmn}} + W_{8_{pqmn}}}{U_4(p,q,s,t\sin(\Omega)) + U_8(p,q,s,t\sin(\Omega))}$$
$$= \frac{k_{\mathrm{T}_{pq}}^2 Y_{\mathrm{TE}_{mn}}^I}{|\gamma_{pq}^{II}|^2 st \sin(\Omega)}$$
$$\times \begin{cases} \left[-\left(\dfrac{mb}{na}\right) W_{5_{pqmn}} + \left(\dfrac{na}{mb}\right) W_{7_{pqmn}}\right] & \text{if } m \neq 0 \wedge n \neq 0 \\ 0 & \text{if } m = 0 \vee n = 0, \end{cases} \tag{6.D.38}$$

$$W_{e_{mnpq}} = \frac{Y_{\mathrm{TE}_{pq}}^{II}}{Y_{\mathrm{TE}_{mn}}^{I}} \frac{W_{9_{mnpq}} + W_{11_{mnpq}}}{O_1(m,n,a,b) + O_5(m,n,a,b)}$$

$$= 2\frac{Y^{II}_{\mathrm{TE}_{pq}}}{Y^{I}_{\mathrm{TE}_{mn}}}\frac{k^4_{\mathrm{T}_{mn}}}{(\omega\mu_0\mu^I_{\mathrm{r}})^2 ab}(-1)^{m+n}\{W_{1_{pqmn}} + W_{3_{pqmn}}\}$$

$$\times \begin{cases} \left(\dfrac{b}{n\pi}\right)^2 & \text{if } m = 0 \wedge n \neq 0 \\ \left(\dfrac{a}{m\pi}\right)^2 & \text{if } m \neq 0 \wedge n = 0 \\ \dfrac{2}{k^2_{\mathrm{T}_{mn}}} & \text{if } m \neq 0 \wedge n \neq 0, \end{cases} \tag{6.D.39}$$

$$W_{f_{mnpq}} = \frac{Y^{II}_{\mathrm{TM}_{pq}}}{Y^{I}_{\mathrm{TE}_{mn}}}\frac{W_{10_{mnpq}} + W_{12_{mnpq}}}{O_1(m,n,a,b) + O_5(m,n,a,b)}$$

$$= 2\frac{Y^{II}_{\mathrm{TM}_{pq}}}{Y^{I}_{\mathrm{TE}_{mn}}}\left(\frac{\gamma^{II}_{pq}}{\gamma^{II*}_{pq}}\right)\frac{k^4_{\mathrm{T}_{mn}}}{(\omega\mu_0\mu^I_{\mathrm{r}})^2 ab}(-1)^{m+n}\{W_{5_{pqmn}} + W_{7_{pqmn}}\}$$

$$\times \begin{cases} \left(\dfrac{b}{n\pi}\right)^2 & \text{if } m = 0 \wedge n \neq 0 \\ \left(\dfrac{a}{m\pi}\right)^2 & \text{if } m \neq 0 \wedge n = 0 \\ \dfrac{2}{k^2_{\mathrm{T}_{mn}}} & \text{if } m \neq 0 \wedge n \neq 0, \end{cases} \tag{6.D.40}$$

$$W_{g_{mnpq}} = \frac{Y^{II}_{\mathrm{TE}_{pq}}}{Y^{I}_{\mathrm{TM}_{mn}}}\frac{W_{13_{mnpq}} + W_{15_{mnpq}}}{O_4(m,n,a,b) + O_8(m,n,a,b)}$$

$$= 4\frac{Y^{II}_{\mathrm{TE}_{pq}}}{Y^{I}_{\mathrm{TM}_{mn}}}Y^{I*}_{\mathrm{TE}_{mn}}\frac{k^2_{\mathrm{T}_{mn}}}{|\gamma^I_{mn}|^2 ab}(-1)^{m+n}$$

$$\times \left\{-\left(\frac{mb}{na}\right)W_{1_{pqmn}} + \left(\frac{na}{mb}\right)W_{3_{pqmn}}\right\}$$

$$\times \begin{cases} 1 & \text{if } m \neq 0 \wedge n \neq 0 \\ 0 & \text{if } m = 0 \vee n = 0, \end{cases} \tag{6.D.41}$$

$$W_{h_{mnpq}} = \frac{Y^{II}_{\mathrm{TM}_{pq}}}{Y^{I}_{\mathrm{TM}_{mn}}}\frac{W_{14_{mnpq}} + W_{16_{mnpq}}}{O_4(m,n,a,b) + O_8(m,n,a,b)}$$

$$= 4\frac{Y^{II}_{\mathrm{TM}_{pq}}}{Y^{I}_{\mathrm{TM}_{mn}}}Y^{I*}_{\mathrm{TE}_{mn}}\left(\frac{\gamma^{II}_{pq}}{\gamma^{II*}_{pq}}\right)\frac{k^2_{\mathrm{T}_{mn}}}{|\gamma^I_{mn}|^2 ab}(-1)^{m+n}$$

$$\times \left\{-\left(\frac{mb}{na}\right)W_{5_{pqmn}} + \left(\frac{na}{mb}\right)W_{7_{pqmn}}\right\}$$

$$\times \begin{cases} 1 & \text{if } m \neq 0 \wedge n \neq 0 \\ 0 & \text{if } m = 0 \vee n = 0. \end{cases} \tag{6.D.42}$$

REFERENCES

1. R.C. Hansen, *Phased Array Antennas*, John Wiley & Sons, New York, 1998.

2. H.J. Visser and M. Guglielmi, 'CAD of waveguide array antennas based on "filter" concepts', *IEEE Transactions on Antennas and Propagation*, Vol. 47, No. 3, pp. 542–548, March 1999.

3. D. Bakers, *Finite Array Antennas: An Eigencurrent Approach*, PhD thesis, Eindhoven University of Technology, 2004.

4. B. Morsink, *Fast Modeling of Electromagnetic Fields for the Design of Phased Array Antennas in Radar Systems*, PhD thesis, Eindhoven University of Technology, 2005.

5. S. Monni, *Frequency Selective Surfaces Integrated with Phased Array Antennas: Analysis and Design Using Multimode Equivalent Networks*, PhD thesis, Eindhoven University of Technology, 2005.

6. D.J. Bekers, S.J.L. van Eijndhoven, A.A.F. van de Ven, P.-P. Borsboom and A.G. Tijhuis, 'Eigencurrent analysis of resonant behavior in finite array antennas', *IEEE Transactions on Microwave Theory and Techniques*, Vol. 54, No. 6, pp. 2821–2829, June 2006.

7. S. Monni, G. Gerini, A. Neto and A.G. Tijhuis, 'Multimode equivalent networks for the design and analysis of frequency selective surfaces', *IEEE Transactions on Antennas and Propagation*, Vol. 55, No. 10, pp. 2824–2835, October 2007.

8. H.J. van Schaik, *Theory and Performance of a Space-Fed, Planar, Phased Array Antenna with 849 Iris-Loaded Rectangular Waveguide Elements and External Matching Sheet*, PhD thesis, Delft University of Technology, 1979.

9. N. Marcuvitz, *Waveguide Handbook*, McGraw-Hill, New York, 1951.

10. W.P.M.N. Keizer, A.P. de Hek and A.B. Smolders, 'Theoretical and experimental performance of a wideband wide-scan angle rectangular waveguide array antenna', *Proceedings of the IEEE Antennas and Propagation International Symposium, APS1991*, pp. 1724–1727, 24–28 June 1991.

11. G. Gerini, 'Phased arrays of compact waveguide radiators and frequency selective surfaces: A multi-mode equivalent network approach', *Proceedings of the 12th International Conference on Antennas and Propagation*, pp. 454–457, March–April 2003.

12. B.J. Morsink, G.H.C. van Werkhoven, A.G. Tijhuis and S.W. Rienstra, 'An accelerated coupled feed-radiator-frequency selective surface model for the next generation phased array systems', *Proceedings of the IEEE International Symposium on Phased Array Systems and Technology*, pp. 470–475, October 2003.

13. A. Ellgardt, 'Study of rectangular waveguide elements for planar wide-angle-scanning phased array antennas', *Proceedings of the IEEE Antennas and Propagation Society International Symposium*, pp. 815–818, July 2005.

14. E.G. Magill and H.A. Wheeler, 'Wide-angle impedance matching of a planar array antenna by a dielectric sheet', *IEEE Transactions on Antennas and Propagation*, Vol. AP-14, pp. 49–53, January 1966.

15. R. Mittra and S.W. Lee, *Analytical Techniques in the Theory of Guided Waves*, Macmillan, New York, 1971.

16. J.J. Campbell and B.V. Popovich, 'A broad-band wide-angle scan matching technique for large environmentally restricted phased arrays', *IEEE Transactions on Antennas and Propagation*, Vol. AP-20, pp. 421–427, July 1972.

17. S.W. Lee and W.R. Jones, 'On the suppression of radiation nulls and broadband impedance matching of rectangular waveguide phased arrays', *IEEE Transactions on Antennas and Propagation*, Vol. AP-18, pp. 41–51, January 1971.

18. C.C. Chen, 'Broad-band impedance matching of rectangular waveguide phased arrays', *IEEE Transactions on Antennas and Propagation*, Vol. AP-21, pp. 298–302, May 1973.

19. V. Galindo and C.P. Wu, 'Numerical solutions for an infinite phased array of rectangular waveguides with thick walls', *IEEE Transactions on Antennas and Propagation*, Vol. AP-14, pp. 652–654, September 1966.

20. V. Galindo and C.P. Wu, 'Properties of a phased array of rectangular waveguides with thick walls', *IEEE Transactions on Antennas and Propagation*, Vol. AP-14, pp. 163–173, March 1966.

21. G.A. Gesell and I.R. Ciric, 'Recurrence modal analysis for multiple waveguide discontinuities and its application to circular structures', *IEEE Transactions on Microwave Theory and Techniques*, Vol. 41, No. 3, pp. 484–490, March 1993.

22. U. Papziner and F. Arndt, 'Field theoretical computer-aided design of rectangular and circular iris coupled rectangular or circular waveguide cavity filters', *IEEE Transactions on Microwave Theory and Techniques*, Vol. 41, No. 3, pp. 462–471, March 1993.

23. U. Tucholke, F. Arndt and T. Wriedt, 'Field theory design of square waveguide iris polarizers', *IEEE Transactions on Microwave Theory and Techniques*, Vol. 34, No. 1, pp. 156–160, January 1986.

24. G.L. James, 'Analysis and design of TE11-to-HE11 corrugated cylindrical waveguide mode converters', *IEEE Transactions on Microwave Theory and Techniques*, Vol. 29, No. 10, pp. 1059–1066, October 1981.

25. O.P. Franza and W.C.C. Chew, 'Recursive mode matching method for multiple waveguide junction modeling', *IEEE Transactions on Microwave Theory and Techniques*, Vol. 44, No. 1, pp. 87–92, January 1996.

26. H. Patzelt and F. Arndt, 'Double-plane steps in rectangular waveguides and their application for transformers, irises and filters', *IEEE Transactions on Microwave Theory and Techniques*, Vol. 30, No. 5, pp. 771–776, May 1982.

27. P. Foster and S.M. Tun, 'Modelling waveguide components using mode-matching techniques', *Microwave Engineering Europe*, pp. 23–25, April 1996.

28. R.R. Mansour and R.H. Macphie, 'An improved transmission matrix formulation of cascaded discontinuities and its applications to E-plane circuits', *IEEE Transactions on Microwave Theory and Techniques*, Vol. 34, No. 12, pp. 1490–1498, December 1986.

29. A. Moumen and L.P. Ligthart, 'Radiation pattern and aperture admittance of dielectric filled waveguides with rectangular cross section using the mode matching technique', *Proceedings of the 28th European Microwave Conference*, Amsterdam, pp. 587–592, 1998.

30. H.J. Visser, 'Waveguide phased array analysis and synthesis', *Proceedings of RADAR 94*, Paris, pp. 707–712, 1994.

31. K.K. Chan and R.M. Turner, 'Modal design of broadband wide-scan waveguide phased array', *Proceedings of RADAR 94*, Paris, pp. 25–30, 1994.

32. K.K. Chan, H.J. Visser, R.M. Turner and W.P.M.N. Keizer, 'Dual polarized waveguide phased array for SAR applications', *Proceedings of PIERS 94*, The Netherlands, 1994.

33. H.J. van Schaik, 'The performance of an iris-loaded planar phased array antenna of rectangular waveguides with an external dielectric sheet', *IEEE Transactions on Antennas and Propagation*, Vol. AP-26, No. 3, pp. 413–419, May 1978.

34. R.W.P. King, *Transmission Line Theory*, McGraw-Hill, New York, 1955.

35. T.C. Collocot and A.B. Dobson (eds.), *Dictionary of Science and Technology*, revised edition, Chambers, Edinburgh, UK, 1982.

36. R.F. Harrington, *Time-Harmonic Electromagnetic Fields*, John Wiley and Sons, New York, 2001.

37. H.J. Visser, *Array and Phased Array Antenna Basics*, John Wiley & Sons, Chichester, 2005.

38. G.V. Eleftheriades, A.S. Omar, L.P.B. Katehi and G.M. Rebeiz, 'Some important properties of waveguide junction generalized scattering matrices in the context of the mode matching techniques', *IEEE Transactions on Microwave Theory and Techniques*, Vol. 42, No. 10, pp. 1896–1903, October 1994.

39. A.S. Omar and K. Schünemann, 'Transmission matrix representation of finline discontinuities', *IEEE Transactions on Microwave Theory and Techniques*, Vol. MTT-33, No. 9, pp. 765–770, September 1985.

40. F. Alessandri, G. Bartolucci and R. Sorrentino, 'Admittance matrix formulation of waveguide discontinuity problems: Computer-aided design of branch guide directional couplers', *IEEE Transactions on Microwave Theory and Techniques*, Vol. 36, No. 2, pp. 394–403, February 1988.

41. G.H. Golub and C.F. van Loan, *Matrix Computations*, John Hopkins University Press, Baltimore, 1989.

42. T. Itoh (ed.), *Numerical Techniques for Microwave and Millimeter-Wave Passive Structures*, John Wiley & Sons, New York, 1989.

43. M. Leroy, 'On the convergence of the solution of numerical results in modal analysis', *IEEE Transactions on Antennas and Propagation*, Vol. AP-31, pp. 655–659, July 1983.

44. M. Guglielmi, 'Simple CAD procedure for microwave filters and multiplexers', *IEEE Transactions on Microwave Theory and Techniques*, Vol. 42, No. 7, pp. 1347–1352, July 1994.

45. A. Weisshaar, M. Mongiardo and V.K. Tripathi, 'CAD-oriented equivalent circuit modeling of step discontinuities in rectangular waveguides', *IEEE Microwave and Guided Wave Letters*, Vol. 6, No. 4, pp. 171–173, April 1996.

46. D.M. Pozar, 'The active element pattern', *IEEE Transactions on Antennas and Propagation*, Vol. 42, No. 8, pp. 1176–1178, August 1994.

47. B.L. Diamond, 'A generalized approach to the analysis of infinite planar array antennas', *Proceedings of the IEEE*, Vol. 56, No. 11, pp. 1837–1851, November 1968.

48. K.K. Chan, R.M. Turner and K. Chadwick, 'Modal analysis of rectangular waveguide phased arrays', *Proceedings of the IEEE Antennas and Propagation International Symposium*, pp. 1400–1403, June 1995.

7 Summary and Conclusions

Even in this age of powerful computers and numerical methods, a need still exists for approximate antenna models. The models are needed in a two-stage approach to synthesizing antenna designs that also involves the use of full-wave analysis methods, and they have a right to exist in themselves. Approximate antenna models have been developed for five classes of antennas, having specific applications in mind: loops and solenoids for intravascular MR antennas, printed monopoles for integration on printed circuit boards, folded-dipole antennas and arrays thereof for RFID use, microstrip patch antennas for use in rectennas, and infinite planar arrays of open-ended waveguide radiators.

7.1 FULL-WAVE AND APPROXIMATE ANTENNA ANALYSIS

Since the end of World War II, a lot of effort has been put into the development of numerical electromagnetic analysis and, for sure, full-wave numerical solvers have come a long way. Nevertheless, the automated design of integrated antennas based on full-wave analysis is not yet feasible. An automated design would require some form of stochastic optimization, and the time that each full-wave analysis iteration would take is still too long. A two-stage approach involving both approximate and full-wave analysis seems more feasible [1]. In this approach, a stochastic optimization is used in combination with an approximate analysis. Then, the outcome of this optimization is used as input in a line search optimization in combination with full-wave modeling. So, approximate antenna models are needed for automated antenna design. But approximate antenna models also have a right to exist in themselves.

Approximate Antenna Analysis for CAD Hubregt J. Visser

As a recent benchmarking of commercially distributed full-wave analysis programs has shown [2, 3], an antenna designer needs to be careful to select the right full-wave analysis method for the right problem. Although the strong advice to use at least two different full-wave analysis methods is fully supported by the present author, it is also understood that not many companies or institutions can afford to purchase or lease multiple commercially distributed full-wave analysis programs. In fact, for many small companies it may be difficult even to get access to a single full-wave analysis program. The availability of approximate models or a method to develop approximate models may be a good alternative. Approximate models result in fast calculations and have the additional advantage that they provide insight into the physical phenomenon that is the basis of the problem. They relate all relevant parameters and variables and allow frequency and dimensional scaling. The accuracy may be less than that obtained with a full-wave analysis but can be sufficient for design purposes.

The *development* of an approximate antenna model is only beneficial if a full-wave analysis program is not available; it is even more beneficial in such a situation if the antenna to be designed belongs to a class of antennas, i.e. when similar antenna designs are foreseen in the future but for different frequency ranges, materials or environments. When a full-wave analysis program is available and the antenna to be designed is known to be a one-of-a-kind design, an educated software version of a trial-and-error procedure using that program is advised. Both full-wave analysis software and approximate analysis software has to be used by an antenna expert, with carefully considered input of parameter values and interpretation of the analysis results.

In this book, we have described approximate models for five classes of antennas:

1. Loops and solenoids in a conducting medium for intravascular use have been analyzed. These antennas were intended to function as receiving antennas in a magnetic resonance imaging system. Since we were interested in the magnetic field strength close to the antenna, we analyzed the antennas using a quasi-static approximation.

2. Printed monopole antennas for integration on a PCB when limited space is available have been analyzed. An approximate model for this class of antennas was derived using separation of an asymmetrically driven dipole antenna into two grounded monopole antennas. One of the monopole antennas consisted of a microstrip-excited printed monopole antenna, and the other was formed by the ground plane of the microstrip. The two monopole antennas were analyzed using an equivalent-radius dipole antenna with a magnetic covering. To demonstrate the use of a full-wave analysis program, a one-of-a-kind printed UWB antenna was designed.

3. Closely related to PCB antennas are folded-dipole antennas on dielectric slabs. Folded-dipole antennas with additional short circuits in the arms and/or parasitic radiators positioned parallel to the folded-dipole antenna offer the possibility to control the input impedance while maintaining a near-omnidirectional radiation pattern. This makes them very suitable for RFID applications, where the chip impedance in general is anything but 50 Ω and the antenna may be designed to be conjugately matched to the RFID chip. Analytical dipole antenna models and transmission line theory have been applied to model both thin-wire and strip folded-dipole antenna structures and linear arrays of reentrant folded-dipole elements.

4. The concept of designing antennas to be conjugately matched to the RF front end can also be applied to the design of compact, efficient, low-power rectennas, i.e. rectifying antennas. First, the rectifying circuit was modeled using a large-signal equivalent circuit. Next, a modified transmission line model for a rectangular microstrip patch antenna was applied to find the edge feed position that corresponded to an antenna input impedance that was the complex conjugate of the impedance of the rectifying circuit. The concept of conjugate matching, in combination with an even–odd mode analysis, was applied to the analysis of a Wilkinson power combiner where the resistive element was replaced by a rectifying circuit. Thus a system of two microstrip antennas was designed, where the combined RF data input was brought to the output of the power combiner and the mismatch between the two antennas was converted to DC energy.

5. The final class of antennas for which an approximate model has been developed is a little different, in the sense that the radiators were modeled in a full-wave manner and the approximation was in the size of an array consisting of these radiators. A large planar array, consisting of radiators placed in a regular grid, was approximated as being infinite in extent in two directions in the transverse plane. Furthermore, the elements were assumed to be excited uniformly. By making this approximation, the array could now be considered to be periodic and analysis could be restricted to a single unit cell, the analysis of which included all of the information about the mutual coupling with the infinite environment.

7.2 INTRAVASCULAR MR ANTENNAS: LOOPS AND SOLENOIDS

The forming of magnetic resonance images is accomplished by trading off signal-to-noise ratio, imaging speed and spatial resolution. By replacing the receiver coils *outside* the body in a 'standard' MRI setup by *intravascular* coils or antennas, the SNR can be improved considerably, making the visualization of catheter positions and orientations during surgery feasible. Even imaging of the condition of the walls of the large arteries should become feasible. The idea of employing intravascular antennas is not new, and some qualitative comparisons of different intravascular-antenna concepts have even been performed. To perform the first quantitative comparison of various antenna concepts for intravascular catheter tracking and artery wall imaging, an approximate model for intravascular wire antennas has been developed. This quasi-static, approximate model calculates the magnetic field intensity induced by the wire structure. The dimensions of an intravascular antenna and the frequency[1] are such that the artery walls will be positioned in the radiating near field of the antenna, where the field amplitudes locally are inversely proportional to the square of the distance, justifying a quasi-static approximation.

To assess the validity of the quasi-static approximation, the results were compared with analytical expressions for the fields of a small-loop antenna carrying a uniform current. To be

[1] The frequency is the Larmor frequency which, for a static magnetic field of 1.5 T, is about 64 MHz.

able to perform this comparison, the small-loop, uniform-current approximation was first validated. It turns out that we may consider a bare loop antenna immersed in blood and subject to a 1.5 T main magnetic field to carry a uniform current for radii up to 1.7 mm.

Next, the quasi-static model was verified by looking at the dynamic 'sensitivity', which was defined as the transverse magnetic field intensity normalized to the uniform current density. The maximum relative error in the dynamic sensitivity for a single bare loop of radius 0.5 mm immersed in blood was 30% on the axis of the loop and 13% for all other loop orientations, calculated in the region of interest. This region of interest was a horizontal section through the center of the antenna and a coaxial, vertical circular cylinder with a radius between 2 mm and 3 mm, corresponding to the size of the large arteries. Moreover, the sensitivity as a function of distance from the loop center showed a similar appearance for the quasi-static and for the dynamic model. This means that we may employ our quasi-static model for comparison of loop antenna designs.

For a multiturn loop antenna immersed in blood, up to 35 turns may be employed for a radius of the order of 0.5 mm without compromising the model. The uniform current is maintained, provided that a thin insulation layer is applied to the wire. If this insulation layer is not applied, the quasi-static approximation will fail for wire antennas larger than a single loop.

After we had established the validity of the quasi-static approximate model, use of the model revealed that a center return antenna was best suited for tracking purposes. A perpendicular-coils antenna was preferred, however, owing to easier manufacture. This preferred antenna also performed better than the first one when rotated with respect to the main magnetic field.

For imaging purposes, both a dual-opposed-solenoids antenna and a triple-loop antenna were preferred. These exhibit comparable sensitivity profiles, and the manufacturing complexity of the two antennas is expected to be equal. Neither antenna should be rotated in excess of 45° with respect to the main MR magnetic field.

It has been demonstrated that intravascular-antenna designs may be created automatically, employing genetic-algorithm optimization in combination with approximate antenna analysis. These designs, subject to user-defined electromagnetic and geometrical constraints, can be generated within minutes, employing standard office computing equipment. Since the designs thus created are very sensitive to the three-dimensional geometry, precise manufacturing techniques are required for realizing these designs.

Last but not least, there is the issue of patient safety. In the modeling and design, the focus was on the intravascular antennas. The signals received by the antennas need to be transported to the MR hardware, however, and therefore transmission lines are employed. These transmission lines go from the MR hardware, which is exterior to the patient, through the patient's vascular system to the intravascular antennas. A real danger exists that the lengths of these transmission lines may be such that they will be resonant at the Larmor frequency. Radiation will then take place at the tips of the transmission line leads and will be dissipated in the tissue surrounding the lines. The temperature may thus increase up to and beyond a level that will be harmful for the patient. Dissecting the transmission line into sections that are too short to become resonant at the Larmor frequency is recommended as a solution to this problem. A technique involving inductive coupling between transmission line sections as described in [4] could be employed for transferring signals from the antenna to the MR hardware.

7.3 PCB ANTENNAS: PRINTED MONOPOLES

As explained before, approximate antenna models, if not already available, should be developed only for *classes* of antennas. If the design of a certain type of antenna is not to be restricted to a single application but, instead, designs of similar antennas, maybe for different frequency bands, materials or environments, are foreseen for the (near) future, it is worthwhile to invest in the development of an approximate model.

If a one-of-a-kind antenna needs to be designed, or time is critical, an educated trial-and-error design methodology may be followed. The iterative realization and redesign of the antenna may be replaced by the use of a commercially distributed full-wave electromagnetic analysis program. The educated part of the trial-and-error process consists of choosing a starting antenna configuration, the operation of which can be understood on basis of physical reasoning. This process has been demonstrated for a printed UWB antenna, the operation of which may be explained by seeing the antenna as a combination of a 'fat' dipole antenna and two tapered slot antennas. The antenna was upgraded, following the same process, with a slot in one of the radiating arms to block out WLAN frequencies between 5 GHz and 6 GHz. When the original UWB antenna was upgraded for the same reason with a filtering structure in the microstrip transmission line exciting the antenna, it was found that applying an approximate model for the filtering structure sped up the design process.

Next, for the design of non-UWB, planar printed monopole antennas, an approximate model was developed. Printed monopole antennas may be used for various applications, ranging from GSM through GPS, Bluetooth and ZigBee to WLANs, covering frequency bands from around 1 GHz to around 6 GHz. Printed monopole antennas may be incorporated on the rim of FR4-based PCB designs and are especially attractive when not much space is available. The approximate model was based on a model of an equivalent-radius, magnetically coated, circularly cylindrical, wire dipole antenna. This model was applied both to a microstrip-excited monopole antenna and to a monopole antenna that was formed by the ground plane of a microstrip transmission line. If the width of the ground plane is small enough to allow a thin-equivalent-wire approximation and to prevent half-wave resonance effects from the rim of the ground plane, the model may be employed to create (pre)designs of planar printed monopole antennas.

7.4 RFID ANTENNAS: FOLDED DIPOLES

The folded-dipole antenna is an attractive antenna for RFID applications. First, the radiation characteristics of a folded-dipole antenna are identical to those of an ordinary dipole antenna, i.e. the radiation is near-omnidirectional. Then, the geometry of the folded-dipole antenna offers opportunities to tune the input impedance to a desired complex value. This makes it possible – through complex-conjugate matching – to connect the antenna directly to the RFID chip, without having a matching network in between. The antenna may be modeled on the basis of a dipole antenna model and a two-wire transmission line model. The tuning can be achieved by placing additional short circuits in the arms of the folded-dipole antenna, which affect mostly the transmission line characteristics of the antenna and not the dipole characteristics, or by placing parasitic dipoles next to the antenna, which affect mostly the dipole characteristics of the antenna and not the transmission line characteristics, or by a combination of both techniques. These impedance-tuning techniques have been verified when applied to thin-wire folded-dipole antennas. A wire transmission line model for the folded-dipole antenna was applied to these antennas and modified to take account of the tuning mechanisms.

For the more realistic application of a strip folded-dipole antenna on a dielectric slab, a model of an asymmetric strip folded-dipole antenna for an antenna in a uniform medium was modified to take account of a finite-thickness dielectric slab.

To design linear arrays of folded-dipole antennas, which give additional degrees of freedom for obtaining a desired input impedance, an approximate but accurate (wire) model was developed. This model combines closed-form analytical equations for a folded-dipole antenna, a reentrant folded-dipole antenna, a two-wire transmission line, the mutual coupling between two folded-dipole antennas and the mutual coupling between two thin-wire dipole antennas. The model successfully applies separate analysis of the feeding network and of the antennas. A relative error of less than one percent in both the real and the imaginary part of the input impedance was demonstrated over a frequency band of 10%.

7.5 RECTENNAS: MICROSTRIP PATCH ANTENNAS

Wireless batteries, or rectennas – rectifying antennas – are intended for converting wireless RF power into DC power. Although power conversion efficiencies exceeding 80% have been reported for high (20 dBm) input power levels to the rectenna, wireless batteries will be most beneficial at large distances from sources that radiate at power levels limited by national and international regulations. Therefore, the challenge lies in maximizing the power conversion efficiency of wireless batteries for low input power levels, i.e. 0 dBm and below. If a rectifying circuit is conjugately matched directly, to a microstrip patch antenna, the need for a matching network between the two no longer exists. Thus the efficiency of the wireless battery will improve. Moreover, this matching technique automatically suppresses the reradiation of harmonics by the microstrip patch antenna, since the harmonics will be mismatched. Thus, the impedance-matching and filtering network encountered in traditional wireless-battery designs has become obsolete. With the aid of analytical models developed for the antenna and the rectifier, single-layer, internally matched, filtered PCB rectennas have been designed for low input power levels. A large-signal model for the rectifying circuit was employed; this model is analyzed in the time domain and gives, after fast Fourier transformation, the input impedance in the frequency domain. A rectangular microstrip patch antenna was modeled using a modified, multimodal cavity model. After the complex input impedance of the rectifying circuit was determined, the edge feed position on the microstrip patch antenna was determined such that it would give an input impedance of the patch that was the complex conjugate of the impedance of the rectifying circuit. An efficiency of 52% for 0 dBm input power was achieved at 2.45 GHz for a wireless battery constructed on FR4, showing an improvement of more than 10% over a traditional rectenna design, in addition to a reduction in size and complexity. A series connection of these wireless batteries was shown to be able to power a standard household wall clock over a distance of a few meters.

If the resistor in a Wilkinson power combiner connected to two microstrip patch antennas is replaced by a rectifying circuit, the simultaneous reception of data and power becomes feasible, a feature not realizable with a standard rectenna. On the basis of the complex input impedance of the rectifying circuit, a modified Wilkinson power combiner has been designed, based on an even–odd-mode analysis. The calculated unloaded output voltage as a function of frequency remained within 10% of the measured values. A setup for the simultaneous reception of amplitude-modulated data and power was realized. For power generation, an inequality between the two antennas was created by insertion of an additional section of microstrip transmission line between one antenna and the input port of the modified Wilkinson power combiner.

The use of ambient RF power generated by GSM base stations is feasible only if relatively large collecting apertures are employed for the rectenna. For distances ranging from 25 m to 100 m from such a station, power density levels ranging from 0.1 mW m^{-2} to 1.0 mW m^{-2} may be expected at single frequencies. For the total GSM downlink frequency bands, these levels may be elevated by a factor of between one and three, depending on the traffic density. Initial measurements in a WLAN environment indicate power density levels that are at least one order of magnitude lower. A single GSM telephone has been demonstrated to deliver enough energy to wirelessly power small applications at distances of a few decimeters.

7.6 LARGE ARRAY ANTENNAS: OPEN-ENDED RECTANGULAR-WAVEGUIDE RADIATORS

The by now classical mode-matching analysis method for rectangular-waveguide structures and infinite arrays of open-ended rectangular-waveguide radiators has been described in this book. Although, since the development of this method about 15 years ago, more efficient analysis methods have been derived, the classical mode-matching approach may still be of use, especially for analyzing structures that are not overcomplicated. Therefore a thorough description of the method that can be straightforwardly implemented into software may be of value to engineers in the field. As said, the analysis of waveguides and waveguide discontinuities in this method is based on mode matching. For every discontinuity, waveguide-to-waveguide and waveguide-to-unit-cell, a generalized scattering matrix is formed. Generalized scattering matrices of discontinuities are cascaded to create an overall generalized scattering matrix of the complete structure to be analyzed.

The code implemented has been thoroughly validated by comparing analysis results with analysis and measurement results reported in the open literature. A mode preselection scheme, a convergence strategy and a Floquet mode selection scheme have been developed and have been applied to the structures analyzed.

REFERENCES

1. A.G. Tijhuis, M.C. van Beurden, B.P. de Hon and H.J. Visser, 'From engineering electromagnetics to electromagnetic engineering: Using computational electromagnetics

for synthesis problems', *Turkish Journal of Electrical Engineering and Computer Sciences*, Vol. 16, No. 1, pp. 7–19, 2008.

2. A. Vasylchenko, Y. Schols, W. De Raedt and G.A.E. Vandenbosch, 'A benchmarking of six software packages for full-wave analysis of microstrip antennas', *Proceedings of the 2nd European Conference on Antennas and Propagation, EuCAP2007*, Edinburgh, UK, November 2007.

3. A. Vasylchenko, Y. Schols, W. De Raedt and G.A.E. Vandenbosch, 'Challenges in full wave electromagnetic simulation of very compact planar antennas', *Proceedings of the 2nd European Conference on Antennas and Propagation, EuCAP2007*, Edinburgh, UK, November 2007.

4. P. Vernickel, V. Schulz, S. Weiss and B. Gleich, 'A safe transmission line for MRI', *IEEE Transactions on Biomedical Engineering*, Vol. 52, No. 6, pp. 1094–1102, June 2005.

Index

Approximate Antenna Analysis for CAD Hubregt J. Visser